STRIPLINE CIRCULATORS • *Joseph Helszajn*

THE STRIPLINE CIRCULATOR: THEORY AND PRACTICE • *Joseph Hel.*

LOCALIZED WAVES • *Hugo E. Hernández-Figueroa, Michel Zamboni-Rached, and Erasmo Recami (eds.)*

MICROSTRIP FILTERS FOR RF/MICROWAVE APPLICATIONS • *Jia-Sheng Hong and M. J. Lancaster*

MICROWAVE APPROACH TO HIGHLY IRREGULAR FIBER OPTICS • *Huang Hung-Chia*

NONLINEAR OPTICAL COMMUNICATION NETWORKS • *Eugenio Iannone, Francesco Matera, Antonio Mecozzi, and Marina Settembre*

FINITE ELEMENT SOFTWARE FOR MICROWAVE ENGINEERING • *Tatsuo Itoh, Giuseppe Pelosi, and Peter P. Silvester (eds.)*

INFRARED TECHNOLOGY: APPLICATIONS TO ELECTROOPTICS, PHOTONIC DEVICES, AND SENSORS • *A. R. Jha*

SUPERCONDUCTOR TECHNOLOGY: APPLICATIONS TO MICROWAVE, ELECTRO-OPTICS, ELECTRICAL MACHINES, AND PROPULSION SYSTEMS • *A. R. Jha*

OPTICAL COMPUTING: AN INTRODUCTION • *M. A. Karim and A. S. S. Awwal*

INTRODUCTION TO ELECTROMAGNETIC AND MICROWAVE ENGINEERING • *Paul R. Karmel, Gabriel D. Colef, and Raymond L. Camisa*

MILLIMETER WAVE OPTICAL DIELECTRIC INTEGRATED GUIDES AND CIRCUITS • *Shiban K. Koul*

ADVANCED INTEGRATED COMMUNICATION MICROSYSTEMS • *Joy Laskar, Sudipto Chakraborty, Manos Tentzeris, Franklin Bien, and Anh-Vu Pham*

MICROWAVE DEVICES, CIRCUITS AND THEIR INTERACTION • *Charles A. Lee and G. Conrad Dalman*

ADVANCES IN MICROSTRIP AND PRINTED ANTENNAS • *Kai-Fong Lee and Wei Chen (eds.)*

SPHEROIDAL WAVE FUNCTIONS IN ELECTROMAGNETIC THEORY • *Le-Wei Li, Xiao-Kang Kang, and Mook-Seng Leong*

ARITHMETIC AND LOGIC IN COMPUTER SYSTEMS • *Mi Lu*

OPTICAL FILTER DESIGN AND ANALYSIS: A SIGNAL PROCESSING APPROACH • *Christi K. Madsen and Jian H. Zhao*

THEORY AND PRACTICE OF INFRARED TECHNOLOGY FOR NONDESTRUCTIVE TESTING • *Xavier P. V. Maldague*

METAMATERIALS WITH NEGATIVE PARAMETERS: THEORY, DESIGN, AND MICROWAVE APPLICATIONS • Ricardo Marqués, Ferran Martín, and Mario Sorolla

OPTOELECTRONIC PACKAGING • *A. R. Mickelson, N. R. Basavanhally, and Y. C. Lee (eds.)*

OPTICAL CHARACTER RECOGNITION • *Shunji Mori, Hirobumi Nishida, and Hiromitsu Yamada*

ANTENNAS FOR RADAR AND COMMUNICATIONS: A POLARIMETRIC APPROACH • *Harold Mott*

INTEGRATED ACTIVE ANTENNAS AND SPATIAL POWER COMBINING • *Julio A. Navarro and Kai Chang*

ANALYSIS METHODS FOR RF, MICROWAVE, AND MILLIMETER-WAVE PLANAR TRANSMISSION LINE STRUCTURES • *Cam Nguyen*

FREQUENCY CONTROL OF SEMICONDUCTOR LASERS • *Motoichi Ohtsu (ed.)*

WAVELETS IN ELECTROMAGNETICS AND DEVICE MODELING • *George W. Pan*

OPTICAL SWITCHING • *Georgios Papadimitriou, Chrisoula Papazoglou, and Andreas S. Pomportsis*

SOLAR CELLS AND THEIR APPLICATIONS • *Larry D. Partain (ed.)*

PARALLEL SOLUTION OF INTEGRAL EQUATION-BASED EM PROBLEMS IN THE FREQUENCY DOMAIN

PARALLEL SOLUTION OF INTEGRAL EQUATION-BASED EM PROBLEMS IN THE FREQUENCY DOMAIN

YU ZHANG

TAPAN K. SARKAR

With contributions from

Daniel Garcia Doñoro
Hongsik Moon
Mary Taylor
Robert A. van de Geijn

IEEE Press

A JOHN WILEY & SONS, INC., PUBLICATION

Library of Congress Cataloging-in-Publication Data:

Zhang, Yu, 1978 Apr. 20–
  Parallel solution of integral equation based EM problems in the frequency domain / Yu Zhang, Tapan K. Sarkar ; with contributions from Hongsik Moon, Mary Taylor, Robert A. van de Geijn.
    p. cm.
  Includes bibliographical references and index.
  ISBN 978-0-470-40545-1 (cloth)
  1. Electromagnetism—Data processing. 2. Electromagnetic fields—Mathematical models. 3. Electromagnetic waves—Mathematical models. 4. Parallel processing (Electronic computers) 5. Integral domains. 6. Time-domain analysis. I. Sarkar, Tapan (Tapan K.) II. Title.
  QC760.54.Z48 2009
  537.01'5118—dc22   2009001799

Printed in the United States of America.

10 9 8 7 6 5 4 3 2 1

Contents

Preface xiii

Acknowledgments xvii

Acronyms xix

Chapter 1   Introduction 1

     1.0   Summary 1
     1.1   A Brief Review of Parallel CEM 1
        1.1.1 Computational Electromagnetics . . . 1
        1.1.2 Parallel Computation in Electromagnetics . . 3
     1.2   Computer Platforms Accessed in This Book . . 9
     1.3   Parallel Libraries Employed for the Computations . 12
        1.3.1 ScaLAPACK — Scalable Linear Algebra Package 13
        1.3.2 PLAPACK — Parallel Linear Algebra Package . 16
     1.4   Conclusion 19
       References 19

Chapter 2   In-Core and Out-of-Core LU Factorization for
              Solving a Matrix Equation 27

     2.0   Summary 27
     2.1   Matrix Equation from a MoM Code . . . 28
     2.2   An In-Core Matrix Equation Solver . . . 28
     2.3   Parallel Implementation of an In-Core Solver . . 32
        2.3.1 Data Distribution for an LU Algorithm . . 32
        2.3.2 ScaLAPACK: Two-Dimensional Block-Cyclic
            Matrix Distribution 36
        2.3.3 PLAPACK: Physically Based Matrix Distribution 38
        2.3.4 Data Distribution Comparison between
            ScaLAPACK and PLAPACK . . . 40
     2.4   Data Decomposition for an Out-of-Core Solver . . 42
     2.5   Out-of-Core LU Factorization 43
        2.5.1 I/O Analysis of Serial Right-Looking and
            Left-Looking Out-of-Core LU Algorithms . 45
            2.5.1.1   *Right-Looking Algorithm* . . 45
            2.5.1.2   *Left-Looking Algorithm* . . 47

v

2.5.2 Implementation of the Serial Left-Looking
Out-of-Core LU Algorithm   50
2.5.3 Design of a One-Slab Left-Looking Out-of-Core
LU Algorithm   55
2.6   Parallel Implementation of an Out-of-Core LU
Algorithm   61
2.6.1 Parallel Implementation of an Out-of-Core
LU Algorithm Using ScaLAPACK .   .   .   61
2.6.2 Parallel Implementation of an Out-of-Core
LU Algorithm Using PLAPACK   .   .   .   64
2.6.3 Overlapping of the I/O with the Computation   .   65
2.6.4 Checkpointing in an Out-of-Core Solver   .   .   65
2.7   Solving a Matrix Equation Using the Out-of-Core LU
Matrices   66
2.8   Conclusion   69
References   69

Chapter   3   A Parallel MoM Code Using RWG Basis Functions and
ScaLAPACK-Based In-Core and
Out-of-Core Solvers   71

3.0   Summary.   71
3.1   Electric Field Integral Equation (EFIE)   .   .   .   71
3.2   Use of the Piecewise Triangular Patch (RWG) Basis
Functions.   74
3.3   Testing Procedure   76
3.4   Matrix Equation for MoM.   78
3.5   Calculation of the Various Integrals   79
3.5.1 Evaluation of the Fundamental Integrals   .   .   79
3.5.2 Extraction of the Singularity   80
3.6   Calculation of the Fields   81
3.7   Parallel Matrix Filling — In-Core Algorithm   .   .   81
3.8   Parallel Matrix Filling — Out-of-Core Algorithm   .   86
3.9   Numerical Results from a Parallel In-Core MoM Solver   88
3.9.1 Numerical Results Compared with other Methods   88
3.9.1.1 A PEC Cube.   88
3.9.1.2 A Combined Cube-and-Sphere PEC Model   88
3.9.2 Different Metrics Used to Assess the Degree
of Parallel Efficiency   88
3.9.3 Efficiency and Portability of a Parallel MoM
In-Core Solver.   92
3.10 Numerical Results from a Parallel Out-of-Core MoM
Solver   96
3.10.1 Parallel Out-of-Core Solver Can Be as Efficient as
a Parallel In-Core Solver.   96

		3.10.2 Scalability and Portability of the Parallel Out-of-Core Solver	98
	3.11	Conclusion	104
		References	105
Chapter	**4**	**A Parallel MoM Code Using Higher-Order Basis Functions and ScaLAPACK-Based In-Core and Out-of-Core Solvers**	**107**
	4.0	Summary	107
	4.1	Formulation of the Integral Equation for Analysis of Dielectric Structures	107
	4.2	A General Formulation for the Analysis of Composite Metallic and Dielectric Structures	110
	4.3	Geometric Modeling of the Structures . . .	114
		4.3.1 Right-Truncated Cone to Model Wire Structures	114
		4.3.2 Bilinear Surface for Modeling Arbitrarily Shaped Surfaces	116
	4.4	Higher-Order Basis Functions	117
		4.4.1 Current Expansion along a Thin PEC Wire .	117
		4.4.2 Current Expansion over a Bilinear Surface .	119
	4.5	Testing Procedure	124
		4.5.1 Testing Procedure for Thin PEC Wires . .	124
		4.5.2 Testing Procedure for Bilinear Surfaces . .	128
	4.6	Parallel In-Core and Out-of-Core Matrix Filling Schemes	131
		4.6.1 Parallel In-Core Matrix Filling Scheme . .	132
		4.6.2 Parallel Out-of-Core Matrix Filling Scheme .	134
	4.7	Numerical Results Computed on Different Platforms .	136
		4.7.1 Performance Analysis for the Parallel In-Core Integral Equation Solver . . .	136
		4.7.1.1 Comparison of Numerical Results Obtained on Single-Core and Multicore Platforms	136
		4.7.1.2 Numerical Results Obtained on Single-Core Platforms	141
		4.7.1.2.1 Radiation from a Vivaldi Antenna Array . . .	141
		4.7.1.2.2 Scattering from a Full-Size Airplane	144
		4.7.1.3 Numerical Results Obtained on Multicore Platforms	146
		4.7.2 Performance Analysis for the Parallel Out-of-Core Integral Equation Solver	147
		4.7.2.1 Vivaldi Antenna Array — a Large Problem Solved on Small Computer Platforms .	147

4.7.2.2 Solution for a Full-Size Airplane — Parallel Out-of-Core Solver Can Be as Efficient as the Parallel In-Core 149

4.7.2.3 Solution for a Full-Size Airplane — Scalability and Portability of the Parallel Out-of-Core Solver 150

4.7.2.4 Solution for a Full- Size Airplane — a Very Large Problem Solved on Nine Nodes of CEM-4. 153

4.8 Conclusion 155

References 155

Chapter 5   Tuning the Performance of a Parallel Integral Equation Solver **157**

5.0 Summary. 157

5.1 Anatomy of a Parallel Out-of-Core Integral Equation Solver 157

5.1.1 Various Components of a Parallel Out-of-Core Solver that Can Be Observed through Ganglia and Tuned 158

5.1.2 CPU Times of Parallel In-Core and Out-of-Core Integral Equation Solvers 161

5.1.3 Performance of a Code Varies with the Amount of Storage Used on the Hard Disk . . . 165

5.2 Block Size 170

5.3 Shape of the Process Grid 173

5.4 Size of the In-Core Buffer Allocated to Each Process . 176

5.4.1 Optimizing IASIZE for a Parallel MoM Code Using Higher-Order Basis Functions. . . 177

5.4.1.1 Case A: Available 2 GB of RAM/Core . 177

5.4.1.1.1 Overview of Wall Time with Different IASIZE . . . 177

5.4.1.1.2 Details on Matrix Filling and Matrix Solving . . . 181

5.4.1.2 Case B: Available 4 GB of RAM/Core . 188

5.4.2 Optimizing IASIZE for a Parallel MoM Code Using RWG Basis Functions 190

5.4.3 Influence of Physical RAM Size on Performance 194

5.5 Relationship between Shape of the Process Grid and In-Core Buffer Size 197

5.6 Overall Performance of a Parallel Out-of-Core Solver on HPC Clusters 201

5.7 Conclusion 205

References 205

Chapter 6   **Refinement of the Solution Using the Iterative**
        Conjugate Gradient Method **207**

     6.0   Summary   207
     6.1   Development of the Conjugate Gradient Method   .   207
     6.2   The Iterative Solution of a Matrix Equation   .   .   212
     6.3   Parallel Implementation of the CG Algorithm   .   .   213
     6.4   A Parallel Combined LU-CG Scheme to Refine the
         LU Solution   215
     6.5   Conclusion   216
         References   217

Chapter 7   **A Parallel MoM Code Using Higher-Order Basis**
        Functions and PLAPACK-Based In-Core and
        Out-of-Core Solvers **219**

     7.0   Summary   219
     7.1   Introduction   219
     7.2   Factors that Affect a Parallel In-Core and Out-of-Core
         Matrix Filling Algorithm   220
     7.3   Numerical Results   224
         7.3.1 Radiation from an Array of Vivaldi Antennas   .   224
         7.3.2 Scattering from an Electrically Large Aircraft .   228
         7.3.3 Discussion of the Computational FLOPS
             Achieved   230
     7.4   Conclusion   231
         References   231

Chapter 8   **Applications of the Parallel Frequency-Domain**
        Integral Equation Solver — TIDES   .   .   .   **233**

     8.0   Summary   233
     8.1   Performance Comparison between TIDES and a
         Commercial EM Analysis Software   .   .   234
         8.1.1 Analysis of a Scattering Problem   .   .   234
         8.1.2 Analysis of a Radiation Problem   :   .   237
         8.1.3 Analysis of a Coupling Problem   .   .   241
     8.2   EMC Prediction for Multiple Antennas Mounted
         on an Electrically Large Platform   .   .   .   243
     8.3   Analysis of Complex Composite Antenna Array   .   248
     8.4   Array Calibration for Direction-of-Arrival Estimation   249
     8.5   Radar Cross Section (RCS) Calculation of Complex
         Targets   252
         8.5.1 RCS Calculation of a Squadron of Tanks   .   252
         8.5.2 RCS of the Tanks inside a Forest Environment .   254

8.5.3 RCS from an Aircraft and a Formation of Aircraft   257
8.5.4 RCS Simulation with Million Level Unknowns .   259
8.5.5 RCS of an Aircraft Carrier   260
8.6 Analysis of Radiation Patterns of Antennas Operating
     Inside a Radome Along with the Platform on
     Which It Is Mounted   264
8.7 Electromagnetic Interference (EMI) Analysis of a
     Communication System   268
8.8 Comparison between Computations Using TIDES and
     Measurement data for Complex Composite Structures.   271
8.9 Conclusion   273
     References   273

Appendix A   **A Summary of the Computer Platforms Used in This**
              Book **275**

A.0 Summary.   275
A.1 Description of the Platforms Used in This Book.   .   275
A.2 Conclusion   284
     References   285

Appendix B   **An Efficient Cross-Platform Compilation of the**
              ScaLAPACK and PLAPACK Routines .   .   .   **287**

B.0 Summary.   287
B.1 Tools for Compiling both ScaLAPACK and
     PLAPACK   287
B.2 Generating the ScaLAPACK Library   .   .   .   288
     B.2.1 Source Codes for Compiling ScaLAPACK   .   288
     B.2.2 Steps for Compiling ScaLAPACK   .   .   .   288
     B.2.3 Script Files for 32-bit Windows Operating System   289
          B.2.3.1 Script Files for BLAS   289
          B.2.3.2 Script Files for BLACS   293
          B.2.3.3 Script Files for ScaLAPACK .   .   .   296
     B.2.4 Script Files for 64-bit Windows Operating System   297
B.3 Generating the PLAPACK Library   298
     B.3.1 Source Codes for Compiling PLAPACK .   .   298
     B.3.2 Script Files for PLAPACK   298
B.4 Tuning the Performance by Turning on Proper Flags .   300
B.5 Conclusion   301
     References   301

**Appendix C   An Example of a Parallel MoM Source Code for
                Analysis of 2D EM Scattering** **303**

        C.0   Summary.   303
        C.1   Introduction of MoM   303
        C.2   Solution of a Two-Dimensional Scattering Problem .   305
              C.2.1 Development of the Integral Equation and the
                    MoM Solution   305
              C.2.2 Evaluation of the Parameter of Interest   .   .   308
        C.3   Implementation of a Serial MoM Code   .   .   .   309
              C.3.1 Flowchart and Results of a Serial MoM Code .   309
              C.3.2 A Serial MoM Source Code for the 2D
                    Scattering Problem   312
        C.4   Implementation of a Parallel MoM Code   .   .   .   313
              C.4.1 Flowchart and Results of a Parallel MoM Code   313
              C.4.2 A Parallel MoM Source Code Using ScaLAPACK
                    for the 2D Scattering Problem   318
        C.5   Compilation and Execution of the Parallel Code   .   331
        C.6   Conclusion   333
              References   333

Index **335**

Preface

The future of computational electromagnetics is changing drastically with the new generation of computer processors, which are multicore instead of single-core. Previously, advancements in chip technology meant an increase in clock speed, which was typically a benefit that computational code users could enjoy. This is no longer the case. In the new roadmaps for chip manufacturers, speed has been sacrificed for improved power consumption. The general trend in the processor development has been from dual-, quad-, and eight-core chips to ones with tens or even hundreds of cores.

This change represents something of a cataclysmic shift for software. As a result, parallel programming has suddenly become relevant for all computer systems. The burden now falls on the software programmer to revamp existing codes and add new functionality to enable computational codes to run efficiently on this new generation of multicore processors.

It is interesting to note that parallel processing is a topic that is not new. There was a large initiative of parallel processing around 1994. A brief overview of some of the computer platforms that were available and what type of results one could achieve at that time is reviewed in the following reference: M. Swaminathan and T. K. Sarkar, "A Survey of Various Computer Architectures for Solution of Large Matrix Equations," *International Journal of Numerical Modeling: Electronic Networks, Devices and Fields*, Vol. 8, pp. 153–168, 1995 (invited paper). However, the concept did not take off at that time because the various compilers, particularly FORTRAN, and the various scientific subroutine packages were not readily available for working with parallel hardware. In addition, the computer hardware was fabricated using special-purpose chips and was not off-the-shelf building blocks. However, this time around the environment is quite different and so are the reasons why there is a renewed push for parallel processing.

In this book, a new roadmap for computational electromagnetics (CEM) code designers is provided, demonstrating how to navigate along with the chip designers through the multicore advancements in chip design. Computational electromagnetics covers a wide range of numerical methods. This book will focus only on the *method of moments* (MoM) — a method that has remained to be one of the most popular numerical methods since 1967 and will continue to be popular in the future. The methodology presented here will provide a guide to the shortest route of developing highly efficient parallel integral equation–based MoM codes in frequency domain.

One objective of this book is to provide solutions to extend the capability of MoM without resulting in accuracy problems as incurred by fast algorithms and hybrid methods. As is well known, MoM needs to deal with huge dense matrices for very complex problems. This leads to a requirement of a large amount of RAM and thus most of computers on hand may not easily satisfy this criterion. To break such limitation of RAM, out-of-core solvers are introduced to extend the capability of MoM by taking full advantage of the mass storage space of the hard disk.

Assume that enough RAM and hard disk are available to store a large dense complex matrix. The simulation efficiency of a serial MoM code is then constrained by the computing capability of an ordinary computer system. To improve the efficiency of MoM and to adapt to the trend of CPU development, parallel algorithm design is the inevitable way out, which is our major concern in this book. Furthermore, parallelized out-of-core solutions provide tools to generate numerically accurate results within a reasonable time frame by using high-performance computers.

A number of scientific publications covering parallel electromagnetic simulation techniques already exist, but they do not give the complete and exhaustive coverage of parallel techniques for the MoM that will be given in this book. This book collects the state of the art in the parallel solution of the integral equations. Codes using the techniques described here bring computing solutions based on an out-of-core processing kernel and parallel solving capability to the EM engineer's desktop. Here is a thumbnail outline of what is unique in this book:

1. Both the piecewise triangular patch (RWG: Rao–Wilton–Glisson) and higher-order basis functions will be discussed, in relation to the electric field integral equation (EFIE) and the PMCHW (Poggio–Miller–Chang–Harrington–Wu) formulations.

2. Both parallel in-core and out-of-core EM integral equation solvers will be developed to completely utilize computer resources. Methodologies developed will be general enough for all types of MoM problems.

3. Both ScaLAPACK and PLAPACK methodologies will be studied in depth. ScaLAPACK is widely used and has been well commercialized. PLAPACK has its own unique user-friendly interface.

4. Both the parallel direct matrix equation solver using LU factorization and the parallel iterative solver using the conjugate gradient (CG) method will be discussed.

5. Both the design methodology and engineering application of the parallel code are emphasized. Codes developed in this book are tested on a variety of platforms.

Another important objective of this book is to provide a reference text on parallel techniques in computational electromagnetics for academic research scientists, postgraduate students in the field of electromagnetic computation, and in particular, for those undertaking numerical analysis and MoM-related electromagnetic problems.

To be a good, step-by-step guide for new researchers to the methodology in parallel code design for the method of moments, this book will provide all the details of the mathematical formulation through an example source code using a parallel in-core MoM methodology suitable for the analysis of 2D problems.

The content of this book is organized as follows:

A background introduction is presented in Chapter 1. The hardware and the software, with which the research was done, are briefly explained.

Chapter 2 gives a more detailed discussion of parallel LU (lower/upper) factorization for solving the matrix equation of MoM. The data distribution of ScaLAPACK and PLAPACK is explained in detail since this is the key to understanding how the MoM matrix is distributed among many processes. Then, the design of out-of-core algorithms and their parallel implementations are discussed.

On the basis of the matrix equation solver developed in Chapter 2, Chapter 3 presents the parallel design of MoM code with the RWG basis functions. Step by step, this chapter tells how to design an efficient parallel matrix filling scheme for MoM with RWG basis function, which includes strategies for both parallel in-core and parallel out-of-core matrix filling. Then the performance of the code is evaluated on seven different computer platforms.

Since MoM using RWG basis functions needs more RAM when dealing with electrically large objects, higher-order basis functions are introduced in Chapter 4. Performance of the parallel in-core and out-of-core codes using higher-order basis functions are evaluated on ten different computer platforms.

Even though the parallel codes have been successfully executed on so many different platforms in Chapters 3 and 4, we easily realize that we may have a hard time obtaining excellent parallel efficiency if we don't optimize relevant parameters. In Chapter 5, the discussion focuses on the analysis and tuning of the performance of parallel codes on different platforms with special attention placed on parallel out-of-core solvers.

After the discussion on solving matrix equations by a direct LU factorization, the accuracy of the computed results will be described. For the solution of very large matrix equations, the conjugate gradient (CG) method, in addition to LU factorization, can guarantee the accuracy of results to a predefined degree. Therefore, in Chapter 6, the parallel CG method is discussed. In addition, one can use the parallel CG method directly as a solver, and also refine the solution after one solves the matrix equation by LU factorization.

As an alternative to ScaLAPACK, in Chapter 7, PLAPACK-based in-core and out-of-core parallel MoM solvers using the high-order basis functions are presented. Parallel codes are developed and their performances are discussed.

In Chapter 8, the applications of the code [*t*arget *ide*ntification *s*oftware (TIDES)] developed in this book, are demonstrated. The new results presented in this chapter convincingly verify TIDES as an excellent tool for solving challenging real-world electromagnetic problems. Several examples are also presented to explain the functional differences between using the subdomain piecewise basis functions and the higher-order basis functions defined over larger domains. These examples demonstrate that using higher-order basis functions is more efficient for analyzing electrically large problems. Comparison between numerical results and measurement data is also provided in this chapter to illustrate the accuracy of TIDES.

These eight chapters are followed by three appendices. To ease the reader referring to the various hardware used in this book, detailed information of the various computer platforms used for this research is provided in Appendix A.

To help the user perform parallel computation using MoM on different platforms, SCons, which is an open-source software construction tool, is introduced, and a demonstration on how to compile ScaLAPACK and PLAPACK in the Windows operating system is presented in Appendix B.

Appendix C presents an example for the implementation of parallel MoM. A demonstration code is provided to solve the 2D electromagnetic scattering problems. This demo code can be considered as a reference code so that the reader only needs to modify corresponding parts to get their parallel MoM code for some other particular applications.

It is hoped that this book will provide an insight into the various implementations of parallel MoM for solution of EM problems in the frequency domain.

Every attempt has been made to guarantee the accuracy of the materials presented in the book. We would, however, appreciate readers bringing to our attention any errors that may have appeared in this version. Errors and/or any comments may be e-mailed to any of the authors.

Acknowledgments

We would like to thank Oliver E. Allen, John S. Asvestas and Saad N. Tabet from Naval Air Systems Command, Patuxent River, Maryland for their continued support and interest of this topic. Grateful acknowledgement is also made to Gerard J. Genello, Michael C. Wicks and Jeffrey T. Carlo of Air Force Research Laboratory, Rome, New York, and Eric L. Mokole from Naval Research Laboratory for their initial interest in developing parallel solvers for solution of challenging electromagnetic field problems. We also would like to thank Mark Povinelli of Lockheed Martin Corporation, Don McPherson and Harvey Schuman of Syracuse Research Corporation for their continued interest in this work.

We gratefully acknowledge Professor Magdalena Salazar Palma (Universidad Carlos III de Madrid, Spain) and Ms. Yuan Yuan (Xidian University, Xi'an, China) for their continued support in this endeavor. We also gratefully acknowledge Professors Carlos Hartmann (Syracuse University, New York, USA), Changhong Liang (Xidian University, Xi'an, China), Luis Emilio Garcia Castillo (Universidad Carlos III de Madrid, Spain), and Dr. Mengtao Yuan (Syracuse University, New York, USA) and Dr. Sioweng Ting (University of Macau) for their continued support.

We would like to express sincere thanks to Joseph F. Skovira (Cornell University, Ithaca, NY, USA), John Porter (Hewlett-Packard, Cupertino, CA, USA), and Xunwang Zhao (Xidian University, Xi'an, China) for their help with benchmarks of our codes. We would also like to express sincere thanks to Damien Hocking for his help with the compilation of ScaLAPACK.

Thanks are also due to Ms. Robbin Mocarski, Ms. Brenda Flowers, Ms. Maureen Marano and Ms. Christine Sauve for their help on formatting the manuscript. We would also express our thanks to Dr. Jie Yang, Dr. Nuri Yilmazer, Santana Burintramart, La Toya Kendra Brown, Arijit De, Weixin Zhao, Zicong Mei and Jianwei Yuan for their help.

Yu Zhang (yzhang39@syr.edu)
Tapan K. Sarkar (tksarkar@syr.edu)
Syracuse, New York
March 2009

Acronyms

ACML	AMD Core Math Library
AMD	Advanced Micro Devices
BLACS	basic linear algebra communication subprograms
BLAS	basic linear algebra subprograms (PBLAS = parallel BLAS)
CEM	computational electromagnetics
CG	conjugate gradient
CMKL	Cluster Math Kernel Library (Intel)
DDR	double data rate
DOA	direction of arrival
EFIE	electric field integral equation
EMC	electromagnetic compatibility
EMI	electromagnetic interference
FDTD	finite-difference time-domain (method)
FEM	finite element method
FFT	fast Fourier transform
FMM	fast multipole method
FSB	frontside bus
HPC	high performance computing
IASIZE	in-core array size
IE	integral equation (SIE = surface integral equation)
IRA	impulse radiating antenna
LAPACK	linear algebra package (ScaLAPACK = scalable LAPACK; PLAPACK = parallel LAPACK)
MLFMA	multilevel fast multipole algorithm
MoM	method of moments
MOT	marching-on-in-time (method) (MOD = marching-on-in-degree)
MPI	message passing interface (MMPI = managed MPI)
MVP	matrix–vector product
NUN	number of unknowns
OCS	out-of-core solver
PBMD	physically based matrix distribution
PEC	perfect electric conductor
PLA	parallel linear algebra
PMCHW	Poggio–Miller–Chang–Harrington–Wu
PO	physical optics
PVM	parallel virtual machine
RAID	redundant array of inexpensive (*or* independent) disks
RCS	radar cross section
RHS	right-hand side (of equation)
RWG	Rao–Wilton–Glisson

SAS	serial attached SCSI
SATA	serial advanced technology attachment
SCSI	small computer system interface
SFS	scalable file share
TIDES	target identification software
TM	transverse magnetic
UTD	uniform-geometrical theory of diffraction

1

INTRODUCTION

1.0   SUMMARY

This chapter begins with a brief review of the several widely used numerical methods utilized in the field of parallel computational electromagnetics (CEM). This book is intended to provide a parallel frequency-domain solution to radiation and scattering for electromagnetic (EM) field problems. It is important to ensure that the parallel EM codes developed are compatible with all the typical computer platforms that are currently available. Therefore, computers with different architectures and operating systems have been used as benchmark platforms in this book. A brief description of the hardware used for this work is provided for easy reference. ScaLAPACK and PLAPACK, the two software library packages that have been employed for implementing parallel computations, are also introduced in this chapter.

1.1   A BRIEF REVIEW OF PARALLEL CEM

1.1.1   Computational Electromagnetics

Electromagnetic devices and systems, ranging from everyday office appliances to cellular phones, have become an integral part of modern life. The continued development of new technology greatly depends on the engineering analysis and synthesis of electromagnetic systems. These are based on obtaining accurate solutions of Maxwell's equations for the system of interest. The increase of research and development in a wide variety of applications, including antenna design, microwave circuits, photonics, ray tracing, wireless communication, electromagnetic compatibility/interference (EMC/EMI), and so on, have led to systems becoming more and more complex. In many cases, a single device has become a very complicated structure that includes a number of conductors, dielectric, and semiconductors of arbitrary shapes and of a complex physical nature. The expensive fabrication technologies preclude the possibility of modifying a device if its performance is not within the specifications of the

1

designer. Therefore, methods capable of extremely accurate characterization of systems are required to analyze the structures of interest. It is challenging to obtain such methods because of the complex nature of any of the modern electromagnetic devices. The analysis of these devices does not lead to closed-form expressions.

Analytical solutions in closed form are known for only a limited number of special cases, which are hardly ever directly applicable to real-world applications. The failure to derive closed form solutions from Maxwell's equations for real-world applications has led to intensive research on numerical techniques. The application of numerical methods to electromagnetic field problems is known as *computational electromagnetics* (CEM). It is a branch of electromagnetics that deals with computational methods and is a natural extension of the analytical approach to solving the Maxwell equations [1,2].

For a student of engineering electromagnetics or a researcher just starting to work in this area, few textbooks can furnish initial guidance to CEM. So, a short bibliography of the most widely used numerical methods is reviewed in this chapter. Among the methods to be described, the method of moments (MoM) is considered to be one of the most popular integral equation–based numerical methods. It coincides formally with the weighted-residual method because the sources (unknowns) are expanded by a sum of certain basis functions multiplied by unknown coefficients. The residual of the integral equation is weighted using a suitable inner product and a set of weighting functions. This results in a set of linear equations, which can be solved in the usual way. Regarding MoM, R. F. Harrington worked out a systematic, functional-space description of electromagnetic problems, starting from integral equations and using the reaction concept. Harrington summarized his work in a book published in 1968, which is still considered to be a classic textbook even today [3]. For analysis of radiation from conducting structures, it is difficult to surpass the accuracy and efficiency of an integral equation–based computational methodology. For composite structures containing both conductors and dielectrics, where the dielectric inhomogenity is not severe, the integral equation methodology is still an excellent numerical procedure. Besides MoM, a number of variations of the integral equation–based numerical solution procedure have been developed. For example, the *fast multipole method* (FMM) and *multilevel fast multipole algorithm* (MLFMA) have been proposed to accelerate the matrix–vector product that arises in the iterative solution of the MoM equations [4].

The *finite element method* (FEM) is a numerical method that is used to solve boundary-value problems characterized by a partial differential equation and a set of boundary conditions. The first book completely committed to finite elements in electromagnetics was by P. P. Silvester and R. L. Ferrari and was published in 1983 [5]. The introductory text by J. M. Jin in 1993 was particularly directed to scattering and antenna problems [6]. The book by M. Salazar-Palma et al. [7] introduced an iterative and self-adaptive finite-element technique for electromagnetic modeling in 1998, with a review of the history of numerical methods in electromagnetics and a categorization of these methods as presented in the first chapter [7]. A selected bibliography for FEM in microwaves was also provided by R. Coccioli et al. [8].

The *finite-difference time-domain* (FDTD) method is a finite-difference solution for electromagnetic problems. The method was developed by K. S. Yee in 1966 [9]. With the advent of low-cost, powerful computers and improvement of the method itself, FDTD has become one of the most widely used methods for solving Maxwell's equations for scientific and engineering applications. K. L. Shlager and J. B. Schneider published a selective survey of the FDTD literature in 1995 [10]. In the same year, A. Taflove authored a comprehensive textbook that he and S. C. Hagness subsequently expanded and updated to a third edition in 2005 [11].

MoM, FEM, and FDTD, the methods mentioned so far, are the most frequently used algorithms in CEM. Many more books are also available for these and other numerical methods, which will not be discussed further.

As S. M. Rao states: "CEM has evolved rapidly during the past decade to a point where extremely accurate predictions can be made for very general scattering and antenna structures" [12]. Numerical simulation technology is the most cost-effective means of meeting many technical challenges in the areas of electromagnetic signature processing, antenna design, electromagnetic coupling, microwave device design and assembly, and so on. With the rapid increase in the performance of desktop computers, it has become feasible for designers to apply computational electromagnetics to determine the electromagnetic behavior of many systems. However, when performing an electromagnetic simulation associated with a large-scale configuration, it is difficult to achieve a high level of accuracy when using conventional computing platforms. The numerical capability of CEM can be enhanced substantially by using scalable parallel computer systems, which will be discussed next.

1.1.2 Parallel Computation in Electromagnetics

Large-scale CEM simulations have encountered computational limitations in terms of physical memory space and central processing unit (CPU) time of the computer. Parallel computing on clusters and/or computers with multicore processors continues to be the method of choice for addressing modern engineering and scientific challenges that arise from the extremely complicated real-life applications.

Parallel computing is a form of computational methodology in which many instructions are carried out simultaneously [13]. Parallel computing operates on the principle that large problems can almost always be divided into smaller ones, which may be carried out concurrently ("in parallel"). In the simplest sense, parallel computing is the simultaneous use of multiple computational resources to solve a numerical problem. In parallel computing, a problem is broken up into discrete parts that can be solved concurrently, and each part is further broken down into a series of instructions, and the instructions from each part are executed simultaneously on different CPUs (or cores). A very good tutorial on the topic of parallel computing is provided online [14].

The primary reasons for using parallel computing are that it saves time (wall clock time), solves larger problems, and provides concurrency (multiple tasks at the same time). A Lawrence Livermore Laboratory report [14] also points out additional reasons such as taking advantage of nonlocal computer resources available on a wide area network, or even the Internet when local resources are scarce. Also, cost saving can be a consideration for using multiple "cheap" computing resources instead of buying computational time on a supercomputer. Another reason for using parallel computing is to overcome storage constraints. For large-scale problems, using the memory and hard disk of multiple computers may overcome the limited resources of a single computer.

Parallel computing has been used for many years, mainly in high performance computing (HPC), but interest has been growing more recently mainly because of physical constraints preventing high-frequency scaling. Parallelism has become the dominant paradigm in computer architecture, mainly in the form of multicore processors. Since 2002 or so, the trends indicated by ever faster networks, distributed systems, multicore CPUs, and multi-processor computer architectures clearly show that parallelism is the future of computing [14].

The last two decades (1990s–2000s) witnessed significant developments in both computer hardware capabilities and implementations of fast and efficient algorithms. The performance of parallel computations on any computer system is closely tied to communication and latency. These factors, particular to each individual system, are introduced by the communication protocol and operating system implementation, both of which have profound influence on the performance of a parallel code. Therefore, a judicious tradeoff between a balanced workload and interprocessor communication is needed to efficiently use distributed-memory, multinode computers. Such a need is intrinsically related to the numerical algorithms and hardware architectures. A synergism of the relatively new numerical procedures and high-performance parallel computing capability opens up a new frontier in electromagnetics research [15].

Table 1.1 lists some selected publications on parallel CEM code development since 1998, from which one can see that parallel computing techniques have penetrated into mainstream numerical methods. Generally speaking, CEM codes are currently run on all typical computer platforms. Note that the works cited in Table 1.1 do not indicate the evolutionary process in parallel CEM, but only show that research has touched on different CEM methods to date.

The subject of reviewing parallel CEM is too broad to be covered extensively in several pages, and therefore only a synoptic view is given here. Even though the topic of this book is parallel frequency-domain MoM, parallel implementations of the other most widely used frequency-domain and time-domain numerical methods, like FEM and FDTD, are also briefly introduced here to provide an overview on the subject.

TABLE 1.1. Several Publications on Parallel CEM

Method	Research Topic	Year
Frequency Domain		
MoM	A parallel implementation of NEC for the analysis of large structures [16]	2003
	Highly efficient parallel schemes using out-of-core solver for MoM [17]	2007
FMM	10 million unknowns: Is it that big? [18]	2003
	Massively parallel fast multipole method solutions of large electromagnetic scattering problems [19]	2007
FEM	Open-region, electromagnetic finite-element scattering calculations in anisotropic media on parallel computers [20]	1999
Time Domain		
TD-FEM	Implementing a finite-element time-domain program in parallel [21]	2000
FDTD	A parallel FDTD algorithm using the MPI library [22]	2001
	Study on the optimum virtual topology for MPI-based parallel conformal FDTD algorithm on PC clusters [23]	2005
	A robust parallel conformal finite-difference time-domain processing package using the MPI [24]	2005
TDIE	A parallel marching-on-in-time solver [25]	2008

MoM is the most well known technique among integral equation (IE)-based methods. As the most popular frequency-domain integral equation method, MoM is an extremely powerful and versatile numerical methodology for discretizing an integral equation to a matrix equation. Because of the many advantages of MoM, the parallel implementation of the method itself, the hybrid methods and fast algorithms related to MoM, have all been significant research topics during the more recent decades and will continue to be popular for years to come.

The parallel MoM code was developed at the beginning of the 1990s [26–28]. J. E. Patterson programmed and executed the numerical electromagnetics code (NEC) [29] in a parallel processing environment, which was developed at Lawrence Livermore National Laboratory, in 1990. T. Cwik, J. Partee, and J. E. Patterson used MoM to solve scattering problems in parallel in 1991. Later, working with Robert A. van de Geijn, the author of PLAPACK [30], T. Cwik, developed a parallel MoM code using the Rao–Wilton–Glisson (RWG) basis function in 1994 [31,32]. In 1998, parallel MoM employing the RWG basis functions, using ScaLAPACK library package [33] as the solver, was implemented for the Cray T3E system [34].

Since the mid-1990s, research has been carried out on parallel implementations of standard production-level MoM codes. One of the frequently studied parallel implementations of existing codes is the widely used NEC. Successfully implemented in 2003, parallel NEC is portable to any platform that supports message-passing parallel environments such as the *message passing interface* (MPI) [35] and the *parallel virtual machine* (PVM) [36]. The code could even be executed on heterogeneous clusters of computers programmed with different operating systems [16].

The bottleneck of this traditional parallel MoM partly comes from the memory storage requirements. One remedy to overcome this disadvantage is to parallelize the hybrid of MoM and such high-frequency methods as *uniform geometrical theory of diffraction* (UTD) [37], *physical optics* (PO) [38], and so on. For example, the hybrid MoM-UTD method, which greatly extends the capability of MoM, is implemented in SuperNEC [39], and is then parallelized. The parallel iterative MoM and PO hybrid solver for arbitrary surfaces using the RWG basis functions also has been implemented [40,41].

Fast algorithms can also be employed to reduce the overall memory and time requirements. Some of the frequently used fast algorithms are the conjugate gradient (CG) fast Fourier transform (CG-FFT) [42], adaptive integral method (AIM) [43], fast multipole method (FMM) [44], and precorrected FFT [45]. The FMM, arguably the most popular of these methods, was even further improved via the multilevel fast multipole algorithm (MLFMA) to further improve the time taken for calculation of a matrix–vector product [46,47]. While difficult to implement, the MLFMA has become the algorithm of choice when solving large-scale scattering problems arising in electromagnetics. Therefore, the parallelization of MLFMA on distributed memory computing clusters [48–56] and shared memory multiprocessors [57] has always been a topic of great interest.

While some success has been demonstrated in parallelizing MLFMA in distributed memory environments [49,50,55,56], it requires sophisticated load distribution strategies involving shared and distributed partitions of the MLFMA tree. Thus, parallel scaling of the MLFMA algorithm has been limited to a handful of processors as addressed in [19]. An FFT extension of the conventional FMM, known as FMM-FFT, lowers the matrix–vector multiplication time requirement of the conventional algorithm, while preserving the propensity for parallel scaling of the single-level FMM (since it does not employ the tree structure of MLFMA). The research published in 2007 has demonstrated that a parallel FMM-FFT algorithm is quite attractive (when compared to the MLFMA) in the context of massively parallel distributed-memory machines [19].

These methods have been implemented mainly in commercially available software [58]. However, the use of hybrid methods and fast algorithms sacrifice accuracy to accommodate the solution of large problems. Also, the disadvantage of the MoM with subdomain basis functions, which is implemented in NEC or in commercial software, is that it generates much more unknowns than the MoM formulation using the entire-domain or larger subdomain basis functions, thus leading to large memory requirements. Although 64-bit computers these days theoretically have "huge" memory capacity, the development and applications of in-core solvers are limited by the amount of RAM available. Using virtual memory of the computer is not a good alternative because the significant degradation in performance results in increased time for generating a solution.

On the other hand, as an example, a 10-node Dell 1855 cluster equipped with 20 CPUs and 80 gigabytes (GB) of memory could complete a problem that occupies all the available memory using an in-core solver for a dense linear matrix equation in less than 3 hours. This suggests that the processing power of high-performance machines is underutilized and much larger problems can be tackled before runtimes become prohibitively long.

For this reason, one optional method to overcome the memory storage bottleneck of MoM while maintaining its accuracy is using an out-of-core solver, rather than using hybrid methods or fast algorithms, which may incur accuracy problems in some cases, e.g., coupling analysis during antenna design. Since the matrix generated by MoM is a full dense matrix, the LU decomposition method used to solve the matrix equation is computationally intensive when compared with the read/write process of the matrix elements from/into the hard disk [59]. Therefore, it is natural to introduce an out-of-core solver to tackle large, dense, linear systems generated using the MoM formulation. Furthermore, to solve the large problem out-of-core as quickly as possible, the code can be designed in a parallel way to run on a cluster.

Another way to overcome the bottleneck in storing the matrix elements in the memory is to apply the higher-order basis functions defined over a large domain to MoM [60–62] as this requires fewer numbers of unknowns (by a factor of ≥ 10 involving canonical planar structures). In two publications in 2007, the authors presented a parallel in-core implementation of MoM using higher-order basis functions [63,64], and the details of this methodology will be further discussed in this book along with the parallel out-of-core implementation [65]. The most important difference between the parallel in-core solver and the parallel out-of-core solver is that the latter has the ability to break the random access memory (RAM) restriction and thus can go far beyond the limit of the physical memory in any computer or cluster to solve a large MoM matrix equation within a reasonable amount of time.

Besides MoM, another well-established frequency-domain technique is the finite element method (FEM), which is an effective means for analyzing electromagnetic problems. The principal attribute of FEM is that it efficiently models highly irregular geometries as well as penetrable and inhomogeneous material media. Publications on parallel implementations of FEM on supercomputers, workstations, and PCs during the period of 1985–1995 were summarized in a paper published in 1996 [66]. In a massively parallel environment, traditional sequential algorithms will not necessarily scale and may lead to very poor utilization of the architecture associated with the multiprocessor. A domain decomposition method based on the finite-element tearing and interconnecting (FETI) algorithm was considered to be more scalable and efficient on parallel platforms than traditional iterative methods such as a preconditioned conjugate gradient algorithm when solving large matrices [67].

Research on parallel implementation of time-domain numerical methods has also been of interest for many years, accompanied by the development in computer technology. Two popular time-domain methods in the analysis of electromagnetic phenomena are the FDTD and FETD (finite-element time-domain) schemes. FDTD is, by nature, essentially data-parallel. A. Taflove et al. started the effort in 1988 on a parallel implementation of FDTD on CM-1 and CM-2 supercomputers [68]. Parallelization of the FDTD method using distributed computing was done in 1994 by V. Varadarajan and R. Mittra [69], who suggested rules and tolerances for implementation on a cluster of computers. They used the parallel virtual machine (PVM) message-passing protocol over TCP/IP on a cluster of eight HP computers. Since then, many problems have been solved using FDTD on various hardware platforms, processor configurations, and software. Z. M. Liu et al. implemented parallel FDTD on a CM-5 parallel computer in 1995 [70]. A. D. Tinniswood et al. ran parallel FDTD on 128-node IBM SP-2 cluster later in 1996 [71]. Note that the reported maximum speedup factor from different publications varies significantly and is not even a relative measure of the performance of the underlying algorithm. The works described so far have been restricted by the available technology, software systems, or factors under examination, as well as differences in the size of the domain of the selected problem [72].

With the development of MPI programming technology, a parallel FDTD implementation using the MPI Cartesian 2D topology to simplify and accelerate the algorithm was presented in 2001 and it allowed noncontiguous locations in memory to be associated with the data type [22]. Furthermore, suggestions on improved domain partitioning conditions based on the estimated processing time for different types of media [dispersive, perfectly matched layer (PML), and other boundary conditions] were given, and good performance results have been obtained. These aspects appear to be evolutionary for the FDTD computation for the following reasons. MPI is a widely embraced standard; the operations occur only at initialization; the methods are scalable and extensible; and there is no impact due to runtime load balancing to yield an improvement in overall performance. In the same year, an MPI Cartesian 3D topology was used for parallel FDTD, and different MPI communication patterns were investigated [73]. Later, the optimum MPI topology for 3D parallel FDTD was studied on a PC cluster [23,74,75]. The influence of different virtual topology schemes on the performance of a parallel FDTD code was discussed in detail, and the general rules were presented on how to obtain the highest efficiency of the parallel FDTD algorithm by optimizing the MPI virtual topology. In 2005, an MPI-based parallel conformal FDTD package was developed with a friendly graphical user interface (GUI) [24]. Parallel implementation of FDTD using MPI is continuing to find even wider use today.

The parallel *finite-element time-domain* (FETD) package using the MPI message-passing standard was developed in 2000 [76]. Different mathematical backgrounds of FETD and FDTD methods have led to different properties. The numerical performance of the distributed implementations of the FDTD and FETD methods was compared in [77] with respect to scalability, load balancing, and speedup.

Another promising time-domain method used to analyze the scattering and radiation from arbitrary structures is the time-domain integral equation (TDIE) based marching-on-in-time (MOT) method or marching-on-in-degree (MOD) method [78,79]. Note that the MOT method may sometimes suffer from its late-time instability. This can be overcome by using the associated Laguerre polynomials for the temporal variation. As such, the time derivatives can be handled analytically and stable numerical results can be obtained even for late times in the MOD method. A parallel time-domain simulator for analyzing the electromagnetic scattering and radiation phenomena has been developed by the authors of this book and partly presented in a previous reference [25].

To summarize, parallel CEM has been a fervently pursued research topic, and what was reviewed in the preceding paragraphs provide just a glimpse of its development. There is much room left for future study and improvement, specifically in the parallel out-of-core implementation of MoM.

In the next section, the computer hardware and software that supports the research on parallel CEM presented in this book are introduced.

1.2   COMPUTER PLATFORMS ACCESSED IN THIS BOOK

Current computers generally have a single processor and use the IA-32 (Intel architecture, 32-bit) [80] instruction set, which is one of the instruction sets of Intel's most successful microprocessor generically termed x86-32. IA-32 is the 32-bit extension of the original Intel x86 processor architecture that defines the instruction set installed in most personal computers (PC) in the world. The limitation of the 32-bit architecture is that it can address only 2^{32} bits at most. This restricts the size of the variable to be stored in the memory, which cannot exceed 4 GB. A typical 2 GB of RAM can store a matrix size of approximately $15,000 \times 15,000$ when the variable is defined in single-precision (complex) arithmetic or a matrix size of approximately $11,000 \times 11,000$ when using double-precision (complex) arithmetic. In other words, one cannot deal with an integral equation solver in a MoM context if the number of unknowns exceeds 15,000 for single precision and 11,000 for double precision using LAPACK [81] for serial codes or ScaLAPACK [82]/PLAPACK [83] for parallel codes.

The IA-32 structure was extended by Advanced Micro Devices (AMD) Corporation in 2003 to 64 bits. The first family of processors that AMD built was the AMD K8 processors, which AMD subsequently named AMD64. This was the first time any company other than Intel Corporation had any significant additions to the 32-bit architecture. Intel Corporation was forced to do something, and they came up with the IA-32e, or the NetBurst family of processors. Later, Intel called them EM64T (extended-memory 64-bit technology). These 64-bit AMD and Intel processors are backward-compatible with the 32-bit code without any loss in the performance.

In addition, there is the IA-64 (Intel architecture, 64-bit), which is a true 64-bit processor architecture developed cooperatively by Intel and Hewlett-Packard and was implemented in the Itanium and the Itanium 2 processors. A 64-bit computer can theoretically address up to $2^{64} = 16.8$ million terabytes (TB) directly, which is sufficient to store a $10^9 \times 10^9$ matrix in single precision or a $0.7 \times 10^9 \times 0.7 \times 10^9$ matrix in double precision, provided enough physical memory and virtual memory are available.

While manufacturing technology continues to improve, breaking the physical limits of semiconductor-based microelectronics has become a major design concern. Some effects of these physical limitations can cause significant heat dissipation and data synchronization problems. The demand for more capable microprocessors causes CPU designers to try various methods of increasing performance. Some instruction-level parallelism (ILP) methods, like superscalar pipelining, are suitable for many applications, but are inefficient for others that tend to contain a difficult-to-predict code. Many applications are better suited for thread-level parallelism (TLP) methods. Utilizing multiple independent CPUs is one common method used to increase a system's overall TLP. A combination of increased available space due to refined manufacturing processes and the demand for increased TLP is the logic behind the creation of multicore CPUs [84].

A multicore CPU [or chip-level multiprocessor (CMP)] combines two or more independent cores into a single package composed of a single integrated circuit (IC), called a *die*, or more dies packaged together. A dual-core processor contains two cores, and a quad-core processor contains four cores. A multicore processor implements multiprocessing in a single physical package. Cores in a multicore device may share a single coherent cache at the highest on-device cache level (e.g., L2 for the Intel core 2) or have separate caches (e.g., current AMD dual-core processors). The processors also share the same interconnect to the rest of the system, like the L2 cache and the interface to the frontside bus (FSB). Each core independently implements optimizations such as superscalar execution, pipelining, and multithreading. Software benefits from multicore architectures where a code can be executed in parallel. Multicore processors can deliver significant benefits in performance for multithreaded software by adding processing power with minimal latency, given the proximity of the processors.

The general trend in processor development has been from multicore to many-core. AMD was the first x86 processor manufacturer to demonstrate a fully functioning dual-core processor on a shipping platform [85]. Intel built a prototype of a processor with 80 cores that delivered TFLOPS (tera-floating-point operations per second) performance in 2006.

The compute resources for parallel computing may include a single computer with multiple processors or one processor with multiple cores, an arbitrary number of computers connected by a network, or a combination of both.

Research efforts in this book involve all three types of computing resources and have employed 18 different computer platforms to observe how the performance of the in-core and out-of-core parallel solvers scales in various hardware and software environments.

Table 1.2 provides a summary of the various computing platforms that the authors have had access to during this research on frequency-domain parallel IE solvers. Here, the term *node* is used to characterize a piece of hardware with a network address. A node may contain multiple cores or CPUs for processing.

The platforms listed in Table 1.2 cover a range of computer architectures, including single-core, dual-core, and quad-core CPUs from both of the primary CPU manufacturers, Intel and AMD. The two common operating systems (OS) for computational platforms, Linux and Windows, and the three system manufacturers, Dell, IBM, and HP, are all represented in this study. Detailed information on the various platforms can be found in Appendix A.

TABLE 1.2. A Summary of the Computer Platforms Used for This Research

Processor Architecture	Platform (Manufacturer)	CPU	Total Nodes	Total Cores	Operating System
IA-32	CEM-1	Intel single-core	1	8	Windows
	CEM-5	Intel single-core	6	6	Linux
EM64T	CEM-2 (Dell)	Intel dual-core	1	4	Windows
	CEM-4 (Dell)	Intel single-core	10	20	Linux
	CEM-7 (Dell)	Intel quad-core	2	16	Linux
	CEM-8 (Dell)	Intel quad-core	1	8	Windows
	CEM-9 (IBM)	Intel quad-core	2	16	Linux
	CEM-10	Intel dual-core	1	2	Windows
	Cluster-1 (HP)	Intel dual-core	40	160	Linux
	Cluster-2 (HP)	Intel quad-core	65	520	Linux
	Cluster-3 (HP)	Intel quad-core	20	160	Linux
	Cluster-4 (HP)	Intel quad-core	20	160	Linux
	Cluster-5 (HP)	Intel dual-core	40	160	Linux
	Cluster-6 (HP)	Intel dual-core	60	240	Linux
	Cluster-7 (IBM)	Intel quad-core	14	112	Linux
	Cluster-8 (IBM)	Intel quad-core	8	64	Linux
AMD64	CEM-6 (IBM)	AMD dual-core	2	16	Linux
IA-64	CEM-3 (HP)	Intel single-core	1	4	Windows

The portability of the codes developed in this book has been verified by executing them on the various platforms listed in Table 1.2. Portability is a characteristic of a well-constructed program (code) that is written in such a way that the same code can be recompiled and run successfully in different environments, i.e., different operating systems, different processor architectures, different versions of libraries, and the like. Portability is important to long-term maintenance of the code by ensuring its robustness to different computational platforms.

1.3   PARALLEL LIBRARIES EMPLOYED FOR THE COMPUTATIONS

Choice of the appropriate software (operating system and the relevant scientific subroutine package libraries) is very important to achieve an optimum efficiency for parallel computation. There are several projects for the parallelization of numerical software. An overview of the public-domain libraries for high performance computing can be found at the HPCNetlib homepage at http://www.nhse.org/hpc-netlib. These libraries are generally very specialized and optimized for the particular hardware platform on which the problem is to be executed.

Basically, the libraries in use for parallel computation consist of mathematical subroutines and programs implementing the MPI protocols. Understanding the importance of matching the software with the hardware has led to various computer manufacturers developing their own libraries.

For example, as for math libraries, the AMD Core Math Library (ACML) released by AMD is specifically designed to support multithreading and other key features of AMD's next-generation processors. ACML currently supports OpenMP, while future releases will expand on its support of multiplatform, shared-memory multiprocessing [86]. Intel has come up with the Intel Math Kernel Library (Intel MKL), a math library highly optimized for Intel Itanium, Intel Xeon, Intel Pentium 4, and Intel Core 2 Duo processor-based systems. Intel MKL performance is competitive with that of other math software packages on non-Intel processors [87].

As for the MPI library, the HP-MPI has been developed by the Hewlett-Packard (HP) company for Linux, HP-UX, Tru64 UNIX, Microsoft Windows Compute Cluster Server 2003, and Windows XP Professional systems. It is a high-performance and production quality implementation of the message passing interface (MPI) standard for HP servers and workstations [88]. Intel MPI library from the Intel Company implements the high-performance MPI-2 specification on multiple fabrics. The Intel MPI library enables users to quickly deliver end-user maximum performance even when they change or upgrade to new interconnects, without requiring major changes to the software or to the operating environment. Intel also provides a free runtime environment kit for products developed with the Intel MPI library [89].

For the computations used in this book, Intel MKL is employed on the computational platforms with Intel CPUs, and AMD ACML is used on platforms with AMD CPUs. These math libraries along with ScaLAPACK and PLAPACK, two of the most general parallel library packages based on message passing with MPI, can be used to obtain a satisfactory parallel computational efficiency. While HP-MPI is used for compiling the CEM source codes and launching the parallel jobs on HP clusters with the Linux operating system, Intel MPI is used on other computer platforms also using the Linux operating system. For computer platforms that use the Windows operating system, different MPICH2 packages are downloaded from the Argonne National Laboratory webpage (http://www.mcs.anl.gov/research/projects/mpich2/index.php) and installed according to the system architecture.

In the following paragraphs, ScaLAPACK and PLAPACK will be introduced in some detail.

1.3.1   ScaLAPACK — Scalable Linear Algebra PACKage

ScaLAPACK is the largest and most flexible public-domain library with basic numerical operations for distributed-memory parallel systems to date. It can solve problems associated with systems of linear equations, linear least-squares problems, eigenvalue problems, and singular value decomposition. ScaLAPACK can also handle many associated computations such as matrix factorizations and estimation of the condition number of a matrix.

ScaLAPACK is a parallel version of LAPACK in both function and software design. Like LAPACK, the ScaLAPACK routines are based on block-partitioned algorithms in order to minimize the frequency of data movement between different levels of the memory hierarchy. The fundamental building blocks of the ScaLAPACK library are distributed-memory versions of the level 1, level 2, and level 3 BLAS (*b*asic *l*inear *a*lgebra *s*ubprograms [90]), called the *p*arallel BLAS (PBLAS) [91], and a set of *b*asic *l*inear *a*lgebra *c*ommunication *s*ubprograms (BLACS) [92] for communication tasks that arise frequently in parallel linear algebra computations. In the ScaLAPACK routines, the majority of interprocessor communication occurs within the PBLAS, so the source code of the top software layer of ScaLAPACK resembles that of LAPACK.

Figure 1.1 describes the ScaLAPACK software hierarchy [93]. The components below the dashed line, labeled "Local", are called on a single process, with arguments stored on a single process only. The components above the dashed line, labeled "Global", are synchronous parallel routines, whose arguments include matrices and vectors distributed across multiple processes. The components below the solid line are machine-specific, while those above the solid line are machine-independent. Each component in Figure 1.1 is described in the following with the various acronyms defined.

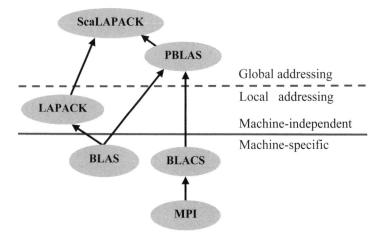

Figure 1.1. ScaLAPACK software hierarchy.

MPI. The message passing interface (MPI) standard is a library specification for message-passing, proposed as a standard by a broadly based committee of vendors, implementers, and users. It is a language-independent communications protocol used to program parallel computers. The MPI interface is meant to provide essential virtual topology, synchronization, and communication functionality between a set of processes (that have been mapped to nodes/servers/computer instances) in a language-independent way, with language-specific syntax (bindings), plus a few features that are language-specific. MPI has such functions included, but are not limited to, point-to-point rendezvous-type send/receive operations, choosing between a Cartesian or graph-like logical process topology, exchanging data between process pairs (send/receive operations), combining partial results of computations (gathering and reduction operations), synchronizing nodes (barrier operations), as well as obtaining network-related information (such as the number of processes), neighboring processes accessible in a logical topology, and so on. MPI programs always work with processes, although usually people talk about processors. When one tries to achieve maximum performance, one process per processor/core is selected as part of the mapping activity. This mapping activity occurs at runtime, through the agent that starts the MPI program, normally called mpirun or mpiexec [94,95].

BLAS. BLAS (*b*asic *l*inear *a*lgebra *s*ubprograms [90]) include subroutines for common linear algebra computations such as dot product, matrix–vector multiplication, and matrix–matrix multiplication. An important aim of the BLAS is to provide a portability layer for computation. As is well known, using matrix–matrix operations (in particular, matrix multiplication) tuned for a particular architecture can mask the effects of the memory hierarchy (cache misses, *t*ranslation *l*ook-aside *b*uffer (TLB) misses, etc.) and permit floating-point operations to be performed near peak speed of the machine.

Optimized versions of the BLAS can be found at http://www.tacc.utexas.edu/resources/software/#blas.

BLACS – BLACS (*b*asic *l*inear *a*lgebra *c*ommunication *s*ubprograms [92]) is a message-passing library designed for linear algebra. An important aim of the BLACS is to provide a portable, linear algebra-specific layer for communication. The computational model consists of a one- or two-dimensional *process grid*, where each process stores pieces of the matrices and vectors. The BLACS include synchronous send/receive routines to communicate a matrix or submatrix from one process to another, to broadcast submatrices to many processes, or to compute global reductions (sums, maxima, and minima). There are also routines to construct, change, or query the process grid. Since several ScaLAPACK algorithms require broadcasts or reductions among different subsets of processes, the BLACS permit a process to be a member of several overlapping or disjointed process grids, each one labeled by a context. Some message-passing systems, such as MPI, also include this context concept; MPI calls this a "communicator". The BLACS provide facilities for safe interoperation of system contexts and BLACS contexts.

LAPACK. LAPACK, or *l*inear *a*lgebra *pack*age [96], is a collection of routines for solving linear systems, least-squares problems, eigenproblems, and singular problems. High performance is attained by using algorithms that do most of their work in calls to the BLAS, with an emphasis on matrix–matrix multiplication. Each routine has one or more *performance tuning parameter*, such as the sizes of the blocks operated on by the BLAS. These parameters are machine-dependent and are obtained from a table defined when the package is installed and referenced at runtime.

The LAPACK routines are written as a single thread of execution. LAPACK can accommodate shared-memory machines, provided parallel BLAS are available (in other words, the only parallelism is implicit in calls to BLAS). More detailed information about LAPACK can be found at http://www.netlib.org/lapack/.

PBLAS. To simplify the design of ScaLAPACK, and because BLAS have proved to be useful tools outside LAPACK, the authors of ScaLAPACK chose to build a parallel set of BLAS, called PBLAS, which perform message passing and whose interface is as similar to the BLAS as possible. This decision has permitted the ScaLAPACK code to be quite similar, and sometimes nearly identical, to the analogous LAPACK code. Further details of PBLAS can be found in reference [91].

ScaLAPACK also contains additional libraries to treat distributed matrices and vectors. One is the tools library, which offers useful routines, for example, to find out which part of the global matrix a local process has in its memory or to identify the global index of a matrix element corresponding to its local index and vice versa.

The research done for this book has found some drawbacks associated with the commercial math library packages. For example, each process cannot address more than 2 GB RAM for LU decomposition when using the earlier version of

Intel Cluster Math Kernel Library (CMKL). Parallel computation on the Itanium 2 platform, using the Windows operating system, is not supported by Intel or HP [97]. Fortunately, these problems have so far been solved during the research work for this book by the authors and/or the vendors.

1.3.2   PLAPACK — Parallel Linear Algebra PACKage

The *p*arallel *l*inear *a*lgebra *pack*age (PLAPACK) is a prototype of a more flexible alternative to ScaLAPACK. Containing a number of parallel linear algebra solvers, PLAPACK is an MPI-based parallel linear algebra package designed to provide a user-friendly infrastructure for building parallel dense linear algebra libraries.

Figure 1.2 gives the layering of the PLAPACK infrastructure with the part concerning *m*anaged *m*essage *p*assing *i*nterface (MMPI) omitted from reference [83] since it is not used by the library employed in the solution of the problems related to this book. The components of PLAPACK shown in the bottom row (below the solid line) in Figure 1.2 are machine-dependent. The components that are above the solid line are machine-independent.

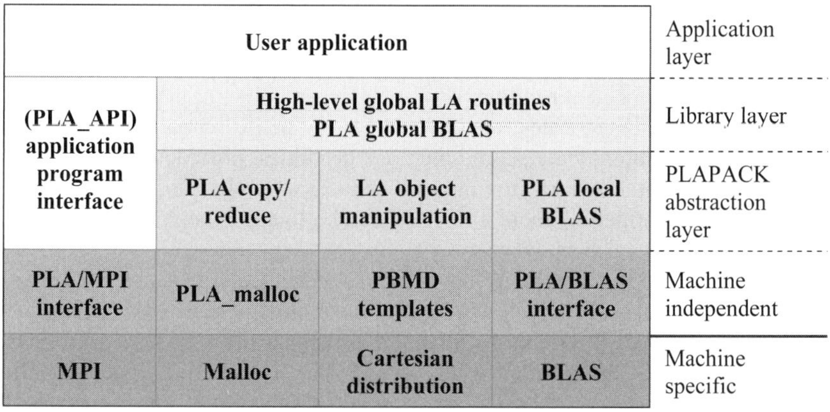

(PLA_API) application program interface	User application			Application layer
	High-level global LA routines PLA global BLAS			Library layer
	PLA copy/ reduce	LA object manipulation	PLA local BLAS	PLAPACK abstraction layer
PLA/MPI interface	PLA_malloc	PBMD templates	PLA/BLAS interface	Machine independent
MPI	Malloc	Cartesian distribution	BLAS	Machine specific

Figure 1.2. Layering of the PLAPACK infrastructure.

To ensure completeness of the book, a very brief introduction is given for each layer of the PLAPACK infrastructure, based on the literature [98] to which the reader may refer to for a detailed description.

Machine/distribution-dependent layer. To ensure portability, PLAPACK uses standardized components, namely, the MPI for communication, the standard memory management routines provided by the C programming language (*malloc*/*calloc*), and the BLAS that is generally optimized by the vendors. The author of PLAPACK also adds Cartesian matrix distributions to this layer, since they provide the most general distribution of matrices commonly used for implementation of parallel dense linear algebra algorithms.

Machine/distribution-independent layer. To achieve machine independence, PLAPACK offers a number of interfaces to the machine-dependent layer. Each interface is briefly described below.

PLAPACK-MPI interface. PLAPACK relies heavily on collective communications like scatter (MPI_scatter), gather (MPI_gather), collect (MPI_allgather), broadcast (MPI_bcast), and others. PLAPACK's developers created an intermediate layer that can be used to pick a particular implementation of such an operation.

PLAPACK-memory management interface. PLAPACK uses dynamic memory allocation to create space for storing data. By creating this extra layer, it provides a means for customizing memory management, including the possibility of allowing the user to provide all space used by PLAPACK.

Physically based matrix distribution (PBMD) and templates. The approach taken to describe Cartesian matrix distributions is fundamental to PLAPACK. In particular, PLAPACK recognizes the important role that vectors play in applications and thus all distribution of data starts with the distribution of the vectors [30]. The details of the distribution are hidden by describing the generic distribution of imaginary vectors and matrices (the template) and indicating how actual matrices and vectors are aligned to the template.

Physically based matrix distribution was proposed by the authors of PLAPACK as a basis for a set of more flexible parallel linear algebra libraries. It was claimed that developers of applications will not have to "unnaturally" modify the way they want to distribute the data in order to fit the distribution required by the format of the respective library. They have the freedom to distribute the vectors (which contain the data of "physical significance") across processors in a way that depends only on the application. The matrix (the operator) is then distributed in a conforming way that, in effect, optimizes matrix−vector products (MVPs) involving the vectors and the matrix.

PLAPACK-BLAS interface. Computation is generally performed locally on each processing node by the BLAS, which have a FORTRAN-compatible calling sequence. Since the interface between C and FORTRAN is not standardized, a PLAPACK-BLAS interface is required to hide these platform specific differences. To make the development of programs easier and to achieve better performance, PLAPACK includes the parallel version of the BLAS routines.

PLAPACK abstraction layer. This layer of PLAPACK provides the abstraction that frees users from details like indexing, communication, and local computation.

Linear algebra objects and their manipulation. All information that describes a linear algebra object, like a vector or a matrix, is encapsulated in an opaque object. This component of PLAPACK allows users to create, initialize, and destroy such objects. In addition, it provides an abstraction that allows users to transparently index into a submatrix or vector. Finally, it provides a mechanism for describing the duplication of the data.

Copy/reduce: duplication and consolidation of data. Communication in PLAPACK is not specified by explicit communication operations. Instead, linear algebra objects are used to describe how data are to be distributed or duplicated. Communication is achieved by copying or reducing from one (duplicated) object to another. This raises the level of abstraction at which communication can be specified.

PLAPACK local BLAS. Since all information about matrices and vectors is hidden in the linear algebra objects, a call to a BLAS routine on a given processor requires extraction of that information. Rather than exposing this, PLAPACK provides routines (the PLAPACK local BLAS) that extract the information and subsequently call the correct sequential BLAS routine on each processor.

Library layer. The primary intent for the PLAPACK infrastructure is to provide the building blocks for creating higher-level libraries. Thus, the library layer for PLAPACK consists of global (parallel) basic linear algebra subprograms and higher-level routines for solving linear systems and algebraic eigenvalue problems.

PLAPACK global BLAS. The primary building blocks provided by PLAPACK are the global (parallel) versions of the BLAS. These allow dense linear algebra algorithms to be implemented quickly without exposing parallelism in any form.

PLAPACK higher-level linear algebra routines. Higher-level algorithms can be easily implemented using the global BLAS. However, to ensure better performance, it is often desirable to implement these higher-level algorithms directly using the abstraction level of PLAPACK. Development of such implementations can often be attained by incrementally replacing calls to global BLAS with calls that explicitly expose parallelism by using objects that are duplicated in nature.

PLAPACK application interface. A highlight of the PLAPACK infrastructure is the inclusion of a set of routines that allow an application to build matrices and vectors in an application friendly manner.

PLAPACK does not offer as many blackbox solvers as ScaLAPACK, but is designed as a parallel infrastructure to develop routines for solving linear algebra problems. With PLAPACK routines, the user can create a global matrix, vectors, and multiscalars, multivectors, and may fill them with values with the help of an API (application programming interface).

Since PLAPACK is not as commercialized as ScaLAPACK, joint efforts have been made by the authors of this book and Robert A. van de Geijn, the author of PLAPACK, to make the library compatible to EM64T systems

[99,100]. The method for generating a Windows version of the PLAPACK library is also provided in Appendix B.

1.4 CONCLUSION

An introduction to parallel CEM is given with a special attention to a parallel MoM methodology to solve EM problems in frequency domain. As will be discussed in the later sections, this methodology can be a powerful tool to solve challenging computational radiation and scattering problems by using parallel out-of-core techniques. A summary of the various computing platforms and the software to parallelize the computations is also provided. The parallel computing platforms involve the most typical architectures available, including IA-32, EM64T, and IA-64 systems. Two popular parallel scientific libraries, ScaLAPACK and PLAPACK, are briefly described to familiarize the reader with them as they will be used later on for implementing parallel solvers for matrix equations.

REFERENCES

[1]   A. E. Bondeson, T. Rylander, and P. Ingelström, *Computational Electromagnetics,* Springer, New York, 2005.

[2]   R. A. Elliott, *Electromagnetics*, McGraw-Hill, New York, 1966.

[3]   R. F. Harrington, *Field Computation by Moment Methods,* Macmillan, New York, 1968.

[4]   W. C. Chew, J. Jin, E. Michielssen, and J. Song, *Fast and Efficient Algorithms in Computational Electromagnetics*, Artech House, Norwood, MA, 2000.

[5]   P. P. Silvester and R. L. Ferrari, *Finite Elements for Electrical Engineers*, Cambridge University Press, New York, 1983.

[6]   J. M. Jin, *The Finite Element Method in Electromagnetics*, Wiley, New York, 1993.

[7]   M. Salazar-Palma, T. K. Sarkar, L. E. Garcia-Castillo, T. Roy, and A. Djordjevic, *Iterative and Self-Adaptive Finite-Elements in Electromagnetic Modeling*, Artech House, Norwood, MA, 1998.

[8]   R. Coccioli, T. Itoh, G. Pelosi, and P. P. Silvester, *Finite Element Methods in Microwaves: A Selected Bibliography*, University of Florence, Department of Electronics and Telecommunciations. Available at: http://ingfi9.die.unifi.it/fem corner/map/cont2_1.htm. Accessed July 2008.

[9]   K. S. Yee, "Numerical Solution of Initial Boundary Value Problems Involving Maxwell's Equations in Isotropic Media," *IEEE Transactions on Antennas and Propagation*, Vol. 14, No. 3, pp. 302–307, May 1966.

[10] K. L. Shlager and J. B. Schneider, "A Selective Survey of the Finite-Difference Time-Domain Literature," *IEEE Antennas and Propagation Magazine*, Vol. 37, No. 4, pp. 39–56, Aug. 1995.

[11] A. Taflove and S. C. Hagness, *Computational Electrodynamics: The Finite-Difference Time-Domain Method,* 3rd ed., Artech House, Norwood, MA, 2005.

[12] S. M. Rao and N. Balakrishnan, "Computational Electromagnetics — A Review,"

       Indian Academy of Science, Current Science Online, 2000. Available at:
       http://www.ias.ac.in/currsci/nov25/articles24.htm. Accessed July 2008.

[13]   G. S. Almasi and A. Gottlieb, *Highly Parallel Computing*, Benjamin-Cummings,
       Redwood City, CA, 1989.

[14]   Lawrence Livermore National Laboratory, "Introduction to Parallel Computing,"
       High Performance Computing Training, 2007. Available at: https://computing
       .llnl.gov/tutorials/parallel_comp/#WhyUse. Accessed July 2008.

[15]   J. S. Shang, "Computational Electromagnetics," *ACM Computing Surveys*, Vol.
       28, No. 1, pp. 97–99, March 1996.

[16]   A. Rubinstein, F. Rachidi, M. Rubinstein, and B. Reusser, "A Parallel
       Implementation of NEC for the Analysis of Large Structures," *IEEE Transactions
       on Electromagnetic Compatibility*, Vol. 45, No. 2, pp. 177–188, May 2003.

[17]   Y. Zhang, T. K. Sarkar, P. Ghosh, M. Taylor, and A. De, "Highly Efficient
       Parallel Schemes Using Out-of-Core Solver for MoM," *IEEE Applied
       Electromagnetics Conference–AEMC*, Kolkata, India, Dec. 2007.

[18]   S. Velamparambil, W. C. Chew, and J. Song, "10 Million Unknowns: Is It That
       Big?" *IEEE Antennas and Propagation Magazine*, Vol. 45, No. 2, pp. 43–58,
       April 2003.

[19]   C. Waltz, K. Sertel, M. A. Carr, B. C. Usner, and J. L. Volakis, "Massively
       Parallel Fast Multipole Method Solutions of Large Electromagnetic Scattering
       Problems," *IEEE Transactions on Antennas and Propagation,* Vol. 55, No. 6, pp.
       1820–1816, June 2007.

[20]   G. Hennigan and S. Castillo, "Open-region, Electromagnetic Finite-element
       Scattering Calculations in Anisotropic Media on Parallel Computers," *IEEE
       Antennas and Propagation Society International Symposium,* Orlando, FL, 1999.

[21]   D. H. Malan and A. C. Metaxas, "Implementing a Finite Element Time Domain
       Program in Parallel," *IEEE Antennas and Propagation Magazine*, Vol. 42, No. 1,
       pp. 105–109, 2000.

[22]   C. Guiffaut and K. Mahdjoubi, "A Parallel FDTD Algorithm Using the MPI
       Library," *IEEE Antennas and Propagation Magazine*, Vol. 43, No. 2, pp. 94–103,
       2001.

[23]   Y. Zhang, W. Ding, and C. H. Liang, "Study on the Optimum Virtual Topology
       for MPI Based Parallel Conformal FDTD Algorithm on PC Clusters," *Journal of
       Electromagnetic Waves and Applications*, Vol. 19, No. 13, pp. 1817–1831, Oct.
       2005.

[24]   W. Yu, Y. Liu, T. Su, N. Huang, and R. Mittra, "A Robust Parallelized Conformal
       Finite Difference Time Domain Field Solver Package Using the MPI Library,"
       IEEE Antennas and Propagation Magazine, Vol. 47, No. 3, 2005.

[25]   Y. Zhang, A. De, B. H. Jung, and T. K. Sarkar, "A Parallel Marching-on-in-Time
       Solver," *IEEE AP-S & USNC/URSI Symposium*, San Diego, CA, July 2008.

[26]   J. E. Patterson, T. Cwik, R. D. Ferraro, N. Jacobi , P. C. Liewer, T. G. Lockhart,
       G. A. Lyzenga, J. W. Parker, and D. A. Simoni, "Parallel Computation Applied to
       Electromagnetic Scattering and Radiation Analysis," *Electromagnetics*, Vol. 10,
       No. 1–2, pp. 21–39, Jan. –June 1990.

[27]   T. Cwik, J. Partee, and J. E. Patterson, "Method of Moment Solutions to
       Scattering Problems in a Parallel Processing Environment," *IEEE Transactions on
       Magnetics*, Vol. 27, No. 5, pp. 3837–3840, Sept. 1991.

[28]   D. B. Davidson, "Large Parallel Processing Revisited: A Second Tutorial," *IEEE
       Antennas and Propagation Magazine*, Vol. 34, No. 5, pp. 9–21, Oct. 1992.

[29]   T. Marshall, "Numerical Electromagnetics Code (Method of Moments),"
       Numerical Electromagnetics Code NEC2 Unofficial Home Page, 2002. Available

at: http://www.nec2.org/. Accessed Aug. 2008.

[30] R. A. van de Geijn, *Using PLAPACK— Parallel Linear Algebra Package*, MIT Press, Cambridge, MA, 1997.

[31] S. M. Rao, D. R. Wilton, and A. W. Glisson, "Electromagnetic Scattering by Surfaces of Arbitrary Shape," *IEEE Transactions on Antennas and Propagation*, Vol. 30, No. 3, pp. 409–418, May 1982.

[32] T. Cwik, R. A. van de Geijn, and J. E. Patterson, "The Application of Parallel Computation to Integral Equation Models of Electromagnetic Scattering," *Journal of the Optical Society of America*, Vol. 11, No. 4, pp. 1538–1545, April 1994.

[33] L. S. Blackford, J. Choi, A. Cleary, E. D'Azevedo, J. Demmel, I. Dhillon, J. Dongarra, S. Hammarling, G. Henry, A. Petitet, K. Stanley, D. Walker, and R. C. Whaley, *ScaLAPACK User's Guide*, Society for Industrial Mathematics, Philadelphia, PA, 1997.

[34] U. Jakobus, "Parallel Computation of Electromagnetic Fields Based on Integral Equations," *Penn State Citeseer*, 1999. Available at: http://citeseer.ist.psu.edu /cachedpage/438895/9. Accessed Aug. 2008.

[35] Lawrence Livermore National Laboratory, "Message Passing Interface (MPI)," *High Performance Computer Training*. Available at: https://computing.llnl.gov /tutorials/mpi/. Accessed Aug. 2008.

[36] Oak Ridge National Laboratory, "PVM: Parallel Virtual Machine," *Computer Science and Mathematics*. Available at: http://www.csm.ornl.gov/pvm/. Accessed Aug. 2008.

[37] E. F. Knott and T. B. A. Senior, "Comparison of Three High-Frequency Diffraction Techniques," *Proceedings of the IEEE*, Vol. 62, No. 11, pp. 1468–1474, Nov. 1974.

[38] J. S. Asvestas. "The Physical Optics Method in Electromagnetic Scattering," *Journal of Mathematical Physics*, Vol. 21, No. 2, pp. 290–299, Feb. 1980.

[39] SuperNEC-EM Antenna Simulation Software, *SuperNEC–Parallel MoM User Reference Manual. Version 2.7*. Available at: http://www.supernec.com /manuals/snparurm.htm. Accessed Aug. 2008.

[40] J. Edlund, *A Parallel, Iterative Method of Moments and Physical Optics Hybrid Solver for Arbitrary Surfaces*, licentiate thesis, Uppsala University, 2001.

[41] Y. Zhang, X. W. Zhao, M. Chen, and C. H. Liang, "An Efficient MPI Virtual Topology Based Parallel, Iterative MoM-PO Hybrid Method on PC Clusters," *Journal of Electromagnetic Waves and Applications*, Vol. 20, No. 5, pp. 661–676, 2006.

[42] T. K. Sarkar, E. Arvas, and S. M. Rao, "Application of FFT and the Conjugate Gradient Method for the Solution of Electromagnetic Radiation from Electrically Large and Small Conducting Bodies," *IEEE Transactions on Antennas and Propagation*, Vol. 34, No. 5, pp. 635–640, May 1986.

[43] E. Bleszynski, M. Bleszynski, and T. Jaroszewicz, "AIM: Adaptive Integral Method for Solving Large-Scale Electromagnetic Scattering and Radiation Problems," *Radio Science*, Vol. 31, No. 5, pp. 1225–1251, 1996.

[44] V. Rokhlin, "Rapid Solution of Integral Equations of Scattering Theory in Two Dimensions," *Journal of Computational Phyics*, Vol. 86, pp. 414–439, 1990.

[45] X. C. Nie, L. W. Li, and N. Yuan, "Precorrected-FFT Algorithm for Solving Combined Field Integral Equations in Electromagnetic Scattering," *IEEE Antennas and Propagation Society International Symposium*, San Antonio, TX, Vol. 3, pp. 574–577, June 2002.

[46] J. M. Song, C. Lu, and W. Chew, "Multilevel Fast Multipole Algorithm for

Electromagnetic Scattering by Large Complex Objects," *IEEE Transactions on Antennas and Propagation*, Vol. 45, No. 10, pp. 1488–1493, Oct. 1997.

[47]   J. Song, C. Lu, W. Chew, and S. Lee, "Fast Illinois Solver Code (FISC)," *IEEE Antennas and Propagation Magazine,* Vol. 40, No. 3, pp. 27–34, June 1998.

[48]   S. Velamparambil, J. Schutt-Aine, J. Nickel, J. Song, and W. Chew, "Solving Large Scale Electromagnetic Problems Using a Linux Cluster and Parallel MLFMA," *IEEE Antennas and Propagation Society International Symposium,* Vol. 1, pp. 636–639, Orlando, FL, July 1999.

[49]   S. Velamparambil and W. Chew, "Analysis and Performance of a Distributed Memory Multilevel Fast Multipole Algorithm," *IEEE Transactions on Antennas and Propagation*, Vol. 53, No. 8, pp. 2719–2727, Aug. 2005.

[50]   S. Velamparambil, J. Song, and W. C. Chew, "Parallelization of Multilevel Fast Multipole Algorithm on Distributed Memory Computers in Fast and Efficient Algorithms," *Computational Electromagnetics*, Chap. 4, Artech House, Boston, MA, 2001.

[51]   E. J. Lu and D. Okunbor, "A Massively Parallel Fast Multipole Algorithm in Three Dimensions," *Proceedings of the 5th IEEE International Symposium on High Performance Distributed Computing,* pp. 40–48, Aug. 1996.

[52]   E. J. Lu and D. Okunbor, "Parallel Implementation of 3-D FMA Using MPI," *Proceedings of 2nd MPI Developer's Conference, IEEE Computer Society Technical Committee on Distributed Processing*, pp. 119–124, Notre Dame, IN., July 1996.

[53]   P. Have, "A Parallel Implementation of the Fast Multipole Method for Maxwell's Equations," *International Journal for Numerical Methods in Fluids*, Vol. 43, No. 8, pp. 839–864, Nov. 2003.

[54]   M. A. Stalzer, "A Parallel Fast Multipole Method for the Helmholtz Equation," *Parallel Processing Letters*, Vol. 5, No. 2, pp. 263–274, June 1995.

[55]   S. Velamparambil, W. C. Chew, and M. L. Hastriter, "Scalable Electromagnetic Scattering Computations," *IEEE Antennas and Propagation Society International Symposium, 2002,* Vol. 3, pp. 176–179, San Antonio, TX, June 2002.

[56]   S. Velamparambil, J. Song, W. Chew, and K. Gallivan, "ScaleME: A Portable Scaleable Multipole Engine for Electromagnetic and Acoustic Integral Equation Solvers," *IEEE Antennas and Propagation Society International Symposium, 1998*, Vol. 3, pp. 1774–1777, June 1998.

[57]   M. Nilsson, *A Parallel Shared Memory Implementation of the Fast Multipole Method for Electromagnetics*, Technical Report 2003-049, Department of Information Technology, Scientific Computing, Uppsala University, Uppsala, Sweden, Oct. 2003.

[58]   FEKO, "Comprehensive EM Solutions Field Computations Involving Objects of Arbitrary Shape," Available at: http://www.feko.info/. Accessed Aug. 2008.

[59]   E. D'Azevedo and J. J. Dongarra, *The Design and Implementation of the Parallel Out-of-Core ScaLAPACK LU, QR and Cholesky Factorization Routines*, University of Tennessee, Knoxville, Citeseer, 1997. Available at: http://citeseer. ist.psu.edu/article/dazevedo97design.html. Accessed Aug. 2008.

[60]   B. M. Kolundzija , J. S. Ognjanovic T. K. Sarkar, and R. F. Harrington, *WIPL: Electromagnetic Modeling of Composite Wire and Plate Structures — Software and User's Manual,* Artech House, Norwood, MA, 1995.

[61]   B. M. Kolundzija, J. S. Ognjanovic, and T. K. Sarkar, *WIPL-D: Electromagnetic Modeling of Composite Metallic and Dielectric Structures, Software and User Manual*, Artech House, Norwood, MA, 2000.

[62]   B. M. Kolundzija, J. S. Ognjanovic, and T. K. Sarkar, "Analysis of Composite

Metallic and Dielectric Structures — WIPL-D Code," *Proceedings of 17th Applied Computational Electromagnetics Conference*, Monterey, CA, pp. 246–253, March 2001.

[63] Y. Zhang, T. K. Sarkar, H. Moon, A. De, and M. C. Taylor, "Solution of Large Complex Problems in Computational Electromagnetics Using Higher Order Basis in MoM with Parallel Solvers," *IEEE Antennas and Propagation Society International Symposium*, pp. 5620–5623, Honolulu, HI, June 2007.

[64] Y. Zhang, T. K. Sarkar, H. Moon, M. Taylor, and M. Yuan, "The Future of Computational Electromagnetics for Solving Large Complex Problems: Parallel In-Core Integral Equation Solvers," *EMTS 2007 International URSI Commission B — Electromagnetic Theory Symposium*, Ottawa, ON, Canada, July 2007.

[65] Y. Zhang, J. Porter, M. Taylor, and T. K. Sarkar, "Solving Challenging Electromagnetic Problems Using MoM and a Parallel Out-of-Core Solver on High Performance Clusters," *2008 IEEE AP-S & USNC/URSI Symposium*, San Diego, CA, July 2008.

[66] J. Mackerle, "Implementing Finite Element Methods on Supercomputers, Workstations and PCs a Bibliography (1985−1995)," *Engineering Computations*, Vol. 13 No. 1, pp. 33–85, 1996. Available at: http://www.emeraldin sight.com/Insight/ViewContentServlet?Filename=/published/emeraldabstractonly article/pdf/1820130103.pdf. Accessed Aug. 2008.

[67] C. T. Wolfe, U. Navsariwala, and S. D. Gedney, "A Parallel Finite-element Tearing and Interconnecting Algorithm for Solution of the Vector Wave Equation with PML Absorbing Medium," *IEEE Transactions on Antennas and Propagation*, Vol. 48, No. 2, pp. 278–284, Feb. 2000.

[68] W. P. Pala, A. Taflove, M. J. Piket, and R. M. Joseph, "Parallel Finite Difference-Time Domain Calculations," *International Conference on Computation in Electromagnetics*, pp. 83–85, Nov. 1991.

[69] V. Varadarajan and R. Mittra, "Finite-Difference Time-Domain Analysis Using Distributed Computing," *IEEE Microwave and Guided Wave Letters*, Vol. 4, pp. 144–145, 1994.

[70] Z. M. Liu, A. S. Mohan, T. A. Aubrey, and W. R Belcher, "Techniques for Implementation of the FDTD Method on a CM-5 Parallel Computer," *IEEE Antennas and Propagation Magazine,* Vol. 37, No. 5, pp. 64–71, 1995.

[71] A. D. Tinniswood, P. S. Excell, M. Hargreaves, S. Whittle, and D. Spicer, "Parallel Computation of Large-Scale FDTD Problems," *Computation in Electromagnetics, 3rd International Conference, (Conf. Publ. No. 420)*, Vol. 1, pp. 7–12, April 1996.

[72] H. Simon, *Parallel Rigorous Electrodynamic Simulation using the Finite-Difference Time Domain Method*, National Energy Research Scientific Computing (NERSC) Center, Computational Research Division. Available at: http://www.nersc.gov/~simon/cs267/hw0/hafeman/. Accessed Aug. 2008.

[73] U. Andersson, *Time-Domain Methods for the Maxwell Equations*, PhD thesis, KTH, Sweden, 2001.

[74] Y. Zhang, W. Ding, and C. H. Liang, "Analysis of Parallel Performance of MPI Based Parallel FDTD on PC Clusters," *Microwave Conference Proceedings, 2005, APMC 2005, Asia-Pacific Conference Proceedings*, Vol. 4, pp. 3, Dec. 2005.

[75] Y. Zhang, *Parallel Computation in Electromagnetics*, Xidian University Press, Xi'an, China, 2006.

[76] H. Malan and A. C. Metaxas, "Implementing a Finite Element Time Domain

Program in Parallel," *IEEE Antennas and Propagation Magazine*, Vol. 42, No. 1, pp. 105–109, Feb. 2000.

[77]   B. Butrylo, C. Vollaire, and L. Nicolas, "Parallel Implementation of the Vector Finite Element and Finite Difference Time Domain Methods," *Parallel Computing in Electrical Engineering, 2002. PARELEC '02, Proceedings. International Conference on*, pp. 347–352, Sept. 2002.

[78]   S. M. Rao and T. K. Sarkar, "Numerical Solution of Time Domain Integral Equations for Arbitrarily Shaped Conductor/Dielectric Composite Bodies," *IEEE Transactions on Antennas and Propagation*, Vol. 50, No. 12, pp. 1831–1837, Dec. 2002.

[79]   B. H. Jung, Z. Ji, T. K. Sarkar, M. Salazar-Palma and M. Yuan, "A Comparison of Marching-on-in-time Method with Marching-on-in-degree Method for the TDIE Solver," *Progress in Electromagnetics Research*, Vol. 70, pp. 281–296, 2007.

[80]   Wikipedia contributors, "IA-32," *Wikipedia, The Free Encyclopedia,* July 2007. Available at: http://en.wikipedia.org/w/index.php?title=IA-32&oldid=146752470. Accessed Oct. 2007.

[81]   Netlib Repository at UTK and ORNL, "LAPACK — Linear Algebra PACKage." Available at: http://www.netlib.org/lapack/. Accessed Aug. 2008.

[82]   Netlib Repository at UTK and ORNL, *The ScaLAPACK Project*. Available at: http://www.netlib.org/scalapack/. Accessed Aug. 2008.

[83]   G. Morrow and R. A. van de Geijn, *Half-day Tutorial on: Using PLAPACK: Parallel Linear Algebra PACKage*, Texas Institute for Computational and Applied Mathematics, University of Texas at Austin, Austin, TX. Available at: http://www.cs.utexas.edu/users/plapack/tutorial/SC98/. Accessed Aug. 2008.

[84]   Wikipedia contributors, "Multicore," *Wikipedia, The Free Encyclopedia.* Available at: http://en.wikipedia.org/wiki/Multi-core_(computing)#Advantages. Accessed Aug. 2008.

[85]   Advanced Micro Devices, Inc., *AMD Multi-core White Paper.* Available at: http://www.sun.com/emrkt/innercircle/newsletter/0505multicorewp.pdf. Accessed Aug. 2008.

[86]   Advanced Micro Devices, Inc., *AMD Core Math Library (ACML)*, AMD Developer Central. Available at: http://developer.amd.com/cpu/libraries/ acml/Pages/default.aspx. Accessed Aug. 2008.

[87]   Intel®Software Network, *Intel®Math Kernel Library 10.0 Overview.* Available at: http://www.intel.com/cd/software/products/asmo-na/eng/307757.htm. Accessed Aug. 2008.

[88]   HP Technical Documentation, *HP-MPI User's Guide.* Available at: http://docs.hp.com/en/B6060-96022/index.html. Accessed Aug. 2008.

[89]   Intel®Software Network, *Intel®MPI Library 3.1 for Linux or Microsoft Windows Compute Cluster Server (CCS).* Available at: http://www.intel.com/ cd/software/products/asmo-na/eng/cluster/mpi/308295.htm

[90]   Netlib Repository at UTK and ORNL, *Basic Linear Algebra Subprograms (BLAS).* Available at: http://www.netlib.org/blas/. Accessed Aug. 2008.

[91]   Netlib Repository at UTK and ORNL, The ScalAPACK Project, *Parallel Basic Linear Algebra Subprograms (PBLAS) Home Page.* Available at: http://www.netlib.org/scalapack/pblas_qref.html. Accessed Aug. 2008.

[92]   Netlib Repository at UTK and ORNL, *Basic Linear Algebra Subprograms (BLACS).* Available at: http://www.netlib.org/blacs/. Accessed Aug. 2008.

[93]   L. S. Blackford, *ScaLAPACK Tutorial*, Available at: http://www.netlib.org/ scalapack/tutorial/sld053.htm. Accessed Aug. 2008.

[94]   Wikipedia contributors, "Message Passing Interface," *Wikipedia, The Free*

Encyclopedia. Available at: http://en.wikipedia.org/wiki/Message_Passing_ Interface. Accessed Aug. 2008.

[95] Message Passing Interface Forum, *MPI Documents.* Available at: http://www.mpi-forum.org/docs/docs.html. Accessed Aug. 2008.

[96] Netlib Repository at UTK and ORNL, *LAPACK — Linear Algebra PACKage.* Available at: http://www.netlib.org/lapack/. Accessed Aug. 2008.

[97] Y. Zhang, T. K Sarkar, A. De, N. Yilmazer, S. Burintramart, and M. Taylor, "A Cross-Platform Parallel MoM Code with ScaLAPACK Solver," *IEEE Antennas and Propagation Society International Symposium,* pp. 2797–2800, Honolulu, HI, June 2007.

[98] P. Alpatov, G. Baker, H. C. Edwards, J. Gunnels, G. Morrow, J. Overfelt, and R. A. van de Geijn, "PLAPACK Parallel Linear Algebra Package Design Overview," *Supercomputing, ACM/IEEE 1997 Conference*, Nov. 1997.

[99] Y. Zhang, T. K. Sarkar, R. A. van de Geijn and M. C. Taylor, "Parallel MoM Using Higher Order Basis Function and PLAPACK In-Core and Out-of-Core Solvers for Challenging EM Simulations," *IEEE AP-S & USNC/URSI Symposium*, San Diego, CA, July 2008.

[100] M. C. Taylor, Y. Zhang, T. K. Sarkar, and R. A. van de Geijn, "Parallel MoM Using Higher Order Basis Functions and PLAPACK Out-of-Core Solver for a Challenging Vivaldi Array," *XXIX URSI General Assembly*, Chicago, IL, Aug. 2008.

2

IN-CORE AND OUT-OF-CORE LU FACTORIZATION FOR SOLVING A MATRIX EQUATION

2.0   SUMMARY

The parallel implementation of a MoM code involves parallel matrix filling followed by a parallel matrix solution. The approach used to fill the matrix in parallel is largely influenced by the type of matrix equation solver being used. To efficiently fill the matrix for a MoM code in parallel, it is important to understand how the data are distributed over different processors (cores) among the different computer nodes.

On the basis of this consideration, this chapter explores the details of solving the dense complex matrix equation for MoM using the LU factorization method rather than describing the MoM procedure itself. After introducing the matrix equation for MoM, the serial versions of the LU decomposition method are investigated. Then, the discussion moves to the parallel implementation of the LU algorithm. After the data distribution patterns used by the ScaLAPACK (*sca*lable *l*inear *a*lgebra *pack*age) and PLAPACK (*p*arallel *l*inear *a*lgebra *pack*age) library packages are explained, they are compared in terms of flexibility of use.

When the matrix from a MoM code is too large to fit into the RAM of the computer at hand, the space on the hard disk should be used to extend the storage capability. An out-of-core solver in addition to the RAM uses the space on the hard disk, whereas an in-core solver uses only the RAM. Both the in-core and the out-of-core LU factorization algorithms are described in this chapter. A comparison is made between the input/output (I/O) operation of the left-looking and right-looking out-of-core algorithms, followed by a one-slab left-looking out-of-core algorithm and its corresponding parallel implementation using ScaLAPACK and PLAPACK. With the LU factors and forward/backward substitution algorithm, the matrix equation obtained from a MoM code can be efficiently solved in parallel.

2.1   MATRIX EQUATION FROM A MOM CODE

Following the MoM procedure to solve an integral equation, a matrix equation can be obtained as

$$[Z][I] = [V],$$ (2.1)

where $[Z]$ is an $M \times M$ matrix in which M unknowns are used to represent the continuous electric and/or the magnetic current over the surface of interest, $[I]$ is an $M \times 1$ coefficient vector of the unknown current coefficients to be solved for, and $[V]$ is an $M \times 1$ excitation vector. Here, $[Z]$ is often referred to as the *impedance matrix*. $[V]$ is termed as the *voltage* or the *excitation matrix*, and $[I]$ is referred to as the matrix containing the unknown current coefficients. Detailed steps for computing the elements of the matrices $[Z]$ and $[V]$, for different basis functions, will be given in the subsequent chapters. The following discussion is focused on solving Eq. (2.1).

The numerical solution of Eq. (2.1) generally proceeds by factoring the impedance matrix $[Z]$ into a lower and an upper triangular matrix, called the *LU factorization* or *decomposition*. This LU-factored matrix is then used to solve for the induced current given an excitation. When a shared-memory machine is used, the impedance matrix is stored in the memory and operated on. When a distributed memory machine is used instead, the matrix must be broken up into pieces and distributed across the memory of all the processes involved [1−3]. The algorithms will be explained in detail in the following section.

2.2   AN IN-CORE MATRIX EQUATION SOLVER

For descriptive convenience, let us rewrite the dense linear system, Eq. (2.1), as

$$[A][X] = [B],$$ (2.2)

where $[A] \in C^{M \times M}$, that is, elements of the matrix are complex quantities. If the use of pivoting during the LU decomposition for stability of the solution is ignored for the moment, it is customary to first perform the LU factorization as follows

$$[A] = [L][U],$$ (2.3)

where $[L] \in C^{M \times M}$ is a lower triangular matrix with unity elements on the diagonal, and $[U] \in C^{M \times M}$ is an upper triangular matrix. The system given by Eq. (2.3) is then solved using a two-step procedure. First solve for $[Y]$ by using

$$[L][Y]=[B]$$ (2.4)

and then solve for $[X]$ using the knowledge of $[Y]$ from

$$[U][X]=[Y].$$ (2.5)

In the past, the solution of the matrix equation could have been easily accomplished by calling the appropriate LINPACK (*lin*ear algebra *pack*age) [4] library routines. However, on a typical processor with hierarchical memory, the brute-force use of LINPACK is inefficient because its memory access patterns disregard the multi-layered memory hierarchies of the computing machines, and thereby spending too much time transferring data instead of carrying out useful floating-point operations. In essence, $O(M)$ [$O(\bullet)$: represents of the order of] operations are performed on $O(M)$ data, and the processor cache, which allows high data access rates, cannot be optimally utilized [5].

In computer science, a cache (http://en.wikipedia.org/wiki/Cache) is a collection of data duplicating original values stored elsewhere or computed earlier, where the original data are expensive to fetch (owing to longer access time) or to compute, compared to the cost of reading the cache. In other words, a cache is a temporary storage area where frequently accessed data can be stored for rapid access. Once the data are stored in the cache, future use can be made by accessing the cached copy rather than refetching or recomputing the original data, so that the average access time is shorter. A cache has proved to be extremely effective in many areas of computing because access patterns in typical computer applications have locality of reference. There are several kinds of locality, but here, we deal primarily with data that are accessed close together in time (temporal locality). The data also might or might not be located physically close to each other (spatial locality). Cache, therefore, helps to expedite data access that the CPU would otherwise need to fetch from the RAM [6–8]. Cache thus has the capability to speed up the execution of the installed software on the system because of a concept in program execution known as *locality* [9]. Locality assumes that if a piece of data has been accessed, it is highly likely that it will be accessed again soon (temporal locality) and that its neighbors in the main memory will also be accessed soon (spatial locality). If codes exhibit locality, then resulting applications will run faster because the data will be in the nearby fast cache.

Before explaining how to optimally use caches, one needs to understand the memory hierarchy — the hierarchical arrangement of storage in current computer architectures. The memory hierarchy is a major subsystem that contributes significantly to the overall system performance. Since the early 1980s, the yearly increase in the computational speed of processors has outpaced the increase in the rate of data transfer from the main memory (RAM). A central processing unit (CPU) in a modern computer is small, fast, and expensive, while the size of the RAM is large, cheap, and comparatively slower. Computer architects have remedied this situation by creating a hierarchy of memories: faster memory near the CPU, and slower memory near the main memory. Figure 2.1 shows a typical memory hierarchy. The faster memory near the CPU (called *cache*) is smaller in capacity and more expensive, while the main memory is slightly slower, is less expensive, and has a higher capacity. It is desirable to keep data, which is likely to be needed next by the CPU, in cache where it can be accessed more quickly than if it had been in the RAM [6–9].

In Figure 2.1, the size of the different types of cache increases from left to right, while the speed decreases. In other words, as the capacity increases, it takes longer to move the data in and out.

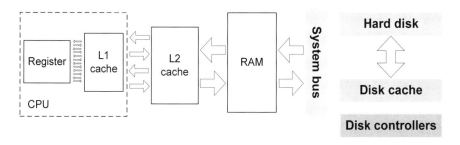

Figure 2.1. A typical hierarchy in the architecture of the memory.

In today's typical microprocessors, there are two cache areas: L1 (level 1) and L2 (level 2). L1 cache is also known as *on-die cache*, *on-chip cache*, or *primary cache*. L1 cache usually rests directly on the microprocessor and has a smaller memory area than the L2 cache. L1 cache is the first area that the microprocessor checks when looking for data stored in the memory. In general, L2 cache memory, also called the *secondary cache*, resides on a chip separate from the microprocessor chip, although more and more microprocessors are including L2 caches in their architectures. L2 cache is the second area that the microprocessor checks for data stored in the memory. It attempts to compensate for the speed mismatch between RAM and the processor. Also, L2 cache will always have more storage capacity as it stores much more data than the L1 cache. The amount of L2 cache in various computer systems has increased during the past five years. With a large amount of cache, the performance of the processor has improved since the processor does not have to request information from the RAM as frequently.

Often, the RAM is not large enough to store all the temporary values that the CPU needs for processing. One remedy is to use the space on the hard disk. In Figure 2.1, the hard drive is also added to the memory hierarchy, and then the RAM becomes the "cache" for the hard drive. The hard drive is much slower in speed, but the cache misses in the RAM are much less frequent since large memory pages are brought in from the disk each time a miss occurs. The term *cache miss* implies that the CPU cannot find the data it is looking for in the cache and it needs to find a place to store that item. The difference in speed between the hard drive and the main memory is so dramatic that often an additional cache is added in the disk controller.

As pointed out earlier, LINPACK is inefficient because its memory access pattern disregards the multilayered memory hierarchies of the machines. Actually, LINPACK has been largely superseded by LAPACK [10], which has been designed to run efficiently on shared-memory, vector supercomputers. LAPACK addresses the problem of spending too much time transferring data instead of doing useful floating-point operations by reorganizing the algorithms to use block matrix operations, such as matrix multiplication, in the innermost loops. These block matrix operations can be optimized to account for the memory hierarchy, and make it possible to achieve high efficiency on diverse modern machines.

In LAPACK, matrix $[A]$, $[L]$ and $[U]$ can be partitioned as

$$\begin{bmatrix}[A_{11}] & [A_{12}] \\ [A_{21}] & [A_{22}]\end{bmatrix} = \begin{bmatrix}[L_{11}] & [0] \\ [L_{21}] & [L_{22}]\end{bmatrix}\begin{bmatrix}[U_{11}] & [U_{12}] \\ [0] & [U_{22}]\end{bmatrix}, \tag{2.6}$$

where $[A_{11}], [L_{11}], [U_{11}] \in C^{k \times k}$ and $[0]$ is the null matrix. This leads to the following set of equations:

$$\begin{bmatrix}[A_{11}] \\ [A_{21}]\end{bmatrix} = \begin{bmatrix}[L_{11}] \\ [L_{21}]\end{bmatrix}[U_{11}] \tag{2.7}$$

$$[A_{12}] = [L_{11}][U_{12}], \tag{2.8}$$

$$[A_{22}] - [L_{21}][U_{12}] = [L_{22}][U_{22}]. \tag{2.9}$$

The LU factorization algorithm can now be formulated by overwriting a panel of matrix elements of width k with its LU factorization [Eq. (2.7)], followed by solution of the triangular system with multiple right-hand sides [Eq. (2.8)]. Then the bulk of the computation is in updating $[A_{22}]$, using a matrix–matrix multiplication, followed by a recursive factorization of the result [Eq. (2.9)]. During the updating of the elements of $[A_{22}]$, $2k(M-k)^2$ operations are performed on the (M^2-k^2) data, which requires $(M-k)k(M-k)$ multiplications plus $(M-k)(k-1)(M-k)$ additions for calculating $[L_{21}][U_{12}]$, plus $(M-k)^2$ subtractions from the elements of $[A_{22}]$. This allows the data to be brought into cache, after which a large amount of computation occurs before it is returned to the memory. Thus, in this procedure, the bottleneck to memory access is overcome [1].

LAPACK obtains its high performance by using algorithms that do most of the work in calls to the BLAS, with the emphasis placed on matrix–matrix multiplications. (BLAS is an application programming interface standard for publishing libraries to perform basic linear algebra operations such as vector and matrix multiplication. They are used to build larger packages such as LAPACK.) Since the mid-1980s, researchers have been pursuing a high-performance implementation of matrix multiplications. One way to accomplish it is to organize the computation around an "inner kernel", $[C] = [A]^T [B] + [C]$, which keeps one of the operands in the L1 cache, while streaming parts of the other operands through that cache. Variants include approaches that extend this principle to multiple levels of cache or that apply the same principle to the L2 cache while essentially ignoring the L1 cache. The purpose of this tuning is to optimally and gradually reduce the time of moving data between memory layers [11].

The approach in Goto BLAS [11–13] is fundamentally different. It is based on the fact that much of the overhead comes from the *translation look-aside buffer* (TLB) table misses in the current generation of architectures. What is most important in the Goto-BLAS approach is the minimization of the TLB misses, though the importance of caches is taken into consideration as well.

2.3   PARALLEL IMPLEMENTATION OF AN IN-CORE SOLVER

2.3.1   Data Distribution for an LU Algorithm

LAPACK provides an approach for a sequential algorithm to solve the matrix equation [Eq.(2.1)]. Knowing this, it is desirable to parallelize LAPACK to solve the matrix equation using high-performance computer clusters. First, it is important to discuss how a user may wish to allocate the assignment of various portions of the matrix to different processes.

Assume that the processes P_{ij} form a logical two dimensional $P_r \times P_c$ array, where i and j refer to the row and the column index of the processes, respectively. Then the simplest assignment of matrix elements to processes is to partition the matrix into the following form

$$[A] = \begin{bmatrix} \begin{bmatrix} \hat{A}_{11} \end{bmatrix} & \cdots & \begin{bmatrix} \hat{A}_{1P_c} \end{bmatrix} \\ \vdots & \vdots & \vdots \\ \begin{bmatrix} \hat{A}_{P_r 1} \end{bmatrix} & \cdots & \begin{bmatrix} \hat{A}_{P_r P_c} \end{bmatrix} \end{bmatrix}, \tag{2.10}$$

where $\begin{bmatrix} \hat{A}_{ij} \end{bmatrix} \in C^{m_i \times n_j}$, $m_i \approx M / P_r$, $n_j \approx M / P_c$. The submatrix $\begin{bmatrix} \hat{A}_{ij} \end{bmatrix}$ is then assigned to process $P_{(i-1)(j-1)}$.

A straightforward parallel LU factorization, again ignoring pivoting, can now proceed as follows. In Eq. (2.6), assume that the column panel consisting of

$[A_{11}]$ and $[A_{21}]$ resides within a single column of processes, while the row panel consisting of $[A_{11}]$ and $[A_{12}]$ resides within a single row of processes. The procedure starts by having the processes that hold portions of the column panel collaborate between each other to factor that panel. Next, the results can be broadcast within the rows of processes to the other columns. The row of processes that holds $[A_{12}]$ can then update this panel in parallel. Finally, $[A_{22}]$ must be overwritten by $[A_{22}] - [L_{21}][U_{12}]$. It is observed that on an arbitrary process P_{ij}, only the portions of $[L_{21}]$ that reside in row i and the portions of $[U_{12}]$ that reside in column j must be present. The former submatrix is already on the process, due to the broadcast of the factored panel within rows of the process grid. The latter can be brought to the appropriate processes by broadcasting portions of $[U_{12}]$ within columns of the process grid.

As an example, consider a matrix partitioned into 3×3 blocks of size $m_i \times n_j$ as shown in Figure 2.2 (a), and the corresponding 3×3 process grid to which the partitioned submatrices are distributed as shown in Figure 2.2 (b).

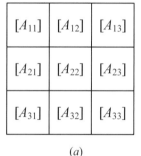   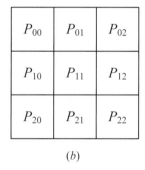

(a)   (b)

Figure 2.2. A matrix and a process grid containing its corresponding blocks: (a) a matrix partitioned into 3×3 blocks; (b) processes owning the corresponding blocks given in (a).

Now observe the computation and the communication involved at each step of factoring this matrix. Suppose that Eq. (2.3) is factored. Then the matrix is block-partitioned in the following form:

$$\begin{bmatrix} [A_{11}] & [A_{12}] & [A_{13}] \\ [A_{21}] & [A_{22}] & [A_{23}] \\ [A_{31}] & [A_{32}] & [A_{33}] \end{bmatrix} = \begin{bmatrix} [L_{11}] & [0] & [0] \\ [L_{21}] & [L_{22}] & [0] \\ [L_{31}] & [L_{32}] & [L_{33}] \end{bmatrix} \begin{bmatrix} [U_{11}] & [U_{12}] & [U_{13}] \\ [0] & [U_{22}] & [U_{23}] \\ [0] & [0] & [U_{33}] \end{bmatrix}. \quad (2.11a)$$

Multiplying $[L]$ with $[U]$ and equating the corresponding terms with respect to $[A]$, Eq. (2.11a) becomes

$$[A_{11}]=[L_{11}][U_{11}]; [A_{12}]=[L_{11}][U_{12}]; \qquad [A_{13}]=[L_{11}][U_{13}];$$
$$[A_{21}]=[L_{21}][U_{11}]; [A_{22}]=[L_{21}][U_{12}]+[L_{22}][U_{22}]; [A_{23}]=[L_{21}][U_{13}]+[L_{22}][U_{23}];$$
$$[A_{31}]=[L_{31}][U_{11}]; [A_{32}]=[L_{31}][U_{12}]+[L_{32}][U_{22}]; [A_{33}]=[L_{31}][U_{13}]+[L_{32}][U_{23}]+[L_{33}][U_{33}].$$

$$(2.11b)$$

With these relations, a recursive algorithm can be developed, which computes one block of row and column at each step and uses them to update the trailing submatrix. The computation and communication at each step are illustrated by Figure 2.3 $(a1)$–$(c2)$.

The arrows in Figure 2.3 $(a1)$, $(b1)$ are used to show the direction of communication between various processes at each step. Recall that the submatrix in each small box belongs to a corresponding process given in Figure 2.2 (b). The computation and the communication involved with this process are now discussed with the aid of Figure 2.3 $(a2)$, $(b2)$, and $(c2)$.

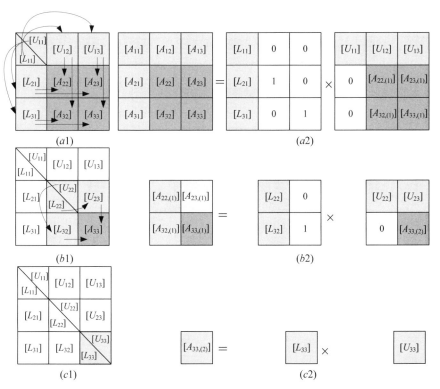

Figure 2.3. Illustration of the three steps for factoring a matrix: $(a1)$–$(a2)$ is carried out at the first step; $(b1)$–$(b2)$ is performed at the second step; $(c1)$–$(c2)$ completes the third step. The computation in the shaded regions will be finished at each step and communicated between processes shown by arrows. The darker shaded regions will then be updated and then processed in the following step.

In the first step, the first column panel is factored by process P_{00}, P_{10}, P_{20}, and the results are broadcast within the rows of processes [in Fig. 2.2 (b)], i.e., from P_{00} to P_{01} and P_{02}, and from P_{10} to P_{11} and P_{12}, and finally from P_{20} to P_{21} and P_{22}. Then the remaining columns of the first row are factored, i.e., $[A_{12}]$ is factored by process P_{01}, and $[A_{13}]$ by P_{02}. Results are broadcast within the columns of the processes [in Fig. 2.2 (b)] from P_{01} to P_{11} and P_{21}, and from P_{02} to P_{12} and P_{22}. Each of the four remaining matrices, $[A_{22}]$, $[A_{23}]$, $[A_{32}]$, and $[A_{33}]$, is updated by a corresponding process. Figure 2.3 ($a1$) indicates the communication involved at the first step and the darker shaded region in it will be factored in the second step. Figure 2.3 ($a2$) indicates the computations involved during the first step, labeled as tasks (1)–(9):

(1) $[A_{11}] \rightarrow [L_{11}][U_{11}]$; (2) $[L_{21}] = [A_{21}]([U_{11}])^{-1}$; (3) $[L_{31}] = [A_{31}]([U_{11}])^{-1}$;

(4) $[U_{12}] = ([L_{11}])^{-1}[A_{12}]$; (5) $[U_{13}] = ([L_{11}])^{-1}[A_{13}]$; (6) $[A_{22}] = [A_{22}] - [L_{21}][U_{12}]$;

(7) $[A_{32}] = [A_{32}] - [L_{31}][U_{12}]$; (8) $[A_{23}] = [A_{23}] - [L_{21}][U_{13}]$; (9) $[A_{33}] = [A_{33}] - [L_{31}][U_{13}]$.

$$(2.12a)$$

At the second step, the second column panel is factored by process P_{11} and P_{21}, and the results are broadcast within rows of processes from P_{11} to P_{12}, and from P_{21} to P_{22}. The remaining column of the second row, the updated $[A_{23}]$, is factored by process P_{12}, and the results are broadcast within columns of processes from P_{12} to P_{22}. The remaining matrix $[A_{33}]$ is updated by process P_{22}. Figure 2.3 ($b1$) indicates the communication at the second step, and the darker shaded region in it will be factored at the third step. Figure 2.3 ($b2$) indicates the computations during the second step, labeled as tasks (10)–(13):

(10) $[A_{22}] \rightarrow [L_{22}][U_{22}]$; (11) $[L_{32}] = [A_{32}]([U_{22}])^{-1}$;

(12) $[U_{23}] = ([L_{22}])^{-1}[A_{23}]$; (13) $[A_{33}] = [A_{33}] - [L_{32}][U_{23}]$.

$$(2.12b)$$

At the third step, the updated $[A_{33}]$ matrix is factored by process P_{22}. Figure 2.3 ($c1$) indicates the final LU results at the third step. Figure 2.3 ($c2$) indicates the computation during the third step, labeled as task (14):

(14) $[A_{33}] \rightarrow [L_{33}][U_{33}]$. $$(2.12c)$$

While this scheme is straightforward, it does not lead to a well-distributed workload among the various processes. As the algorithm recursively proceeds with the updated submatrix $[A_{22}]$ in Eq. (2.6), more and more processes become idle, until eventually only one process is busy. This can be changed by partitioning the matrix into much smaller $m_b \times n_b$ submatrices, and mapping them in two dimensions on the logical process array. In other words, partition the $M \times M$ matrix $[A]$ as

$$[A] = \begin{bmatrix} [A_{11}] & \cdots & [A_{1q}] \\ \vdots & \vdots & \vdots \\ [A_{p1}] & \cdots & [A_{pq}] \end{bmatrix}, \qquad (2.13)$$

where $p \approx M / m_b$ and $q \approx M / n_b$, and each $[A_{ij}]$ is a block of m_b rows and n_b columns, with the possible exception of the blocks in the last column, which may have fewer columns than n_b, and blocks in the last row, which may have fewer rows than m_b. Then block $[A_{ij}]$ is assigned to different processes. The algorithm proceeds as before, but as the problem deflates, processes continue to participate, instead of becoming idle as in the previous scheme. Adding the strategy of pivoting to this approach is relatively straightforward.

Now the problem that is to be treated is the mapping of each block $[A_{ij}]$ to different processes. ScaLAPACK and PLAPACK use different ways to distribute the matrix to different processes [14−16].

2.3.2   ScaLAPACK: Two-Dimensional Block-Cyclic Matrix Distribution

For reasons related to the performance and load balancing, ScaLAPACK uses a two-dimensional, block-cyclic distribution of full matrices (readers may refer to the ScaLAPACK *User's Guide*). First, the matrix is partitioned into blocks of size $m_b \times n_b$. These blocks are then uniformly distributed across the $P_r \times P_c$ process grid in a cyclic manner. As a result, every process owns a collection of blocks, which are contiguously stored in a two-dimensional "column-major" array. This local storage convention allows the ScaLAPACK software to efficiently use local memory by calling BLAS3 routines for submatrices that may be larger than a single $m_b \times n_b$ block.

Here, the parallel matrix distribution of ScaLAPACK, known as *block-cyclic distribution*, is briefly reviewed. The $P_r \times P_c$ processes available to the application are viewed as filling a logical two-dimensional grid of P_r rows and P_c columns. Given the matrix $[A]$ and its partition in Eq. (2.11), the block $[A_{ij}]$ is mapped to the process at the $\{(i−1) \bmod P_r, (j−1) \bmod P_c\}$ position of the process grid, i.e. the process in row $\{(i−1) \bmod P_r\}$ and column $\{(j-1) \bmod P_c\}$, where "mod" represents the modulo operator.

For example, consider a 9×9 matrix that is distributed using a 2×2 block size over six processes, which are mapped into a 2×3 two-dimensional message passing interface (MPI) virtual topology grid with a cyclic boundary condition along its two directions, as shown in Figure 2.4 (*a*). After this arrangement in Figure 2.4 (*a*), the 9×9 matrix is divided into 2×2 blocks (The boxes are marked by solid gray lines. The dashed gray lines associated with the rightmost and the bottommost boxes indicate that these blocks may not be fully filled), which results in six blocks corresponding to six processes as shown in Figure 2.4 (*b*).

Figure 2.4 (*c*) shows another case, in which the 9×9 matrix is divided into blocks of size 3×2, across a 2×2 process grid. Figure 2.4 (*d*) shows the matrix distribution on each process resulting from the distribution in Figure 2.4 (*c*). In Figure 2.4 (*c*), the outermost numbers denote the indices for the process rows and columns. Each solid gray line encloses a block, similar to that shown in Figure 2.4 (*a*), while the block size is changed to be 3×2; the dashed line in the rightmost boxes indicates that those blocks are not fully filled.

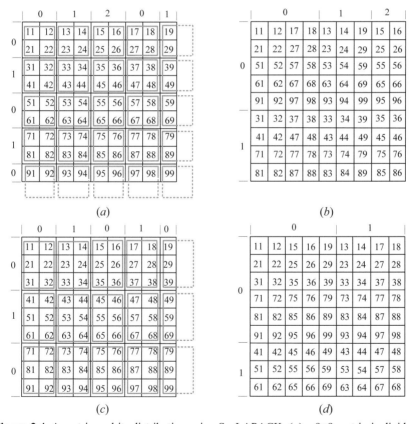

Figure 2.4. A matrix and its distribution using ScaLAPACK: (*a*) a 9×9 matrix is divided into 2×2-sized blocks; (*b*) the rearrangement of the blocks in (*a*) for each process in a 2×3 process grid; (*c*) a 9×9 matrix is divided into 3×2-sized blocks; (*d*) the rearrangement of the blocks in (*c*) for each process in a 2×2 process grid.

2.3.3   PLAPACK: Physically Based Matrix Distribution

The *physically based matrix distribution* (PBMD) was proposed by the authors of PLAPACK as a basis for a more flexible parallel linear algebra library. The application developers will not have to "unnaturally" modify the way they want to distribute data in order to fit the required distribution required by the library. It is possible to distribute the vectors (which contain the data of "physical significance") across processes in a way that depends only on the application. The matrix is then distributed in a conformal way that, in effect, optimizes the computations of the matrix–vector product.

Even though the distribution of the data is automatically done in PLAPACK, it may still be interesting to know how the vectors and the matrices are distributed across the processes for those who want to develop programs using PLAPACK. It is particularly worthwhile to discuss how PLAPACK distributes the data, because PLAPACK adopts a method that is different from the ScaLAPACK library package.

Given a two-dimensional grid of P_r rows and P_c columns of processes, a vector (one-dimensional) is assumed to be uniformly distributed by blocks of $n_b \times 1$ components across the processes of the entire grid, in a column-major order. Assume that $[X]$ and $[B]$ are vectors of length M that are related by $[A][X]=[B]$ (where $[A]$ is an $M \times M$ matrix). Then PLAPACK requires that $[X]$ and $[B]$ be distributed in a similar fashion, so that $[A]$ can be partitioned into square blocks of order n_b (note that in PLAPACK $m_b = n_b$), and distributed according to the following rule: A block of $[A]$ is in the same column of the process grid as the block of $[X]$ that it multiplies and the same row of the process grid as the block of $[B]$ that it updates. In general, if $[X_i]$ is the *i*th block of $[X]$ (i = 1,2,3,...), and row($[X_i]$) and col($[X_i]$) are the row and column coordinates of the process grid to which $[X_i]$ is mapped, then

$$\text{row}([X_i]) = (i-1)\bmod P_r, \quad \text{col}([X_i]) = \left[floor\left(\frac{i-1}{P_r}\right)\right]\bmod P_c, \qquad (2.14)$$

where the *floor* function returns the highest integer less than or equal to the number, and "mod" represents the modulo operator. Thus, according to the PBMD distribution rule, block $[A_{ij}]$ should be mapped to the process grid in {row($[B_i]$), col($[X_j]$)} position, i.e., the { $(i-1)\bmod P_r$, $\left[floor((j-1)/P_r)\right]\bmod P_c$ } position. To make it clear, the application of PBMD is demonstrated using a small matrix, where a 7×7 matrix is distributed using a 1×1 block size over six processes.

Assume that a 2×3 process grid is chosen; Figure 2.5 (*a*) illustrates the process coordinates of one possible mapping of the six processes to the grid in Figure 2.5 (*b*).

(0,0)	(0,1)	(0,2)
(1,0)	(1,1)	(1,2)

(a)

0	2	4
1	3	5

(b)

Figure 2.5. A two-dimensional grid filled with six processes in a column-major order: (*a*) process coordinates; (*b*) process rank.

The physically based distribution of a 7×7 matrix, using a block size $n_b = 1$, to the 2×3 process grid given in Figure 2.5 is illustrated in Figure 2.6. Figure 2.6 (*a*) shows the 7×7 matrix and illustrates the mapping procedure. Figure 2.6 (*b*) shows the block matrices in every process after distribution.

In Figure 2.6 (*a*), the numbers in the dashed box are coordinates of each process that owns a portion of $[X]$ and $[B]$. The arrows between the blocks of the vector and the dashed-line boxes illustrate the relationship between each block vector and each process coordinate. The numbers below the horizontal dashed-line box indicate the process column coordinates, which correspond to the column coordinates of the processes used for mapping the elements of the $[X]$ vector, as shown by the arrows pointing down from the horizontal dashed-line box. The numbers to the left of the vertical dashed-line box indicate the process row coordinates, which correspond to the row coordinates of the processes used for mapping the elements of the $[B]$ vector, as shown by the arrows pointing to the left from the vertical dashed-line box.

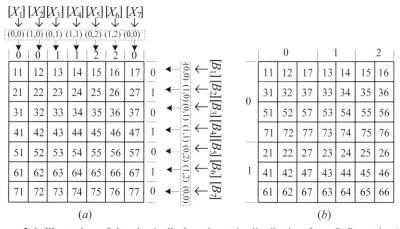

Figure 2.6. Illustration of the physically based matrix distribution for a 7×7 matrix: (*a*) mapping of the vectors for distributing a matrix; (*b*) rearrangement of the blocks in (*a*) for each process in a 2×3 process grid.

As shown in Figure 2.6 (*b*), process (0, 0) has the elements from rows (1,3,5,7) and columns (1,2,7). This is because elements 1, 3, 5, and 7 of vector [*B*] are assigned to all the processes with the row coordinate in the process grid being 0, and elements 1, 2, and 7 of vector [*X*] are assigned to all the processes, where the column coordinate in the process grid is 0.

2.3.4 Data Distribution Comparison between ScaLAPACK and PLAPACK

Now the distributions of the elements of the same matrix using the ScaLAPACK and PLAPACK library packages are presented and compared.

As an example, consider a matrix $[A]$ of Figure 2.7 (*a*), which is distributed to different processes in the 2×3 process grid illustrated in Figure 2.5. Figure 2.7 (*b*) shows which process the blocks of $[A]$ are distributed to using ScaLAPACK's distribution methodology.

In Figure 2.7 (*a*), the outermost numbers denote the row and column indices of the process coordinates. The top and bottom number in any block of Figure 2.7 (*b*) denotes the process ID (rank) and the process coordinates of a certain process, respectively, corresponding to the block of the matrix shown in Figure 2.7 (*a*). By varying the dimensions of the blocks of $[A]$ and those of the process grid, different mappings can be obtained.

(a)

	0	1	2	0	1	2
0	11	12	13	14	15	16
1	21	22	23	24	25	26
0	31	32	33	34	35	36
1	41	42	43	44	45	46
0	51	52	53	54	55	56
1	61	62	63	64	65	66

(b)

0 (0,0)	2 (0,1)	4 (0,2)	0 (0,0)	2 (0,1)	4 (0,2)
1 (1,0)	3 (1,1)	5 (1,2)	1 (1,0)	3 (1,1)	5 (1,2)
0 (0,0)	2 (0,1)	4 (0,2)	0 (0,0)	2 (0,1)	4 (0,2)
1 (1,0)	3 (1,1)	5 (1,2)	1 (1,0)	3 (1,1)	5 (1,2)
0 (0,0)	2 (0,1)	4 (0,2)	0 (0,0)	2 (0,1)	4 (0,2)
1 (1,0)	3 (1,1)	5 (1,2)	1 (1,0)	3 (1,1)	5 (1,2)

Figure 2.7. Block-cyclic distribution of a matrix as performed by ScaLAPACK: (*a*) a matrix consisting of 6×6 blocks of size $m_b \times n_b$ (here $m_b = n_b$); (*b*) rank and coordinates of each process owning the corresponding blocks in (*a*).

Given the same process grid in Figure 2.5 and the block size of $m_b \times n_b$ ($m_b = n_b$), the PBMD of PLAPACK partitions the matrix uniformly into square blocks, and distributes the blocks to the processes, as shown in Figure 2.8. The number of contiguous blocks in a row that are mapped to one process is equal to the number of rows in the grid of the processes.

Figure 2.8 (a) shows the distribution of $[X]$ and $[B]$, in the equation $[A][X]=[B]$, to different processes with block size $n_b \times 1$. The implications of the dashed-line box and the arrows in Figure 2.8 (a) are similar to the one discussed for Figure 2.6 (a). Figure 2.8 (b) shows where each block of $[A]$ [in Fig. 2.8 (a)] resides, i.e., the distribution of the processes and their corresponding process coordinates denoted by the top and bottom number in each box, respectively. In PLAPACK, $[A_{11}]$ and $[A_{12}]$ must reside on process (0, 0), i.e., process 0 in the grid; $[A_{13}]$ and $[A_{14}]$ must reside on process (0, 1), i.e., process 2, and so on.

For both ScaLAPACK and PLAPACK, one needs to distribute the matrix over a two-dimensional process grid, but it can be done in different ways and the users can, more or less, influence the way in which the data related to the matrix are distributed. Users of PLAPACK, who only want to use the routines from the library package, do not need to be concerned about the way data are distributed. The distribution can be done automatically. To use ScaLAPACK, however, the user has to create and fill the local parts of the global matrix manually.

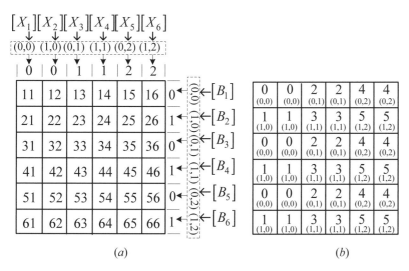

(a)   (b)

Figure 2.8. PBMD distribution of a matrix by PLAPACK: (a) mapping vectors for distributing a 6×6 matrix blocks of size $m_b \times n_b$ (here $m_b = n_b$); (b) rank and coordinates of the processes owning the corresponding blocks in (a).

2.4   DATA DECOMPOSITION FOR AN OUT-OF-CORE SOLVER

The reason for developing an out-of-core algorithm is that the matrix is too large to be stored in the main memory of the system. When the matrices involved are too large to fit in the main memory of the computer, they must be stored on the hard disks. When filling the matrix, one portion of the matrix is computed at a time, and that portion is written to the hard disk, then another portion of the matrix is computed and written, and this procedure is repeated until the complete matrix is evaluated. For the in-core matrix filling algorithm, the entire matrix is filled in one step, but need not be written out. The main idea of designing an out-of-core filling algorithm is to modify the in-core filling algorithm structure and fill a portion of the matrix at a time instead of the whole matrix. When performing an out-of-core LU factorization, each portion of the matrix is read into the RAM and the LU decomposition is started. On completion, the result of the LU factorization of this portion of the matrix is written back to the hard disk. The code then proceeds with the next portion of the matrix until the entire matrix is LU factored. This is the basic mechanism of an out-of-core LU factorization, which will be further discussed in the following sections.

An important point for an out-of-core algorithm is that during the evaluation of the elements of the matrix and its solution through the LU decomposition, the original matrix is decomposed into a set of smaller matrices that can be fitted in-core. To do this, the following assumptions are necessary. The (double-precision complex) matrix has M rows and columns and requires $N_{storage} = M \times M \times 16$ bytes of hard-disk storage. The computer system has M_{RAM} bytes of in-core buffer available to each process. To complete the decomposition of the matrix, each process will handle a specific number of portions I_{slab}, which must satisfy the equation

$$I_{slab} = ceiling \left\{ \frac{N_{storage}}{pM_{RAM}} \right\}, \qquad (2.15)$$

where p is the total number of processes, and the *ceiling* function returns the smallest integer greater than or equal to the number.

When the decomposition is oriented along the column, each in-core filling is a slab of the matrix, as shown in Figure 2.9, consisting of all the M rows of the matrix by the number of columns that will fit in the total amount of the in-core buffer available pM_{RAM}. Note that the value of M_{RAM} for the in-core buffer memory is not required to be the same as the physical memory available to each process on the computer. To obtain the best performance of an out-of-core code, the value of M_{RAM} should always be less than the size of physical memory for each process. This is because the operating system needs some memory resource, and if the out-of-core code uses up all of the physical memory, the system will start to use virtual memory. Using virtual memory of the computer degrades the performance, which will be discussed further in Chapter 5.

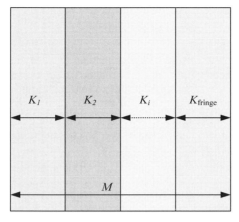

Figure 2.9. Data decomposition for storing an out-of-core matrix.

The width of the ith out-of-core slab K_i, as illustrated in Figure 2.9, is bounded by

$$M = \sum_{i=1}^{I_{\text{slab}}} K_i .$$
(2.16a)

At the last out-of-core fill ($i = I_{\text{slab}}$), the number of unfilled columns is

$$K_{\text{fringe}} = M - \sum_{i=1}^{I_{\text{slab}}-1} K_i .$$
(2.16b)

When only one process is used, the out-of-core matrix filling can easily be done with a slight modification to the serial code. However, when more processes are used, the assignment to fill each slab is distributed over p processes by using the block distribution scheme of ScaLAPACK and PLAPACK. Thus, the matrix filling schemes need to be designed in such a way so as to avoid redundant calculation for better parallel efficiency.

The parallel out-of-core matrix filling algorithm for MoM, using different types of basis functions, will be further explained in the following chapters based on the data distribution given here.

2.5 OUT-OF-CORE LU FACTORIZATION

Assuming that the out-of-core filling of the matrix is completed, the next step involves performing the out-of-core LU factorization. It is well known that high performance can be achieved for an algorithm portable to different computer platforms by designing the algorithms in terms of matrix–matrix multiplications. Such algorithms are called *blocked* algorithms.

With the relationships presented in Eq. (2.11), variants can be developed by postponing the formation of certain components and manipulating the order in which they are formed. Two natural variants occur: the left-looking and the right-looking algorithms. The terms "left" and "right" refer to the regions of the data access, as shown in Figure 2.10.

The shaded parts in Figure 2.10 indicate the matrix elements accessed when forming a block row or column. The darker shaded parts indicate the block row or column being modified. Assuming the matrix is partitioned into 3×3 blocks, the left-looking variant, shown in Figure 2.10 (*a*), computes one block column at a time by using previously computed columns, while the right-looking variant, shown in Figure 2.10 (*b*), computes one block row and one column at each step and uses them to update the trailing submatrix [the trailing submatrix is shown as $\left[\hat{A}_{33} \right]$ in Fig. 2.10 (*b*)]. The submatrices in the darker shaded parts represent the LU factors that will be obtained at the current step. By repartitioning the matrix after each step, the calculation can continue until the whole matrix is factorized.

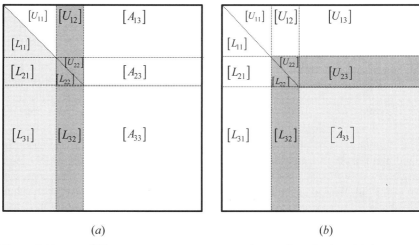

$$(a) \qquad\qquad\qquad\qquad (b)$$

Figure 2.10. Two different memory access patterns for LU factorization: (*a*) data access in the left-looking LU algorithm; (*b*) data access in the right-looking LU algorithm.

In the following section, we will briefly review the analysis of the I/O (input/output) of the two blocked LU factorization algorithms with partial pivoting, the right-looking algorithm and the left-looking algorithm [17,18], which are relevant to an out-of-core implementation.

2.5.1   I/O Analysis of Serial Right-Looking and Left-Looking Out-of-Core LU Algorithms

2.5.1.1 Right-Looking Algorithm. Consider an $M \times M$ matrix $[A]$. Factor it into:

$$[P][A] = [L][U], \qquad (2.17)$$

where $[P]$ is a permutation matrix representing the effects of pivoting performed to improve the numerical stability of the algorithm and involves the interchange of matrix rows; $[L]$ and $[U]$ are the lower and upper triangular factors, respectively.

Assume that a partial factorization of $[A]$ has been obtained so that the first n_b columns of $[L]$ and the first n_b ($m_b = n_b$) rows of $[U]$ have been evaluated. Then the partial factorization may be written in block partitioned form, with square blocks along the leading diagonal, as

$$[P][A] = \begin{bmatrix} [L_{11}] & [0] & [0] \\ [L_{21}] & [I] & [0] \\ [L_{31}] & [0] & [I] \end{bmatrix} \begin{bmatrix} [U_{11}] & [U_{12}] & [U_{13}] \\ [0] & [\hat{A}_{22}] & [\hat{A}_{23}] \\ [0] & [\hat{A}_{32}] & [\hat{A}_{33}] \end{bmatrix}, \qquad (2.18)$$

where $[L_{11}]$ and $[U_{11}]$ are $n_b \times n_b$ matrices.

In Eq. (2.18) the block matrices, $[L]$ and $[U]$, to the left and above the dotted line, have already been evaluated. The blocks labeled $[\hat{A}_{ij}]$ in Eq. (2.18) are the portions of $[A]$ that were updated but have not yet been factored, and will be referred to as the *active submatrix*.

To continue the factorization, the next block column of $[L]$ and the next block row of $[U]$ need to be evaluated, so that

$$\begin{bmatrix} [I] & [0] \\ [0] & [P_2] \end{bmatrix} [P][A] = \begin{bmatrix} [L_{11}] & [0] & [0] \\ [L_{21}] & [L_{22}] & [0] \\ [L_{31}] & [L_{32}] & [I] \end{bmatrix} \begin{bmatrix} [U_{11}] & [U_{12}] & [U_{13}] \\ [0] & [U_{22}] & [U_{23}] \\ [0] & [0] & [\hat{A}_{33}] \end{bmatrix}, \qquad (2.19)$$

$$[P_2] \begin{bmatrix} [\hat{A}_{22}] \\ [\hat{A}_{32}] \end{bmatrix} = \begin{bmatrix} [L_{22}] \\ [L_{32}] \end{bmatrix} [U_{22}], \qquad (2.20)$$

where $[P_2]$ is a permutation matrix of order $M - n_b$. In Eq. (2.19), the block matrices, $[L]$ and $[U]$, lying between the two dotted lines, need to be evaluated. The first thing to do is to factor the first block column of the active submatrix, which will be referred to as the *current column.* This can be observed by comparing Eqs. (2.18) and (2.19). This step results in the block column of $[L]$.

Then the active submatrices, to the right of the current column, and the partial $[L]$ matrix, to the left of the current column, are pivoted

$$\begin{bmatrix} [\hat{A}_{23}] \\ [\hat{A}_{33}] \end{bmatrix} \Leftarrow [P_2] \begin{bmatrix} [\hat{A}_{23}] \\ [\hat{A}_{33}] \end{bmatrix}, \qquad \begin{bmatrix} [L_{21}] \\ [L_{31}] \end{bmatrix} \Leftarrow [P_2] \begin{bmatrix} [L_{21}] \\ [L_{31}] \end{bmatrix}, \qquad (2.21)$$

and the triangular system is solved using

$$[U_{23}] = [L_{22}]^{-1}[A_{23}] \qquad (2.22)$$

to obtain the next block row of $[U]$. Finally, a matrix–matrix product is performed to update $\left[\hat{A}_{33}\right]$:

$$\left[\hat{A}_{33}\right] \Leftarrow \left[\hat{A}_{33}\right] - [L_{32}][U_{23}]. \qquad (2.23)$$

The blocks only need to be relabeled to advance to the next step.

The main advantage of the block partitioned form of the LU factorization algorithm is that the updating of $\left[\hat{A}_{33}\right]$ in Eq. (2.23) involves a matrix–matrix operation if the block size n_b is greater than 1. Machine-dependent parameters, such as cache size, number of vector registers, and memory bandwidth, will dictate the best choice for the block size n_b. Matrix–matrix operations are generally more efficient than the matrix–vector operations on high-performance computers. However, when the block size n_b is equal to 1, a matrix–vector operation is used to perform an outer product, which is generally the least efficient of the level 2 BLAS since it updates the whole submatrix.

Since $[L]$ is lower triangular and $[U]$ is upper triangular, the original array $[A]$ can be used to store the factorization. Note that in this and all other versions of LU factorization, the additional zeros and ones appearing in the representation do not have to be stored explicitly.

Now the amount of I/O required for transfer of data to and from the disk for the block-partitioned, right-looking LU factorization of a $M \times M$ matrix $[A]$, with a block size of n_b ($m_b = n_b$), can be derived. For descriptive clarity, assume that M is exactly divisible by n_b. The factorization proceeds in M/n_b steps, which are indexed as $k = 1, \ldots, M/n_b$. For the general step k, the active submatrix is the $M_k \times M_k$ matrix in the lower right corner of $[A]$, where $M_k = M - (k-1)n_b$. At step k, it is necessary to both read and write all of the active submatrices, so the total amount of I/O for the right-looking algorithm is

$$(R+W)\sum_{k=1}^{M/n_b}\left[M-(k-1)n_b\right]^2 = \frac{M^3}{3n_b}\left[1+O\left(\frac{n_b}{M}\right)\right](R+W),\qquad(2.24)$$

where R and W are the time required to read and write one matrix element, respectively. It is also assumed that there is no time spent at the startup when performing I/O.

2.5.1.2 Left-Looking Algorithm. In terms of data access, the left-looking variant is better than the right-looking variant. Start with the assumption that

$$[P][A]=\begin{bmatrix}[L_{11}] & [0] & [0]\\ [L_{21}] & [I] & [0]\\ [L_{31}] & [0] & [I]\end{bmatrix}\begin{bmatrix}[U_{11}] & [A_{12}] & [A_{13}]\\ [0] & [A_{22}] & [A_{23}]\\ [0] & [A_{32}] & [A_{33}]\end{bmatrix},\qquad(2.25)$$

where the block matrices, $[L]$ and $[U]$, to the left of the dotted line, have already been evaluated. We then proceed to the next step in the factorization procedure to form

$$\begin{bmatrix}[I] & [0]\\ [0] & [P_2]\end{bmatrix}[P][A]=\begin{bmatrix}[L_{11}] & [0] & [0]\\ [L_{21}] & [L_{22}] & [0]\\ [L_{31}] & [L_{32}] & [I]\end{bmatrix}\begin{bmatrix}[U_{11}] & [U_{12}] & [A_{13}]\\ [0] & [U_{22}] & [A_{23}]\\ [0] & [0] & [A_{33}]\end{bmatrix},\qquad(2.26)$$

where the block matrices, $[L]$ and $[U]$, lying between the two dotted lines, need to be evaluated. By comparing Eqs. (2.25) and (2.26), it is seen that the factorization proceeds by first solving the triangular system

$$[U_{12}]=[L_{11}]^{-1}[A_{12}],\qquad(2.27)$$

and then performing a matrix–matrix product to update the rest of the middle block column of $[A]$:

$$\begin{bmatrix} \begin{bmatrix} \hat{A}_{22} \end{bmatrix} \\ \begin{bmatrix} \hat{A}_{32} \end{bmatrix} \end{bmatrix} \Leftarrow \begin{bmatrix} \begin{bmatrix} A_{22} \end{bmatrix} \\ \begin{bmatrix} A_{32} \end{bmatrix} \end{bmatrix} - \begin{bmatrix} \begin{bmatrix} L_{21} \end{bmatrix} \\ \begin{bmatrix} L_{31} \end{bmatrix} \end{bmatrix} \begin{bmatrix} U_{12} \end{bmatrix} . \tag{2.28}$$

Next, perform the factorization:

$$\begin{bmatrix} P_2 \end{bmatrix} \begin{bmatrix} \begin{bmatrix} \hat{A}_{22} \end{bmatrix} \\ \begin{bmatrix} \hat{A}_{32} \end{bmatrix} \end{bmatrix} \Leftarrow \begin{bmatrix} \begin{bmatrix} L_{22} \end{bmatrix} \\ \begin{bmatrix} L_{32} \end{bmatrix} \end{bmatrix} \begin{bmatrix} U_{22} \end{bmatrix} , \tag{2.29}$$

and finally introduce the pivoting:

$$\begin{bmatrix} \begin{bmatrix} A_{23} \end{bmatrix} \\ \begin{bmatrix} A_{33} \end{bmatrix} \end{bmatrix} \Leftarrow \begin{bmatrix} P_2 \end{bmatrix} \begin{bmatrix} \begin{bmatrix} A_{23} \end{bmatrix} \\ \begin{bmatrix} A_{33} \end{bmatrix} \end{bmatrix} ; \qquad \begin{bmatrix} \begin{bmatrix} L_{21} \end{bmatrix} \\ \begin{bmatrix} L_{31} \end{bmatrix} \end{bmatrix} \Leftarrow \begin{bmatrix} P_2 \end{bmatrix} \begin{bmatrix} \begin{bmatrix} L_{21} \end{bmatrix} \\ \begin{bmatrix} L_{31} \end{bmatrix} \end{bmatrix} . \tag{2.30}$$

Note that all the data accesses occur to the left of and within the block column being updated. Matrix elements to the right are referenced only for pivoting purposes, and even this procedure may be postponed until needed by rearranging the above operations.

The amount of I/O for the left-looking out-of-core LU factorization algorithm may be evaluated using two versions of the left-looking algorithm. In the first version, the matrix is always stored on the hard disk in a nonpivoted form, except in the last step of the algorithm, where the whole matrix is written out in pivoted form. In this case, pivoting has to be done "on the fly" when matrix blocks are read in from the hard disk. In the second version of the algorithm, the matrix is stored on the hard disk in a pivoted form.

Consider the first version in which the matrix is stored in a nonpivoted form. Whenever a block is read into RAM, the whole $M \times n_b$ block must be read so that it can be pivoted. On the completion of each step, the latest factored block is the only block that is written to the hard disk, except in the last step where all blocks are written out in a pivoted form so that the complete matrix stored on the hard disk is pivoted. Note that in some cases these writings may be omitted if a nonpivoted matrix is called because the pivots are stored in the pivot vector and can always be applied later. At step k of the algorithm, the amount of I/O required is

$$(R + W) M n_b + R M n_b (k - 1) , \tag{2.31}$$

where the first term represents reading and writing the block that is to be factored at this step and the second term represents reading in the blocks to the left. The sum over k, plus the time to write out all pivoted blocks in the last step, gives the total amount of I/O for this version of the left-looking algorithm as

$$\frac{M^3}{2n_b}\left[1+O\left(\frac{n_b}{M}\right)\right]R + 2M^2\left[1+O\left(\frac{n_b}{M}\right)\right]W . \qquad (2.32)$$

The time required for carrying out the read and writes, of order n_b/M , can be ignored. If the readings and writings are assumed to take approximately the same time, i.e., $R \approx W$, then, when compared with Eq. (2.24), this version of the left-looking algorithm is seen to perform less I/O than the right-looking algorithm.

Now consider the other version where blocks are always stored on the hard disk in a pivoted form. Although it is no longer necessary to read in all rows of a $M \times n_b$ block in this case, it is still necessary to write out partial blocks in each step. This is because the same pivoting carried out when factoring the block column needs to be applied to the blocks to the left, which must then be written to the hard disk. At any step k, all the blocks to be updated must be read in and written out.

As indicated in Figure 2.11 (a), the shaded parts of the blocks, to the left of the dark-shaded column, which must be read from the disk at step $k = 5$, form a stepped trapezoidal shape.

As indicated in Figure 2.11 (b), the shaded parts of the blocks, to the left of the dark-shaded column, which must be written out after applying the pivots for this step ($k = 5$), form a rectangle shape.

The diagonal dashed lines in Figure 2.11 are visual aids to show the trapezoidal shape and the rectangle shape. Note that here k starts from 1.

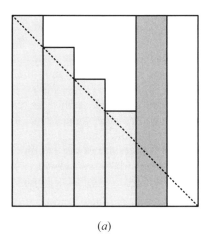
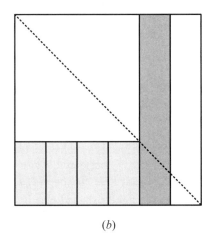

(a) (b)

Figure 2.11. The left-looking LU factorization algorithm that writes the matrix in a pivoted form: (a) data read from the hard disk; (b) data written to the hard disk.

For any step $k > 1$, the amount of I/O required is

$$(R+W)\,Mn_b + Rn_b \sum_{i=1}^{k-1} [M - (i-1)\,n_b] + Wn_b\,[M - (k-1)\,n_b](k-1)\,, \qquad (2.33)$$

and at step $k = 1$ the amount of I/O required is $(R+W)Mn_b$. The total amount of I/O required for the left-looking version of the algorithm is

$$\frac{M^3}{3\,n_b}\left[1+O\left(\frac{n_b}{M}\right)\right]R + \frac{M^3}{6\,n_b}\left[1+O\left(\frac{n_b}{M}\right)\right]W\,. \qquad (2.34)$$

If the read and write to the hard disk use the same amount of time, then the two left-looking versions of the algorithm have the same total amount of I/O, and they both perform less I/O than the right-looking algorithm. Therefore, a left-looking algorithm would be better than a right-looking algorithm for an out-of-core LU factorization technique [17,18].

2.5.2   Implementation of the Serial Left-Looking Out-of-Core LU Algorithm

In this section the implementation of the sequential left-looking out-of-core LU factorization will be discussed. In the left-looking out-of-core algorithm, there are only two block columns of width n_b that may be in-core at any time. One of these is the block column being updated and factored, which is referred to as the *active block*. The other is one of the block columns lying to the left of the active block column, which is referred to as the *temporary block*.

As can be seen in Section 2.5.1.2, the three main computational tasks in a step of the left-looking algorithm are the solution of a triangular system of Eq. (2.27), a matrix–matrix multiplication as given by Eq. (2.28), and an LU factorization of the system of Eq. (2.29).

In the left-looking out-of-core algorithm, a temporary block needs to be read just once to be used in both the solution of the triangular system and the matrix–matrix multiplication, since the two steps are intermingled.

To clarify this, a matrix consisting of 6×6 submatrices as shown in Figure 2.12 (*a*) is considered. The size of each submatrix is considered to be $n_b \times n_b$. The role that the *current temporary block column* (block column 3, marked as the lighter shaded part in Fig. 2.12) plays in the factorization of the *active block column* (block column 5, marked as the darker shaded part in Fig. 2.12) will be illustrated.

Figure 2.12 (*b*) indicates that the factorization for block columns, which are to the left of the current active block column, has already been completed. In other words, the LU submatrices in block columns 1−4 have already been computed.

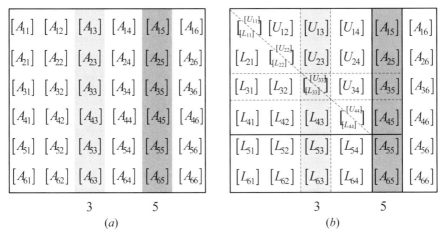

Figure 2.12. Temporary block column 3 and active block column 5: (*a*) a matrix consisting of 6×6 submatrices; (*b*) the block columns that are left to the active block column 5 have been LU-factorized.

In Figure 2.12 (*b*), the diagonal dashed line is used to show the lower and upper parts of the LU factors. The two vertical dashed lines form the boundary of the current temporary block column. The two vertical solid lines are the boundary of the active block column. The LU factors to the left of the left boundary of the active block column have been evaluated. The two horizontal dashed lines delineate the boundary of the third block row (transpose position of the current temporary block column: block column 3). The significance of the solid horizontal line will be explained later, together with the relevant computations.

The computations, using block columns 1−3, for factoring and updating the active block column (block column 5) consist of

$$[U_{15}] \Leftarrow ([L_{11}])^{-1} [A_{15}], \tag{2.35a}$$

$$[U_{25}] \Leftarrow ([L_{22}])^{-1} ([A_{25}] - [L_{21}][U_{15}]), \tag{2.35b}$$

$$[U_{35}] \Leftarrow ([L_{33}])^{-1} ([A_{35}] - [L_{31}][U_{15}] - [L_{32}][U_{25}]), \tag{2.36a}$$

$$[A_{45}] \Leftarrow [A_{45}] - [L_{41}][U_{15}] - [L_{42}][U_{25}] - [L_{43}][U_{35}], \tag{2.36b}$$

$$[A_{55}] \Leftarrow [A_{55}] - [L_{51}][U_{15}] - [L_{52}][U_{25}] - [L_{53}][U_{35}], \tag{2.36c}$$

$$[A_{65}] \Leftarrow [A_{65}] - [L_{61}][U_{15}] - [L_{62}][U_{25}] - [L_{63}][U_{35}]. \tag{2.36d}$$

Evaluation of $[U_{15}]$ and $[U_{25}]$ in Eq. (2.35) can be done by using the earlier temporary block columns, block columns 1 and 2, which are to the left of the current temporary block column. Updates of $[A_{35}]$, $[A_{45}]$, $[A_{55}]$, and $[A_{65}]$ [which are within the boundary of block column 5 and below the topmost horizontal dashed line as shown in Fig. 2.12 (b)], which are indicated in the right side of Eqs. (2.36a)–(2.36d), can be carried out by using the earlier temporary block column with $[U_{15}]$ and $[U_{25}]$. Therefore, the first two blocks, $[A_{13}]$ and $[A_{23}]$, of the current temporary block column plays no role in the computation of the active block column. The submatrix $[U_{35}]$ can be obtained from Eq. (2.36a) by using the level 3 BLAS routines _TRSM, and $[A_{45}]$, $[A_{55}]$, and $[A_{65}]$ can then be further updated using Eqs. (2.36b)–(2.36d) by using the level 3 BLAS routines _GEMM.

After processing of all the block columns to the left of block column 5 (block columns 1–4) has been completed, matrices $[A_{55}]$ and $[A_{65}]$ [below the horizontal solid line as shown in Fig. 2.12 (b)] are factored using the LAPACK routine _GETRF.

For the general case, consider the role that block column i plays in the factorization of block column k (where $i < k$), as shown in Figure 2.13.

In Figure 2.13, the first $(i-1)n_b$ rows of the block column i, labeled $[T]$, are not involved in the factoring of the block column k. The next n_b rows of the block column i are labeled $[T_0]$, and the next $(k-1-i)n_b$ rows are labeled $[T_1]$. The last $M-(k-1)n_b$ rows are labeled $[T_2]$. The corresponding portions of the block column k are labeled $[C]$, $[C_0]$, $[C_1]$, and $[C_2]$.

The function of the dashed and solid lines shown in Figure 2.13 is similar to the ones of Figure 2.12 (b). Here they are used to outline the diagonal, the boundary of the temporary block column, and the boundary of the active block column.

Then the role played by block column i in factoring block column k can be expressed in the following operations

$$[C_0] \Leftarrow (tril[T_0])^{-1}[C_0], \qquad (2.37)$$

$$\begin{bmatrix}[C_1]\\[C_2]\end{bmatrix} \Leftarrow \begin{bmatrix}[C_1]\\[C_2]\end{bmatrix} - \begin{bmatrix}[T_1]\\[T_2]\end{bmatrix}[C_0], \qquad (2.38)$$

where $[C_0]$, which is computed in Eq. (2.37), is used in Eq. (2.38). The notation tril in Eq. (2.37) denotes the lower triangular part of a matrix with diagonal elements replaced by ones.

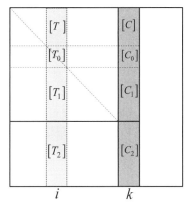

Figure 2.13. The temporary block column i and the active block column k.

In updating the block column k, the out-of-core algorithm sweeps over all block columns to the left of block column k and perform, for each block column, the solution of the triangular system of Eq. (2.37) and the matrix–matrix multiplication in Eq. (2.38). After processing of all the block columns to the left of block column k has been completed, the matrix $[C_2]$ (below the solid horizontal line in Fig. 2.13) is factored. Pivoting can also be carried out as discussed next.

In the version of the algorithm where the matrix is stored on the hard disk in a pivoted form, it is necessary to read in only those parts of the temporary blocks that play a role in the computation for the current active block column. When a partial temporary block column is read into the RAM, the pivots obtained when factoring $[C_2]$ in the previous step must be applied before using it, and it must then be written back out to the disk, as indicated in Figure 2.11. The pivots from factoring $[C_2]$ in the current step will be applied to all temporary blocks (before using them) that play a role in factoring the next active block column.

If the matrix is stored on the hard disk in a nonpivoted form, then whenever a block column is read into the RAM, all the elements of the pivot computed up to that point must be applied. Also, after updating and factoring the active block, the pivots must be applied to it in reverse order to undo the effect of pivoting before writing the block column to the disk.

Figure 2.14 provides the pseudocode of a version of the left-looking algorithm, in which the matrix is stored in a nonpivoted form. Note that a complete block column is always read or written in this version. Since the information about the pivots is maintained in-core, the factored matrix can always be read in later to be pivoted. In the pseudocode of Figure 2.14, it is assumed that the matrix is of size $M \times M$ and M is divisible by the block size n_b ($m_b = n_b$). Fortunately, the general case is only slightly more complicated. Note that it is necessary to position the file pointer (at the start of the file) only once during each pass through the outer loop.

An Out-of-Core Left-Looking LU Factorization Algorithm:

loop over block column, $k = 1, \ldots, M / n_b$;

    read block column k into active block;

    apply pivots to active block;

    go to the start of the file;

        loop over block column left to k, $i=1, \ldots, k\text{-}1$;

            read block column i into temporary block;

            apply pivots to temporary block;

            triangular solve;

            matrix multiply;

        end loop;

    factor matrix $\left[C_2 \right]$;

    undo the effect of pivoting for active block;

    write active block;

end loop;

Figure 2.14. An out-of-core left-looking LU factorization algorithm.

It is seen in the previous section that the left-looking LU factorization routine requires less I/O than the right-looking variant. However, the left-looking algorithm has less inherent parallelism since it acts only on each individual block column. To obtain better performance, the actual implementation of the left-looking factorization should use two full in-core column panels [17], as shown in Figure 2.15. The lightly shaded part of this figure represents temporary block columns and the darker-shaded part represents the active block columns. The dashed and the solid lines have similar implications as illustrated in Figures 2.12 (*b*) and 2.13. Note that the widths of the two panels are different.

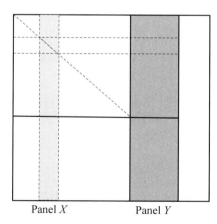

Panel *X*   Panel *Y*

Figure 2.15. The algorithm requires two panels.

The benefit of having two different widths is as follows. When each panel is updated and factored, all the elements to its left must be read from the disk. Thus, the wider the panel width of Y (here panel Y actually refers to slab Y), the less is the I/O required for the complete factorization. By reading the matrix from the left, one panel X at a time, the in-core memory can be mostly dedicated to a single (panel) slab Y, thus nearly maximizing the width of panel Y. This will be referred to as a *one-slab solver* in the following sections.

2.5.3   Design of a One-Slab Left-Looking Out-of-Core LU Algorithm

Given a general $M \times N$ matrix $[A]$, its LU factorization, when employing partial pivoting, is given by

$$[P][A] = [L][U],   \tag{2.39}$$

where $[P]$ is a permutation matrix of order M. The matrix $[L]$ is an $M \times N$ lower triangular matrix, and $[U]$ is an $N \times N$ upper triangular matrix. Denote the computation of $[P]$, $[L]$, and $[U]$ by

$$[A, p] := \left[\{L \setminus U\}, p \right] = LU([A]),   \tag{2.40}$$

where $\{L \setminus U\}$ is the matrix whose strictly lower and upper triangular parts are $[L]$ and $[U]$, respectively. Here it is recognized that $[L]$ has ones on the diagonal, which need not be stored, so the diagonal elements of $\{L \setminus U\}$ belong to $[U]$. The factors $[L]$ and $[U]$ can be stored by overwriting the original contents of $[A]$. The permutation matrix is generally stored in a vector $[p]$ of M integers.

The term *slab solver* is derived from the fact that the matrix is viewed as a collection of slabs consisting of K adjacent columns. The first slab is brought from the hard disk (out-of-core) into the memory (in-core). It is first factored and written back to the hard disk. The second slab is brought into memory and its forward substitution with the first slab starts, requiring the first slab to also be loaded into the memory, although not all at once, as will be seen next. Once updated, the second slab is factored and written back to the disk. The procedure continues with all the remaining slabs. When slab j is reached, it is loaded into the memory. Forward substitution with slabs $1, \ldots, j-1$, continues in order, after which slab j is factored and written back to the disk.

The primary reason that the left-looking algorithm is chosen at the top level is that the required I/O is less than that of a right-looking algorithm. The bulk of the I/O lies with the reading of the prior slabs (the part of the matrix to the left of the current slab, and these data need only to be read) during the forward substitution stage.

The complete left-looking out-of-core algorithm is given in Figure 2.16. The following notation is used to characterize the following matrices: $[A_{TL}]$ and $[A_{BL}]$ are the top left and the bottom left partitions of $[A]$, respectively; $[A_{TR}]$ and $[A_{BR}]$ are the top right and the bottom right partitions, respectively; $[p_T]$ and $[p_B]$ are the top and bottom partitions of the pivot vector $[p]$, respectively; $[B_T]$ is the top partition and $[B_B]$ is the bottom partition of $[B]$; and $n([A_{TL}])$ and $n([A])$ are the column dimensions of the matrices $[A_{TL}]$ and $[A]$, respectively. K is called the *block size* in the algorithm, and corresponds to the width of the slab. The thick and thin lines inside the matrices and vectors are used to explain how the matrix has been partitioned. At the end of each iteration of the repartitioning loop for the matrix, the thick line will move to the position of the thin line, as will be illustrated next.

In the rest of this chapter, the row and column indices of matrices start from 0 rather than 1 (which was used in the discussions above) for an easier C-language implementation of the algorithm. The left-looking out-of-core algorithm is described as follows (ignoring pivoting for now).

Partition the matrices as

$$[A] \rightarrow \begin{bmatrix} [A_{00}] & [A_{01}] & [A_{02}] \\ [A_{10}] & [A_{11}] & [A_{12}] \\ [A_{20}] & [A_{21}] & [A_{22}] \end{bmatrix}, \tag{2.41a}$$

$$[L] \rightarrow \begin{bmatrix} [L_{00}] & [0] & [0] \\ [L_{10}] & [L_{11}] & [0] \\ [L_{20}] & [L_{21}] & [L_{22}] \end{bmatrix}, \tag{2.41b}$$

$$[U] \rightarrow \begin{bmatrix} [U_{00}] & [U_{01}] & [U_{02}] \\ [0] & [U_{11}] & [U_{12}] \\ [0] & [0] & [U_{22}] \end{bmatrix}. \tag{2.41c}$$

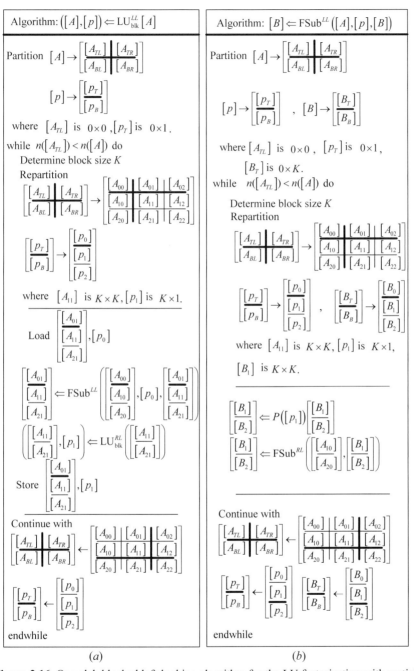

Figure 2.16. One-slab blocked left-looking algorithm for the LU factorization with partial pivoting: (*a*) one-slab left-looking LU factorization algorithm; (*b*) forward substitution algorithm in the left-looking LU algorithm.

We then obtain

$$
\begin{bmatrix}
[A_{00}] & [A_{01}] & [A_{02}] \\
\hline
[A_{10}] & [A_{11}] & [A_{12}] \\
\hline
[A_{20}] & [A_{21}] & [A_{22}]
\end{bmatrix} =
$$

$$
\begin{bmatrix}
[L_{00}][U_{00}] & [L_{00}][U_{01}] & [L_{00}][U_{02}] \\
[L_{10}][U_{00}] & [L_{10}][U_{01}]+[L_{11}][U_{11}] & [L_{10}][U_{02}]+[L_{11}][U_{12}] \\
[L_{20}][U_{00}] & [L_{20}][U_{01}]+[L_{21}][U_{11}] & [L_{20}][U_{02}]+[L_{21}][U_{12}]+[L_{22}][U_{22}]
\end{bmatrix}. \quad (2.42)
$$

Now, assume that before and after each iteration of the loop, the submatrices to the left of the thick line have been overwritten with the corresponding submatrices $[L]$ and $[U]$ [see Fig. 2.16 (*a*), below the line marked "Repartition"]. In other words, $[A_{00}]$, $[A_{10}]$ and $[A_{20}]$ have been overwritten by $\{L\backslash U\}_{00}$, $[L_{10}]$ and $[L_{20}]$, respectively:

$$
[A_{00}] \Leftarrow \{L\backslash U\}_{00}, \qquad [A_{10}] \Leftarrow [L_{10}], \qquad [A_{20}] \Leftarrow [L_{20}].
$$

To continue with the solution process, $[U_{01}]$, $\{L\backslash U\}_{11}$, and $[L_{21}]$ must overwrite the corresponding parts of $[A]$. Equation (2.42) outlines the necessary computations required. The current panel of rows must first be updated with the computation that precedes it through:

$$
\begin{aligned}
[A_{01}] &\Leftarrow [U_{01}] \Leftarrow ([L_{00}])^{-1}[A_{01}] \Leftarrow (tril[A_{00}])^{-1}[A_{01}], \\
[A_{11}] &\Leftarrow [A_{11}] - [A_{10}][A_{01}], \\
[A_{21}] &\Leftarrow [A_{21}] - [A_{20}][A_{01}].
\end{aligned}
$$

The notation *tril* denotes the lower triangular part of a matrix with the diagonal elements replaced by ones, such that it becomes a unit lower triangular matrix, as required for matrix $[L]$.

For the case where pivoting occurs in the computation that precedes the current panel, the computation is similar as in the forward substitution, shown in the operation FSub^{LL} of Figure 2.16 (*b*), where the computation is interleaved with the application of permutations. Subsequently, the updated matrix: $\{[A_{11}][A_{21}]\}^{T}$, is factored and the thick lines are moved since another panel of columns has now been completed [see Fig. 2.16 (*a*), below "Continue with"]. Here $\{\ \}^{T}$ implies that the submatrices are in the transpose position.

During the forward substitution with FSub^{LL}, the computation is interleaved with the periodic application of row exchanges [the multiplication by the permutation matrix $P([p_1])$]. Therefore, the forward substitution routine in Figure 2.16 (*b*) corresponds to that in the right-looking algorithm in Figure 2.17 (*b*) assuming that pivoting has been applied before the computation. The lines inside the matrices and vectors in Figure 2.17 are used to show the partitioning of the matrix, which is similar to the lines given in Figure 2.16.

For the factorization of the slab, once the slab has been updated, it is better to use a right-looking algorithm. The blocked, in-core, right-looking algorithm for LU factorization, with partial pivoting, is shown in Figure 2.17 (*a*). The algorithm assumes that at the start of the loop, $[A_{TL}]$ and $[A_{TR}]$ have been overwritten by $\{L\backslash U\}_{TL}$ and $[U_{TR}]$, respectively, and $[A_{BL}]$ by $[L_{BL}]$. The matrix $[P_T]$ has been obtained, and $[A_{BR}]$ has been updated to where its LU factorization still needs to be computed. The algorithm proceeds by computing an LU factorization with partial pivoting of "the current panel", overwriting $[A_{11}]$ and $[A_{21}]$ with $\{L\backslash U\}_{11}$ and $[L_{21}]$, respectively. It then applies the row-swaps to the remainder of the rows and updates $[A_{12}]$ with

$$[U_{12}] = ([L_{11}])^{-1}[A_{12}].$$

Finally, $[A_{22}]$ is updated with

$$[A_{22}] - [L_{21}][U_{12}],$$

after which $[A_{22}]$ must be factored during future iterations.

The reason for using the right-looking algorithm in the one-slab left-looking algorithm is somewhat subtle. On a sequential architecture, the right-looking algorithm involves most computation in terms of the update

$$[A_{22}] \Leftarrow [A_{22}] - [A_{21}][A_{12}],$$

where $[A_{21}]$ consists of b columns. This particular shape of matrix–matrix multiplication inherently can achieve better performance than the shape that is encountered in the left-looking algorithm. On a parallel architecture, either shared memory or distributed memory, the shape of the matrix–matrix multiplication encountered by the right-looking algorithm parallelizes more naturally, since it requires communication between parts of $[A_{21}]$ and $[A_{12}]$ among processes after which nonoverlapping parts of $[A_{22}]$ can be updated in parallel.

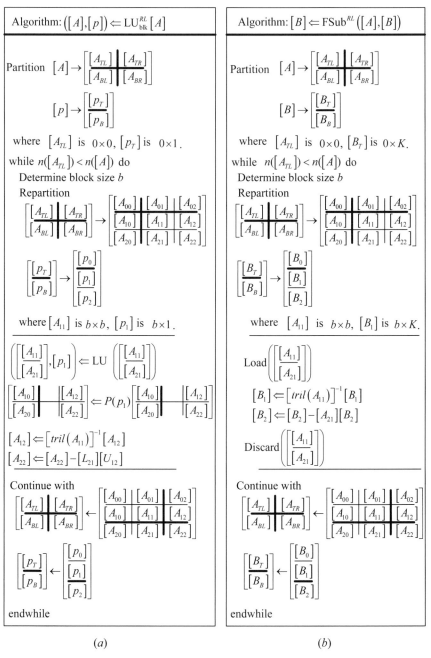

(a)   (b)

Figure 2.17. Blocked right-looking algorithm for the LU factorization with partial pivoting: (a) right-looking LU factorization algorithm; (b) forward substitution algorithm in the right-looking LU algorithm.

What makes the solution procedure in Figure 2.16 a *one-slab* solver is the observation that when slab j is being updated by slab i ($i < j$), it is not necessary to bring slab i completely into the memory. Instead, that slab is brought in by one panel (b columns) at a time.

2.6 PARALLEL IMPLEMENTATION OF AN OUT-OF-CORE LU ALGORITHM

A distributed memory parallel implementation of the algorithm presented in the previous section proceeds exactly as described, except that the matrix to be factored is distributed among the processes and parallel implementations of the various in-core operations are employed.

The two most widely used parallel in-core linear algebra library packages are ScaLAPACK and PLAPACK, the details of which have already been discussed. Both use block distribution to map an in-core matrix (in RAM) onto nodes, although there are some differences. The same block distribution of matrices used in ScaLAPACK or PLAPACK can also map an out-of-core matrix (on hard disk) onto nodes. When bringing a slab or a panel into memory, by design, it is guaranteed that the in-core copy of the submatrix is on the same node as the out-of-core copy so that reading or writing need to only access the local disk attached to that process.

2.6.1 Parallel Implementation of an Out-of-Core LU Algorithm Using ScaLAPACK

In this section, a parallel left-looking algorithm for performing an out-of-core LU factorization of dense matrices is described. Use of out-of-core storage adds an extra layer to the hierarchical memory. In order to have flexible and efficient access to this extra layer of memory, an additional level of partitioning over matrix columns has been introduced into the standard ScaLAPACK algorithm [17,18].

The matrix is now partitioned into blocks at two levels. Each out-of-core matrix is associated with a device unit number, much like the familiar FORTRAN I/O subsystem. Each I/O operation is record-oriented, where each record is conceptually an $mm_b \times nn_b$ ScaLAPACK block-cyclic distributed matrix. Moreover, if this record/matrix is distributed with (m_b, n_b) as the block size on a $P_r \times P_c$ processes grid, then

$$mm_b \bmod \left(m_b \times P_r \right) = 0,$$

$$nn_b \bmod \left(n_b \times P_c \right) = 0,$$

i.e., mm_b (and nn_b) are exact multiples of $m_b \times P_r$ (and $n_b \times P_c$). Data to be transferred are first copied or assembled into an internal temporary buffer (record). This arrangement reduces the number of calls for seeking data and encourages large contiguous block transfers, but incurs some overhead when performing memory-to-memory copies. All processes are involved during each record transfer. Individually, each process writes out a (mm_b / P_r) by (nn_b / P_c) matrix block. The values of mm_b and nn_b can be adjusted to achieve good I/O performance, with large contiguous block transfers, or to match a RAID disk stripe size [17,18]. RAID [redundant array of inexpensive (or independent) disks] is a technology that employs the simultaneous use of two or more hard-disk drives to achieve greater levels of performance, reliability, and/or larger data volume size.

In Figure 2.15, panel X is nn_b columns wide and panel Y (here it is also referred to as slab Y) occupies the remaining memory but should be at least nn_b columns wide. Panel X acts as a buffer to hold and apply previously computed factors to panel Y. Once all the updates are performed, panel Y is factored using an in-core ScaLAPACK algorithm. The results in panel Y are then written to the disk.

The implementation of an LU factorization scheme will be described in more detail below. To illustrate this, consider the block-partitioned matrix $[A]$ as

$$[A] \rightarrow \begin{bmatrix} [A_{00}] & [A_{01}] \\ [A_{10}] & [A_{11}] \end{bmatrix} = \begin{bmatrix} [L_{00}] & [0] \\ [L_{10}] & [I_{11}] \end{bmatrix} \begin{bmatrix} [U_{00}] & [U_{01}] \\ [0] & [U_{11}] \end{bmatrix}, \tag{2.43}$$

where $[A_{00}]$ is a square $K \times K$ submatrix.

The out-of-core LU factorization, carried out through PFxGETRF, involves the following operations:

1. If no updates are required in factoring the first panel, all available storage is used as one panel:

    (i)   LAREAD: reads in part of the original matrix.
    (ii) PxGETRF: performs in-core factorization using ScaLAPACK through

    $$\begin{bmatrix} [L_{00}] \\ [L_{10}] \end{bmatrix} [U_{00}] \Leftarrow [P_0] \begin{bmatrix} [A_{00}] \\ [A_{10}] \end{bmatrix}. \tag{2.44}$$

    (iii) LAWRITE: writes out the factors.

    Otherwise, partition the storage into panels X and Y.

2. Compute updates into panel Y by reading in the previous factors (nn_b columns at a time) into panel X. If panel Y contain $\begin{bmatrix} [A_{01}] \\ [A_{11}] \end{bmatrix}$, then

   (i)   LAREAD: reads in part of the factor into panel X.
   (ii) LAPIV: physically exchanges rows in panel Y to match the permuted ordering in panel X:

$$\begin{bmatrix} [\hat{A}_{01}] \\ [\hat{A}_{11}] \end{bmatrix} \Leftarrow [P_0] \begin{bmatrix} [A_{01}] \\ [A_{11}] \end{bmatrix}. \tag{2.45}$$

   (iii) PxTRSM: computes the upper triangular factor:

$$[U_{01}] \Leftarrow [L_{00}]^{-1} [\hat{A}_{01}]. \tag{2.46}$$

   (iv) PxGEMM: updates the remaining lower part of panel Y:

$$[\hat{A}_{11}] \Leftarrow [\hat{A}_{11}] - [L_{10}][U_{01}]. \tag{2.47}$$

3. Once all the previous updates are performed, apply the in-core ScaLAPACK subroutine PxGETRF to compute the LU factors of panel Y:

$$[L_{11}][U_{11}] \Leftarrow [P_1][\hat{A}_{11}]. \tag{2.48}$$

   The results are then written back to the hard disk.

4. A final extra pass over the computed lower triangular $[L]$ matrix may be required to rearrange the factors in the final permutation order:

$$[\hat{L}_{01}] \Leftarrow [P_1][L_{01}]. \tag{2.49}$$

In this four-step procedure, the various function calls are given in front of the colon; and the purpose of applying it is given after each colon. Take task (iii) of step 2 as an example, where PxTRSM is the function call, and the purpose of applying this function is that it "computes the upper triangular factor".

   Note that although PFxGETRF can accept a general rectangular matrix, a column-oriented algorithm is used. The pivot vector is held in the memory and not written to the hard disk. During the factorization, factored panels are stored on the hard disk with only partially or "incompletely" pivoted row data, whereas factored panels are stored in the original nonpivoted form described in

Section 2.5.2 and repivoted "on the fly". The current scheme is more complex to implement but reduces the number of row exchanges required.

2.6.2   Parallel Implementation of an Out-of-Core LU Algorithm Using PLAPACK

Matrix $[A]$ of Eq. (2.43) is chosen as an example in this section to illustrate the methodology using PLAPACK. In Figure 2.15, panel X is b columns and b should be exact multiples of $n_b \times P_c$ (similar to nn_b in the parallel out-of-core implementation using ScaLAPACK). Panel Y is a slab, which occupies the remaining memory but should be at least b columns wide. The out-of-core LU factorization involves the following operations:

1. If no updates are required in factoring the first panel, all available storage is used as one panel:

    (i)   PLA_Copy: reads in part of the original matrix.
    (ii)   PLA_LU: performs in-core factorization of Eq. (2.44) using PLAPACK.
    (iii) PLA_Copy: writes out the factors.

    Otherwise, partition storage into A_L and A_R (using PLA_Part_1x2). Here A_L and A_R track the part of the matrix that has already been factored and the part yet to be factored, respectively.

2. Compute updates into panel Y by reading in the previous factors A_L (b columns at a time) into panel X. Let Panel Y hold $\begin{bmatrix} [A_{01}] \\ [A_{11}] \end{bmatrix}$:

    (i)   PLA_Copy: reads in part of factor A_L into panel X.
    (ii)   PLA_Apply_pivots_to_rows: physically exchanges rows in panel Y to match permuted ordering in panel X as in Eq. (2.45).
    (iii) PLA_TRSM: computes the upper triangular factor as in Eq. (2.46).
    (iv)   PLA_GEMM: updates the remaining lower part of panel Y as in Eq. (2.47).

3. Once all the previous updates are performed, apply in-core PLAPACK subroutine PLA_LU to compute the LU factors in panel Y as in Eq. (2.48). The results are then written back to the hard disk.

4. A final extra pass over the computed lower triangular $[L]$ matrix may be required to rearrange the factors in the ultimate permutation order, as in Eq. (2.49).

In the description of the tasks, including the four steps that are involved, the function name is given in front of the colon and the purpose of applying it is given after each colon. Take for example, the task (iv) of step 2. In this case, PLA_GEMM is the function name, and it is used to update the remaining lower part of the panel.

The pivot vector can be held in the memory. Similar to the out-of-core implementation using ScaLAPACK, during the PLAPACK out-of-core factorization, factored panels are stored on the hard disk with only partially pivoted row data to reduce the number of row exchanges required.

2.6.3   Overlapping of the I/O with the Computation

The time spent on I/O can be, to some degree, much smaller in amount with respect to the computation time, and can be hidden in the scheme of the things by overlapping the I/O related to the future computations with the current computational process.

When considering the computations associated with a typical slab j, the bulk of I/O is with the loading of the elements of the matrix to its left. Moreover, the matrix to the left is loaded in panels of width b, each of which tends to be occupied by only a few columns of nodes. The slab, by contrast, is partitioned approximately equally among all the nodes. Thus, the loading of panels in Figure 2.17 (b) is much more time consuming than the loading and storing of the slab. Overlapping the loading of the next slab or the writing of the previous slab would require space for an additional slab in the memory, thereby reducing the slab width K, which directly affects the performance. In contrast, to read the next panel in Figure 2.17 (b) when a current panel is being used to update the slab, only the space for an extra panel is required. The width of a panel is much smaller than that of a slab ($b \le K$). This justifies modifying the implementation so that in Figure 2.17 the next panel is read in, when computation associated with the current panel is continuing.

2.6.4   Checkpointing in an Out-of-Core Solver

The second reason why a left-looking algorithm is preferable at the out-of-core level is that it is easy to *checkpoint* and *restart* the solution procedure after a break. The factorization of a matrix that requires storage on the hard disk can easily take hours, days, or even weeks to complete. Thus, it becomes necessary to develop a plan for recovery of the computational process from a system failure without having to restart the computation from the very beginning.

Left-looking algorithms maintain a clear separation between the part of the matrix that has been worked upon and the remaining part that has yet to be updated. This can facilitate checkpointing, which means periodically saving the state of the intermediate results so that one can roll back the computation to such an intermediate state and restart the computation from that point.

Some precaution must be taken while the current slab, after being updated and factored, is being written back to the hard disk. A simple solution is to create an extra slab on the disk to the left of the first slab. Once the first slab has been factored, it can overwrite this extra slab. Similarly, slab j overwrites slab $j-1$, as shown in Figure 2.18, where the arrow shows that each slab is written to its left.

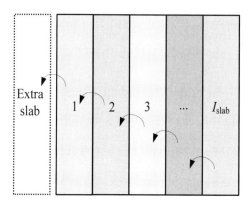

Figure 2.18. Each slab is written to its left neighboring slab after being LU factored.

In this way, if a system failure occurs while slab j is being written to the hard disk, the original data are still available and a restart can commence with the updating and factorization of slab j. The pivoting information can be much more easily maintained on the hard disk since it is a vector and one can simply write it to the hard disk whenever it has been updated.

2.7   SOLVING A MATRIX EQUATION USING THE OUT-OF-CORE LU MATRICES

After the matrix $[A]$ has been factored into the $[L]$ and $[U]$ matrices, the matrix equation given by Eq. (2.2) can be solved by first performing a forward substitution given in Eq. (2.4), and then a backward substitution as specified in Eq. (2.5). The out-of-core algorithm for solving the matrix equation using the LU-factored results is shown in Figure 2.19 (a), where a general matrix equation is assumed to have multiple vectors on the right-hand side. The thick and the thin lines inside the matrices and vectors are used as an aid to explain the partitioning of the matrix. At the end of each iteration of the repartition loop for the matrix, the thick line will move into position of the thin line, as will be discussed next. Figure 2.19 (b) presents one possible backward substitution algorithm used in Figure 2.19 (a). The notation *triu* denotes the upper triangular part of a matrix, and $[X_T]$ and $[X_B]$ are the top and bottom partitions of $[X]$. Other notations for $[A_{TL}]$, $[A_{BL}]$, $[A_{TR}]$, $[A_{BR}]$, $[B_T]$, $[B_B]$ and K are the same as has been discussed in Section 2.5.3.

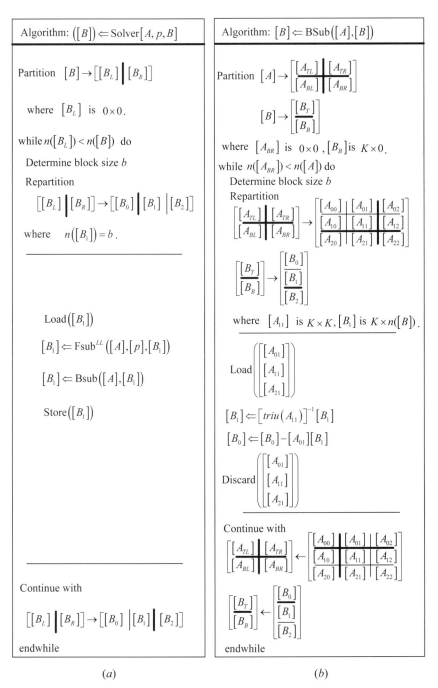

Figure 2.19. An out-of-core matrix equation solver: (*a*) the algorithm for solving the matrix equation; (*b*) backward substitution algorithm.

Since the same FSub^{LL}, as described in Section 2.5.3, can be used as one possible forward substitution procedure, as shown in Figure 2.19, the forward substitution will not be treated further here. The focus will be on the backward substitution procedure [in Fig. 2.19 (b)], which starts from partitioning the matrix as follows:

$$[A] \rightarrow \begin{bmatrix} [A_{00}] & [A_{01}] & [A_{02}] \\ [0] & [A_{11}] & [A_{12}] \\ [0] & [0] & [A_{22}] \end{bmatrix}; \quad [X] \rightarrow \begin{bmatrix} [X_0] \\ [X_1] \\ [X_2] \end{bmatrix}; \quad [B] \rightarrow \begin{bmatrix} [B_0] \\ [B_1] \\ [B_2] \end{bmatrix}. \qquad (2.50)$$

Resulting in the following equation:

$$\begin{bmatrix} [A_{00}] & [A_{01}] & [A_{02}] \\ [0] & [A_{11}] & [A_{12}] \\ [0] & [0] & [A_{22}] \end{bmatrix} \begin{bmatrix} [X_0] \\ [X_1] \\ [X_2] \end{bmatrix} = \begin{bmatrix} [B_0] \\ [B_1] \\ [B_2] \end{bmatrix}. \qquad (2.51)$$

Now, assume that before and after each iteration of the loop, the $[B]$ submatrices below the thick line have been overwritten with the corresponding $[X]$ submatrices. In other words, $[B_2]$ has been overwritten with $[X_2]$. To continue with the solution procedure, $[X_1]$ must overwrite the corresponding parts of $[B_1]$. Equation (2.51) indicates the required computation; blocks $[B_1]$ and $[B_0]$ must first be updated through:

$$[B_1] \Leftarrow [X_1] \Leftarrow (triu[A_{11}])^{-1} [B_1], \text{ and } [B_0] \Leftarrow [B_0] - [A_{01}][B_1].$$

Subsequently, the thick lines are moved upward for $[B]$ and $[X]$, and upward and left for $[A]$, since another $[B]$ submatrix can now be processed.

Note that $[X]$ and its partition, given in Eqs. (2.50) and (2.51), are just employed for a more accessible description of the procedure. However, the vector $[X]$ does not necessarily exist in the code when programming the algorithm, because the memory used for $[B_1]$ is reusable so that it is considered as the corresponding $[X_1]$ at each step.

2.8 CONCLUSION

In this chapter, data distributions of ScaLAPACK and PLAPACK are introduced for factoring a matrix using parallel processing. The data decomposition is then presented for an out-of-core matrix solver, where the storage needed for the matrix exceeds the capacity of the RAM. The one-slab left-looking out-of-core algorithm is described where the right-looking in-core algorithm is used when factoring each slab. Thus, the advantages of both the left-looking and the right-looking algorithms are retained. The in-core and the out-of-core solvers will be used in the following chapters to solve the matrix equations that result from the application of the MoM procedure to an integral equation in the frequency domain.

REFERENCES

[1]   T. Cwik, R. A. van de Geijn, and J. E. Patterson, "The Application of Massively Parallel Computation to Integral Equation Models of Electromagnetic Scattering," *Journal of the Optical Society of America A,* Vol. 11, pp. 1538–1545, 1994.

[2]   T. Cwik and J. E. Patterson, "Computational Electromagnetic and Supercomputer Architecture," *Progress in Electromagnetics Research,* Vol. 7, pp. 23–55, 1993.

[3]   T. Cwik, "Parallel Decomposition Methods for the Solution of Electromagnetic Scattering Problems," *Electromagnetics,* Vol. 12, pp. 343–357, 1992.

[4]   Netlib Repository at UTK and ORNL, *LINPACK, Collection of Fortran Subroutines.* Available at: http://www.netlib.org/linpack/. Accessed Aug. 2008.

[5]   J. J. Dongarra, I. S. Duff, D. C. Sorenson, and H. A. van der Vorst, *Solving Linear Systems on Vector and Shared Memory Computers,* SIAM, Philadelphia, 1991.

[6]   Wikipedia contributors, "Cache," *Wikipedia, The Free Encyclopedia,* Aug. 2008. Available at: http://en.wikipedia.org/wiki/Cache. Accessed Aug. 2008.

[7]   R. A. van der Pas, *Memory Hierarchy in Cache-Based Systems.* Sun Microsystems, CA, Nov. 2002. Available at: http://www.sun.com/blueprints /1102/817-0742.pdf. Accessed Aug. 2008.

[8]   Wikipedia contributors, "Memory Hierarchy," *Wikipedia, The Free Encyclopedia.* Available at: http://en.wikipedia.org/wiki/Memory_hierarchy. Accessed Aug. 2008.

[9]   P. Denning, "The Locality Principle," *Communications of the ACM,* Vol. 48, No. 7, pp. 19–24, 2005.

[10]   Netlib Repository at UTK and ORNL, "LAPACK — Linear Algebra PACKage." Available at: http://www.netlib.org/lapack/. Accessed Aug. 2008.

[11]   Texas Advanced Computing Center, "Goto BLAS," *Software and Tools,* The University of Texas at Austin, 2008. Available at: http://www.tacc.utexas.edu /resources/software/#blas. Accessed Aug. 2008.

[12]   K. Goto and R. A. van de Geijn, "High-Performance Implementation of the Level-3 BLAS," *ACM Transactions on Mathematical Software*, Vol. 35, No. 1, 2008.

[13]   K. Goto and R. A. van de Geijn, "Anatomy of High-Performance Matrix Multiplication," *ACM Transactions on Mathematical Software*, Vol. 34, No. 3, May 2008.

[14]   I. Gutheil, *Data Distribution and Usage of Libraries for Parallel Systems*, Central Institute for Applied Mathematics, John von Neumann Institute for Computing, Research Centre Jülich, Germany. Available at: http://www. fz-juelich.de/jsc/files/docs/vortraege/MPIkursfolien.pdf. Accessed Aug. 2008.

[15]   M. Sidani and B. Harrod, *Parallel Matrix Distributions: Have We Been Doing It All Right?*, Technical Report UT-CS-96-340, University of Tennessee, Knoxville, Nov. 1996.

[16]   V. Eijkhout, J. Langou, and J. Dongarra, *Parallel Linear Algebra Software*, Chap. 13, Netlib Repository at UTK and ORNL, 2006. Available at: http://www.netlib.org/utk/people/JackDongarra/PAPERS/siam-la-sw-survey-chapter13-2006.pdf. Accessed Aug. 2008.

[17]   E. D'Azevedoy and J. Dongarra, "The Design and Implementation of the Parallel Out-of-Core ScaLAPACK LU, QR and Cholesky Factorization Routines," *Concurrency: Practice and Experience*, Vol. 12, pp. 1481–1493, 2000.

[18]   J. J. Dongarra, S. Hammarling and D. W. Walker, "Key Concepts for Parallel Out-of-Core LU Factorization," *Parallel Computing*, Vol. 23, pp. 49–70, 1997.

3

A PARALLEL MOM CODE USING RWG BASIS FUNCTIONS AND SCALAPACK-BASED IN-CORE AND OUT-OF-CORE SOLVERS

3.0   SUMMARY

In this chapter, the solution of the electric field integral equations (EFIE) for perfect electric conductors (PECs) is obtained using the piecewise triangular patch basis functions also known as the Rao–Wilton–Glisson (RWG) basis functions. Using this methodology, an efficient parallel matrix filling scheme for the parallel in-core MoM is discussed in detail starting from a serial code. The parallel out-of-core solver is also employed in this chapter to go beyond the limitation of the RAM of an in-core MoM solver and thus extend its capability. Numerical results computed on several typical computer platforms illustrate that the parallel matrix filling schemes and the matrix equation solvers introduced here are highly efficient and meet the performance benchmarks obtained from the theoretical predictions.

3.1   ELECTRIC FIELD INTEGRAL EQUATION (EFIE)

In this section, the electric field integral equation (EFIE) is introduced for the analysis of radiation and scattering from perfect electric conductors located in free space. The analysis is based on the equivalence principle and on the uniqueness theorem. The EFIE can be used to make an EM analysis of a wide range of arbitrarily shaped objects including open and closed structures; in addition, the objects may contain wire structures. The method can also be extended to handle surfaces with impedance coatings. Scattering and radiation problems are essentially identical; the only difference is in the specification of the excitation. Therefore, only the scattering problem is considered here.

Figure 3.1 shows the scattering from a PEC body of arbitrary shape excited by a plane wave. The surface of the scatterer is denoted by S. It has an outward directed unit normal $\hat{n}$. The scattered fields in the region outside S are of interest.

71

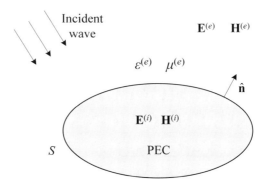

Figure 3.1. Scattering from a PEC body.

Intuitively, an electric current is induced by the incident wave, and generates the scattered field for the outside region. To obtain the induced current quantitatively, the equivalent principle will be applied here.

Inside the PEC body, the electric field $\mathbf{E}^{(i)}$ and the magnetic field $\mathbf{H}^{(i)}$ are equal to zero. According to the equivalent principle, to compensate for the discontinuity of the tangential components of the magnetic fields across the boundary, a sheet of electric currents is placed on the surface of the body. Next, the PEC structure is replaced by the same medium as in the outside region [1,2], since the fields internal to the surface S are zero. Therefore, the problem is reduced to computing the fields external to S with the additional stipulation that the electric currents located on the surface S are now radiating in a homogeneous medium when the structure is excited by a plane wave. The scattered fields can now easily be computed through Maxwell's equations as the equivalent electric currents placed over the virtual surface S, as shown in Figure 3.2, are now radiating in a medium that has the parameters of the external medium.

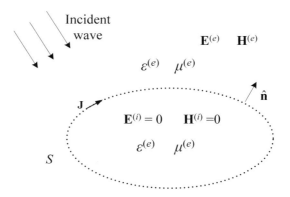

Figure 3.2. Equivalent model of the scattering from a PEC structure.

In addition, the boundary condition on the PEC surface is given by

$$\hat{n} \times (\mathbf{E}_{inc} + \mathbf{E}^{(e)}(\mathbf{J}))\,|_S = 0, \tag{3.1}$$

where $\mathbf{E}_{inc}$ is the incident electric field, and $\mathbf{E}^{(e)}(\mathbf{J})$ is the scattered electric field in the outside region generated by the equivalent electric current $\mathbf{J}$ on the surface.

In this chapter, Eq. (3.1) is employed to find the unknown currents. Because of the presence of a homogenous medium in the equivalent problem, the free space Green's function is used to calculate the potentials and the fields. Using the magnetic vector potential and the electric scalar potential, the scattered field in free space can be written as

$$\mathbf{E}^{(e)}(\mathbf{J}) = -j\omega\mathbf{A}(\mathbf{r}) - \nabla\Phi(\mathbf{r}), \tag{3.2}$$

$$\mathbf{A}(\mathbf{r}) = \mu_0 \int_S \mathbf{J}(\mathbf{r}') \frac{e^{-jkR}}{4\pi R}\, dS' = \mu_0 \int_S \mathbf{J}(\mathbf{r}')\, g(\mathbf{r},\mathbf{r}')\, dS', \tag{3.3}$$

$$\Phi(\mathbf{r}) = \frac{1}{\varepsilon_0} \int_S \sigma(\mathbf{r}') \frac{e^{-jkR}}{4\pi R}\, dS' = \frac{1}{\varepsilon_0} \int_S \sigma(\mathbf{r}')\, g(\mathbf{r},\mathbf{r}')\, dS', \tag{3.4}$$

where ε_0 and μ_0 are the permittivity and permeability for vacuum, respectively, and $g(\mathbf{r},\mathbf{r}') = e^{-jkR}/(4\pi R)$ is the free-space Green's function [2,3]. A time dependence of the mathematical form $\exp(j\omega t)$ is assumed throughout the analysis. In the expression for Green's function, $R = |\mathbf{r} - \mathbf{r}'|$, $\mathbf{r}$ is the position vector of the observation point and $\mathbf{r}'$ is that of the source point defined with respect to the origin of a global coordinate system.

The magnetic vector potential $\mathbf{A}(\mathbf{r})$ is obtained from the surface current distribution, whereas the scalar potential $\Phi(\mathbf{r})$ is evaluated from the free charges residing on the surface of the PEC structure. According to the equation of continuity, the surface charge density σ can be expressed in terms of the current density through the surface divergence as

$$j\omega\sigma = -\nabla_s \cdot \mathbf{J}. \tag{3.5}$$

Therefore

$$\Phi(\mathbf{r}) = \frac{-1}{j\omega\varepsilon_0} \int_S \nabla_s' \cdot \mathbf{J}(\mathbf{r}')\, g(\mathbf{r},\mathbf{r}')\, dS'. \tag{3.6}$$

Substituting Eq. (3.2) into Eq. (3.1), one obtains the electric field integral equation:

$$\hat{\mathbf{n}} \times \mathbf{E}_{\text{inc}} = \hat{\mathbf{n}} \times \left(j\omega\mathbf{A}(\mathbf{r}) + \nabla\Phi(\mathbf{r}) \right), \ \mathbf{r} \text{ is on } S, \tag{3.7}$$

where $\mathbf{A}(\mathbf{r})$ and $\Phi(\mathbf{r})$ are given by Eqs. (3.3) and (3.6).

3.2   USE OF THE PIECEWISE TRIANGULAR PATCH (RWG) BASIS FUNCTIONS

It is very difficult to analytically solve the EFIE. Hence, the problem is solved numerically through the use of the MoM methodology where the operator equation is transformed to a matrix equation through the use of basis and testing functions. The unknown current $\mathbf{J}$ is approximated by the triangular patch basis functions popularly known as the Rao–Wilton–Glisson (RWG) basis functions [3]. The current is expanded in terms of the known vector basis functions $\mathbf{f}_n$ which is multiplied by the scalar coefficients I_n that needs to be solved for. Therefore

$$\mathbf{J}(\mathbf{r}') = \sum_{n=1}^{N} I_n \, \mathbf{f}_n(\mathbf{r}') . \tag{3.8}$$

After substituting Eq. (3.8) into Eq. (3.3), the magnetic vector potential is evaluated as

$$\mathbf{A}(\mathbf{r}) = \sum_{n=1}^{N} \left\{ \mu_0 \int_S \mathbf{f}_n(\mathbf{r}') \, g(\mathbf{r}, \mathbf{r}') \, dS' \right\} I_n . \tag{3.9}$$

Similarly, the scalar potential of Eq. (3.6) is given by

$$\Phi(\mathbf{r}) = \sum_{n=1}^{N} \left\{ \frac{j}{\varepsilon_0 \, \omega} \int_S \nabla_s' \cdot \mathbf{f_n}(\mathbf{r}') \, g(\mathbf{r}, \mathbf{r}') dS' \right\} I_n . \tag{3.10}$$

Once the unknown current $\mathbf{J}$ is obtained, the scattered fields in the outside region can be calculated using the vector and the scalar potential functions.

Next, we review some of the properties of the RWG basis functions. As Figure 3.3 shows, for any triangle pair, T_n^+ and T_n^-, with area A_n^+ and A_n^-, respectively, sharing a common edge l_n, the basis function can be written as

$$\mathbf{f}_n(\mathbf{r}) = \begin{cases} \dfrac{l_n}{2A_n^+}\,\boldsymbol{\rho}_n^+ & \mathbf{r} \in T_n^+ \\[2mm] \dfrac{l_n}{2A_n^-}\,\boldsymbol{\rho}_n^- & \mathbf{r} \in T_n^-\,, \\[2mm] 0 & \text{otherwise} \end{cases} \qquad (3.11)$$

where $\boldsymbol{\rho}_n^+ = \mathbf{r} - \mathbf{r}_n^+$ is the vector directed from the free vertex of the triangle T_n^+ to the observation point, and $\boldsymbol{\rho}_n^- = \mathbf{r}_n^- - \mathbf{r}$ is the vector directed from the observation point to the free vertex of the triangle T_n^-, as shown in Figure 3.3. The basis function is zero outside the two adjacent triangles T_n^+ and T_n^-. The current has no component normal to the boundary (which excludes the common edge) of the surface formed by the triangle pair T_n^+ and T_n^-, and hence there are no line charges along this boundary.

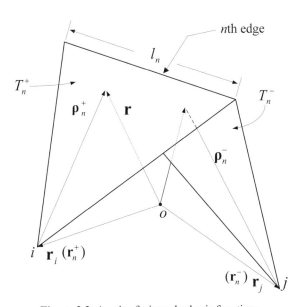

Figure 3.3. A pair of triangular basis functions.

The surface divergence of $\mathbf{f}_n(\mathbf{r})$ is proportional to the surface charge density associated with the basis function, and is given by

$$-j\omega\sigma_n = \nabla_s \cdot \mathbf{f}_n(\mathbf{r}) = \begin{cases} \dfrac{l_n}{A_n^+} & \mathbf{r} \in T_n^+ \\[3mm] -\dfrac{l_n}{A_n^-} & \mathbf{r} \in T_n^- \\[3mm] 0 & \text{otherwise} \end{cases} \tag{3.12}$$

The charge density is constant on each triangle, and the total charge associated with the triangle pair T_n^+ and T_n^- is zero. The current component normal to the nth edge is constant since

$$f_{n,\text{normal}}^+ = \frac{l_n}{2A_n^+} \cdot \frac{A_n^+}{l_n/2} = 1, \tag{3.13a}$$

$$f_{n,\text{normal}}^- = \frac{l_n}{2A_n^-} \cdot \frac{A_n^-}{l_n/2} = 1. \tag{3.13b}$$

The current is also continuous across the common edge as seen in Figure 3.4. It depicts that the normal component of $\boldsymbol{\rho}_n^\pm$ along the nth edge is just the height expressed as $2A_n^\pm/l_n$ of the triangle $T_n^\pm$ with the nth edge as the base.

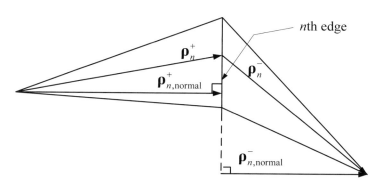

Figure 3.4. The normal components of the current across the nth edge.

3.3   TESTING PROCEDURE

In this section, it is shown how to solve the EFIE of Eq. (3.7) using *Galerkin's method*, where the testing function is chosen to be the same as the basis function. First, an inner product is defined as

$$\langle \mathbf{f}, \mathbf{g} \rangle \overset{\Delta}{=} \int_S \mathbf{f} \cdot \mathbf{g} \ dS. \tag{3.14}$$

Assuming that the testing functions, $\mathbf{f}_m(\mathbf{r})$ ($m = 1, ..., N$), are equal to the basis functions covering the entire surface S, the inner product is carried out by multiplying it on both sides of Eq. (3.7) and then integrating to yield

$$\langle \mathbf{E}_{\text{inc}}, \mathbf{f}_m(\mathbf{r}) \rangle = j\,\omega\,\langle \mathbf{A}, \mathbf{f}_m(\mathbf{r}) \rangle + \langle \nabla\Phi, \mathbf{f}_m(\mathbf{r}) \rangle. \tag{3.15}$$

The last term in this formulation can be written as

$$\langle \nabla\Phi, \mathbf{f}_m(\mathbf{r}) \rangle = \int_S \nabla\Phi \cdot \mathbf{f}_m(\mathbf{r}) \ dS = \int_{T_m^+ + T_m^-} \nabla_s\Phi \cdot \mathbf{f}_m(\mathbf{r}) \ dS. \tag{3.16}$$

Since

$$\nabla_s \cdot \left(\Phi\,\mathbf{f}_m(\mathbf{r})\right) = \Phi\,\nabla_s \cdot \mathbf{f}_m(\mathbf{r}) + \nabla_s\Phi \cdot \mathbf{f}_m(\mathbf{r}), \tag{3.17}$$

Eq. (3.16) can be written as

$$\int_{T_m^+ + T_m^-} \nabla_s\Phi \cdot \mathbf{f}_m(\mathbf{r}) \ dS = \int_{T_m^+ + T_m^-} \nabla_s \cdot \left(\Phi\,\mathbf{f}_m(\mathbf{r})\right) \ dS - \int_{T_m^+ + T_m^-} \Phi\,\nabla_s \cdot \mathbf{f}_m(\mathbf{r}) \ dS. \tag{3.18}$$

Also, note that

$$\int_{T_m^+ + T_m^-} \nabla_s \cdot \left(\Phi\,\mathbf{f}_m(\mathbf{r})\right) \ dS = \oint_C \Phi\,\mathbf{f}_m(\mathbf{r}) \cdot \mathbf{n}_C \ dl, \tag{3.19}$$

where C is the boundary representing the triangle pair $T_m^\pm$, and $\mathbf{n}_C$ is the outward normal to the boundary C, as shown in Figure 3.5.

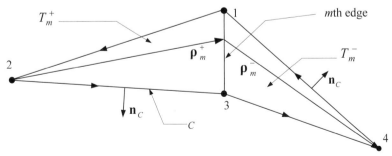

Figure 3.5. A triangle pair $T_m^\pm$.

According to the definition of the RWG basis functions, $\mathbf{f}_m(\mathbf{r})$ becomes zero at vertices 2 and 4, and the direction of the vectors $\boldsymbol{\rho}_m^+$ or $\boldsymbol{\rho}_m^-$ of $\mathbf{f}_m(\mathbf{r})$ must be directed along the boundary of the triangle at vertices 1 and 3. Therefore, the integral over the whole boundary is zero:

$$\oint_C \Phi\, \mathbf{f}_m(\mathbf{r})\bullet \mathbf{n}_c \ dl = 0 . \tag{3.20}$$

Therefore, Eq. (3.18) is simplified to

$$\int_{T_m^+ + T_m^-} \nabla_s\Phi \bullet \mathbf{f}_m(\mathbf{r})\ dS \ = \ - \int_{T_m^+ + T_m^-} \Phi\, \nabla_s \bullet \mathbf{f}_m(\mathbf{r})\ dS , \tag{3.21}$$

and this results in

$$\left\langle \nabla\Phi, \mathbf{f}_m(\mathbf{r}) \right\rangle = -\int_S \Phi\, \nabla_s \bullet \mathbf{f}_m(\mathbf{r})\ dS . \tag{3.22}$$

Finally, one obtains

$$\int_S \mathbf{f}_m(\mathbf{r})\bullet \mathbf{E}_{\text{inc}}\ dS = \mathrm{j}\,\omega\!\int_S \mathbf{f}_m(\mathbf{r})\bullet \mathbf{A}\ dS - \int_S \left(\nabla_s \bullet \mathbf{f}_m(\mathbf{r})\right)\Phi\ dS . \tag{3.23}$$

3.4   MATRIX EQUATION FOR MOM

If the expressions of Eqs. (3.9) and (3.10) are substituted into Eq. (3.23), one obtains a matrix equation that can be written in a compact form as

$$\sum_{n=1}^{N} Z_{mn} I_n = V_m, \ m = 1, ..., N \tag{3.24}$$

where

$$V_m = \int_S \mathbf{f}_m(\mathbf{r})\bullet \mathbf{E}_{\text{inc}}\ dS \tag{3.25}$$

is the "voltage" or the excitation component. The elements of the impedance matrix Z_{mn} are obtained from

$$Z_{mn} = \mathrm{j}\,\omega\,\mu_0 \int_S \int_{S'} \mathbf{f}_m(\mathbf{r})\bullet \mathbf{f}_n(\mathbf{r}')\, g\left(\mathbf{r},\mathbf{r}'\right)\ dS'\ dS$$

$$-\left(\frac{\mathrm{j}}{\varepsilon_0\,\omega}\right)\int_S \int_{S'} \left(\nabla_s \bullet \mathbf{f}_m(\mathbf{r})\right)\left(\nabla_s' \bullet \mathbf{f}_n(\mathbf{r}')\right) g\left(\mathbf{r},\mathbf{r}'\right)\ dS'\ dS. \tag{3.26}$$

The various integrals in this expression are evaluated by the substitution of Eqs. (3.11) and (3.12) into Eq. (3.26) and yields the following expressions for the various integrals as

$$
\iint_{S\ S'} \mathbf{f}_m(\mathbf{r}) \bullet \mathbf{f}_n(\mathbf{r}')\, g(\mathbf{r},\mathbf{r}')\, dS'\, dS = \frac{l_m\, l_n}{4\, A_m^+ A_n^+} \iint_{T_m^+ T_n^+} \left(\boldsymbol{\rho}_m^+ \bullet \boldsymbol{\rho}_n'^+ \right) g(\mathbf{r},\mathbf{r}')\, dS'\, dS
$$

$$
+ \frac{l_m\, l_n}{4\, A_m^+ A_n^-} \iint_{T_m^+ T_n^-} \left(\boldsymbol{\rho}_m^+ \bullet \boldsymbol{\rho}_n'^- \right) g(\mathbf{r},\mathbf{r}')\, dS'\, dS
$$

$$
+ \frac{l_m\, l_n}{4\, A_m^- A_n^+} \iint_{T_m^- T_n^+} \left(\boldsymbol{\rho}_m^- \bullet \boldsymbol{\rho}_n'^+ \right) g(\mathbf{r},\mathbf{r}')\, dS'\, dS
$$

$$
+ \frac{l_m\, l_n}{4\, A_m^- A_n^-} \iint_{T_m^- T_n^-} \left(\boldsymbol{\rho}_m^- \bullet \boldsymbol{\rho}_n'^- \right) g(\mathbf{r},\mathbf{r}')\, dS'\, dS \ ,
$$

$$(3.27)$$

and

$$
\iint_{S\ S'} \left(\nabla_s \bullet \mathbf{f}_m(\mathbf{r}) \right) \left(\nabla_s' \bullet \mathbf{f}_n(\mathbf{r}') \right) g(\mathbf{r},\mathbf{r}')\, dS'\, dS = \frac{l_m\, l_n}{A_m^+ A_n^+} \iint_{T_m^+ T_n^+} g(\mathbf{r},\mathbf{r}')\, dS'\, dS
$$

$$
- \frac{l_m\, l_n}{A_m^+ A_n^-} \iint_{T_m^+ T_n^-} g(\mathbf{r},\mathbf{r}')\, dS'\, dS
$$

$$
- \frac{l_m\, l_n}{A_m^- A_n^+} \iint_{T_m^- T_n^+} g(\mathbf{r},\mathbf{r}')\, dS'\, dS
$$

$$
+ \frac{l_m\, l_n}{A_m^- A_n^-} \iint_{T_m^- T_n^-} g(\mathbf{r},\mathbf{r}')\, dS'\, dS \ .
$$

$$(3.28)$$

In a compact matrix form, Eq. (3.24) can be written as

$$
[Z][I] = [V]. \tag{3.29}
$$

3.5   CALCULATION OF THE VARIOUS INTEGRALS

3.5.1   Evaluation of the Fundamental Integrals

Most of the CPU time required for filling the impedance matrix $[Z]$ for the RWG basis functions is spent on the calculation of the surface integrals presented in Eqs. (3.27) and (3.28). Assuming that all the triangles associated with the structure are numbered by $p = 1,...,P$, then every integral in Eq. (3.27) can be generated from the following form

$$\int_{T_p} \int_{T_q} \left(\boldsymbol{\rho}_i \cdot \boldsymbol{\rho}'_j \right) g(\mathbf{r},\mathbf{r}') \, dS' \, dS, \quad p, q = 1, ..., P.$$

(3.30)

Here, $\boldsymbol{\rho}_i = \mathbf{r} - \mathbf{r}_i$ for any vertex i of patch p, whereas $\boldsymbol{\rho}'_j = \mathbf{r}' - \mathbf{r}_j$ for any vertex j of patch q. Similarly, every integral in Eq. (3.28) can be generated from

$$\int_{T_p} \int_{T_q} g(\mathbf{r},\mathbf{r}') \, dS' \, dS, \quad p, q = 1, ..., P.$$

(3.31)

3.5.2   Extraction of the Singularity

The integrals in Eqs. (3.30) and (3.31) can be calculated in several ways. The singularity of the free-space Green's function is integrable in the double integrals, but the accuracy of the Gaussian quadrature formulas is reduced if this singularity is retained. Therefore, extraction of the singularity is applied to Eqs. (3.30) and (3.31) in the form

$$\int_{T_p} \int_{T_q} \left(\boldsymbol{\rho}_i \cdot \boldsymbol{\rho}'_j \right) g(\mathbf{r},\mathbf{r}') \, dS' \, dS = \int_{T_p} \int_{T_q} \frac{\left(\boldsymbol{\rho}_i \cdot \boldsymbol{\rho}'_j \right)}{4\pi |\mathbf{r} - \mathbf{r}'|} \, dS' \, dS$$

$$+ \int_{T_p} \int_{T_q} \frac{\left(\exp(-jk|\mathbf{r} - \mathbf{r}'|) - 1 \right) \left(\boldsymbol{\rho}_i \cdot \boldsymbol{\rho}'_j \right)}{4\pi |\mathbf{r} - \mathbf{r}'|} \, dS' \, dS,$$

(3.32)

$$\int_{T_p} \int_{T_q} g(\mathbf{r},\mathbf{r}') \, dS' \, dS = \int_{T_p} \int_{T_q} \frac{1}{4\pi |\mathbf{r} - \mathbf{r}'|} \, dS' \, dS$$

$$+ \int_{T_p} \int_{T_q} \frac{\exp(-jk|\mathbf{r} - \mathbf{r}'|) - 1}{4\pi |\mathbf{r} - \mathbf{r}'|} \, dS' \, dS.$$

(3.33)

The first terms on the right-hand sides of Eqs. (3.32) and (3.33) contain a singularity, however, they can be evaluated analytically using the procedure outlined in references [3,4]. Strictly speaking, the integration-by-parts approach of reference [4] handles only the inner of the double integral presented in Eqs. (3.32) and (3.33). The outer integrals are computed numerically using the Gaussian quadratures [5].

Since the focus of this chapter is on the development of parallel MoM codes rather than dealing with the details of calculating the various integrals, it is suggested that one can refer to references [2–4] for a thorough explanation of the procedure used here for the computations.

3.6   CALCULATION OF THE FIELDS

Once the unknown currents using the MoM methodology is known, the scattered (or the radiated) electric field in free space can be obtained from

$$\mathbf{E} = -j\omega\,\mu_0 \sum_{n=1}^{N} \left\{ \int_S \mathbf{f}_n(\mathbf{r}')\,g(\mathbf{r},\mathbf{r}')\,dS' \right\} I_n$$

$$-\frac{j}{\varepsilon_0\,\omega} \sum_{n=1}^{N} \left\{ \int_S \nabla'_s \cdot \mathbf{f}_n(\mathbf{r}')\,\nabla g(\mathbf{r},\mathbf{r}')\,dS' \right\} I_n, \tag{3.34}$$

where I_n are the unknown constants in the MoM formulation for the surface current density. The scattered magnetic field is given by

$$\mathbf{H} = \sum_{n=1}^{N} \left\{ \int_S -\mathbf{f}_n(\mathbf{r}') \times \nabla g(\mathbf{r},\mathbf{r}')\,dS' \right\} I_n. \tag{3.35}$$

3.7   PARALLEL MATRIX FILLING — IN-CORE ALGORITHM

In evaluation of the elements of the impedance matrix, the integrals are performed over a pair of triangles even though the elements of the impedance matrix are associated with the common edge of this triangle pair. Therefore, the code can be designed by looping triangle-to-triangle, rather than edge-to-edge, to calculate the various integrals, in order to avoid redundancy for different edges on a pair of source and field (testing) triangles. The pseudocode of the serial in-core matrix filling is given in Figure 3.6.

As shown in Figure 3.6, the code starts with the outer loops over the triangles, and then performs the integral on each triangle pair, followed by the inner loops over the edges of the triangles. The results for the integral obtained over a triangle pair is accumulated to at most nine related impedance elements, which contributes to one term on the right side of Eqs (3.27) and (3.28).

Serial In-Core Filling Algorithm:

Do q = 1, *NumTriangles*	!loop over source triangles
Do p = 1, *NumTriangles*	!loop over field (testing) triangles
compute_interactions(q,p)	!calculate integral on triangle pair (q,p)
Do e = 1, 3	!loop over edges of the source triangle
nn = edge_num(q,e)	!global index of eth edge on qth source triangle
If (nn.NE.-1) then	!nnth edge is a valid common edge
Do f = 1, 3	!loop over edges of the field triangle
mm = edge_num(p,f)	!global index of fth edge of pth field triangle
If (mm.NE.-1) then	!mmth edge is a valid common edge
Z(mm,nn) = Z(mm,nn) + delta_Z(mm,nn)	!add into the impedance matrix
Endif	
Enddo	!end loop over edges of the field triangle
Endif	
Enddo	!end loop over edges of the source triangle
Enddo	!end loop over field (testing) triangles
Enddo	!end loop over source triangles

Figure 3.6. A serial in-core filling algorithm.

The parallel matrix filling scheme A is obtained from the serial code by choosing the correct process associated with the evaluation of each element of the impedance matrix. The modified code can be obtained by slightly changing the serial code, as demonstrated in Figure 3.7. The differences between the serial code and the parallel matrix filling scheme A are marked by the shaded lines in Figure 3.7.

Parallel Matrix In-Core Filling Algorithm — Scheme A:

Do q = 1, *NumTriangles*	!loop over source triangles
Do p = 1, *NumTriangles*	!loop over field (testing) triangles
compute_interactions(q,p)	!calculate integral on triangle pair (q,p)
Do e = 1, 3	!loop over edges of the source triangle
nn = edge_num(q,e)	!global index of eth edge on qth source triangle
If (nn.NE.-1 .and. nn is on this process) then	!nnth edge is valid and on this process
n = local_Index(nn)	!get local index of the global index nn
Do f = 1, 3	!loop over edges of the field triangle
mm = edge_num(p,f)	!global index of fth edge of pth field triangle
If (mm.NE.-1.and. mm is on this process) then	!mmth edge is valid and on this process
m = local_Index(mm)	!get the local index of the global index mm
Z(m,n) = Z(m,n) + delta_Z(m,n)	!add into the local impedance matrix
Endif	
Enddo	!end loop over edges of the field triangle
Endif	
Enddo	!end loop over edges of the source triangle
Enddo	!end loop over field (testing) triangles
Enddo	!end loop over source triangles

Figure 3.7. A parallel in-core matrix filling algorithm — scheme A.

In scheme A, only a portion of the matrix is placed on each process after the computation so that the RAM required for each process is reduced. However, all the processes have to calculate all the integrals between a pair of the source and the field triangles. This means that the computation for the parallel matrix filling is not reduced when compared with the serial in-core matrix filling scheme, even though only a portion of the matrix is stored on each process. This scheme wastes a lot of computation time and cannot be considered effective.

The parallel matrix filling scheme B is generated from scheme A by moving the integral ["compute_interactions(q,p)" given in the third line below the thick solid line in Fig. 3.7], which compute the interactions between the triangle pair (q,p), inside the inner most loop over the edges of a field patch, as marked in Figure 3.8. The entire procedure of scheme B, shown in Figure 3.8, is now described.

First, the code loops over all the source patches, the field patches, and the edges on a certain source triangle. Then it picks up the corresponding processes that are about to calculate the nth column of the matrix. It then loops over the edges of a field triangle, and then picks up the corresponding processes by which the mth row of the matrix will be calculated.

Only after the correct process has been chosen for the nth column and the mth row, will the integral between the source triangle and the field triangle be evaluated and stored. The matrix filling scheme ensures that only the process that corresponds to the mth row and the nth column will calculate the integral over the surfaces of the source triangle and the field triangle pair. This reduces the redundancy in the calculation of the integral in scheme A.

Parallel Matrix In-Core Filling Algorithm — Scheme B:

Do $q = 1$, *NumTriangles*	!loop over source triangles
Do $p = 1$, *NumTriangles*	!loop over field (testing) triangles
Do $e = 1, 3$	!loop over edges of the source triangle
nn = edge_num(q,e)	!global index of eth edge of qth source triangle
If (nn.NE.-1 .and. nn is on this process) then	!nnth edge is valid and on this process
n = local_ Index(nn)	!get the local index of the global index nn
Do $f = 1, 3$	!loop over edges of the field triangle
mm = edge_num(p,f)	!global index of fth edge of pth field triangle
If (mm.NE.-1.and. mm is on this process) then	!mmth edge is valid and on this process
compute_interactions(q,p)	!calculate integral on triangle pair (q,p)
m = local_Index(mm)	!get the local index of the global index mm
$Z(m,n) = \overline{Z}(m,n) + $ delta_ $Z(m,n)$	!add into the local impedance matrix
Endif	
Enddo	!end loop over edges of the field triangle
Endif	
Enddo	!end loop over edges of the source triangle
Enddo	!end loop over field (testing) triangles
Enddo	!end loop over source triangles

Figure 3.8. A parallel in-core matrix filling algorithm — scheme B.

The easiest way to check the efficiency of a parallel matrix filling procedure in scheme B is to set the number of processes to 1, in which case the code should work in a serial fashion. However, the integral over a triangle pair repeats at most 8 times for each process, since there are at most three edges with each triangle connecting with other triangles. Therefore, this methodology still needs improvement to avoid redundant calculation of the integrals for any given process.

This effort results in scheme C, as shown in Figure 3.9. The differences between scheme B and scheme C are marked by the shaded lines in Figure 3.9. In scheme C, the repeated calculation of the integrals related to the potential functions for each process is avoided by using a flag for each triangle. For each loop of the source triangle, the values of the flags pertaining to all the field triangles are initialized to be false. Once the integral on the given source triangle and the pth field triangle has been performed, the state of the flag corresponding to the field triangle flag(p) is changed to be true. When the loop over the three edges of the source triangle is followed by the loop over the edges of the field triangle, the flag of this field triangle will determine whether to compute the integral. Once the status of the flag is true, the integrals need not be performed again because it is still in the memory.

Parallel Matrix In-Core Filling Algorithm — Scheme C:

Do q = 1, *NumTriangles*	!loop over source triangles
flag(1: *NumTriangles*)=0	!set initial value of flags of integral for all the !field (testing) triangles to be 0
Do p = 1, *NumTriangles*	!loop over field (testing) triangles
Do e = 1, 3	!loop over edges of the source triangle
nn = edge_num(q,e)	!global index of eth edge of qth source triangle
If (nn.NE.-1 .and. nn is on this process) then	!nnth edge is valid and on this process
n = local_ Index(nn)	!get the local index of the global index nn
Do f = 1, 3	!loop over edges of the field triangle
mm = edge_num(p,f)	!global index of fth edge of pth field triangle
If (mm.NE.-1.and. mm is on this process) then	!mmth edge is valid and on this process
If (flag(p)==0) then	
compute_interactions(q,p)	!calculate integral on triangle pair (q,p)
flag(p)=1	!set flag of integral for testing triangle p to be 1
Endif	
m = local_Index(mm)	!get the local index of the global index mm
$Z(m,n) = Z(m,n) +$ delta_ $Z(m,n)$	!add into the local impedance matrix
Endif	
Enddo	!end loop over edges of the field triangle
Endif	
Enddo	! end loop over edges of the source triangle
Enddo	! end loop over field (testing) triangles
Enddo	! end loop over source triangles

Figure 3.9. A parallel in-core matrix filling algorithm — scheme C.

When compared with a sequential code, the only additional overhead in a parallel code is the procedure used by each process to select the edges that contribute to the matrix element belonging to its blocks. The advantage of loop-over-triangles reducing the redundant calculation of the integrals as compared with loop-over-edges still holds true for the parallel code. Thus, one can expect the parallel matrix filling scheme to have a good speedup.

The redundant calculation is efficiently avoided on each individual process in the parallel matrix filling scheme. However, when a large impedance matrix is divided into small blocks, as required by the matrix solution method, there may be some calculations that are repeated. Since each edge of a triangle may be related to different matrix elements mapped onto different processes, the integral over a triangle may still have to be repeated by different processes. To remedy this redundant calculation, one possibility is to use communication between processors to preclude the repeated calculations of the integrals on different processes. This can be achieved by having a process first calculate the integral, and then to broadcast the result to all the other processes that need it. However, on a distributed-memory computer system, communication between nodes can also consume significant time and this could ultimately reduce the efficiency when the number of nodes and/or the problem size is very large.

The tradeoff between redundant calculations on different processes versus increased communication between the nodes has to be carefully considered. In this book, it was decided to carry out the redundant calculations among different processes because of its higher computational efficiency.

The computation for filling the column matrix in the right-hand side of Eq. (3.29) does not take a long time. However, this computation can take significant time when the number of unknowns increase. Once the procedure on how to design a parallel matrix filling algorithm is decided for the computation of an impedance matrix, the right-hand side (RHS) of the matrix equation can be parallelized as shown in Figure 3.10.

Parallel RHS Filling Algorithm:

Do p = 1, NumTriangles	!loop over field (testing) triangles
flag(p)=0	!set initial value of flags for pth field triangle
Do f = 1, 3	!loop over edges of the field triangle
mm = edge_num(p,f)	!global index of fth edge of field triangle p
If (mm.NE.-1 .and. mm is on this process) then	!mmth edge is valid and on this process
If (flag(p)==0) then	
compute_integral_excitation(p)	!calculate integral on pth field triangle
flag(p)=1	!set flag of integral for field triangle p to be 1
Endif	
m = local_index(mm)	!get the local index of the global index mm
$V(m) = V(m)$+ delta_V(m)	!add into the local voltage matrix
Endif	
Enddo	!end loop over edges of the field triangle
Enddo	!end loop over field (testing) triangles

Figure 3.10. A parallel RHS filling algorithm.

It is worth pointing out that when writing a parallel code, subroutine INDXG2P of ScaLAPACK can be used to find a certain position in a process grid given the global row or/and column index of the matrix. Subroutine INDXG2L of ScaLAPACK can be used to find the local index given a global row or/and column index.

3.8   PARALLEL MATRIX FILLING — OUT-OF-CORE ALGORITHM

Solutions of very large problems require storing the impedance matrix out-of-core and this requires an out-of-core matrix equation solver. The reason for developing an out-of-core matrix filling algorithm is that the matrix is too large to be stored in the main memory of the system. We have to fill one portion of the matrix at a time and write it to files on the hard disk, and then fill another portion of the matrix and write it out, and so on. This is different from the in-core matrix filling algorithm, where the matrix is filled once and stored in the RAM. Hence, the main idea of designing an out-of-core algorithm is to modify the in-core filling algorithm structure and fill a portion of the matrix, instead of the whole matrix, and then store this portion on the hard disk.

Similarly, the issues involved with a parallel out-of-core algorithm design are data decomposition and load balancing. As with a parallel in-core filling algorithm, much communication can be avoided in the parallel out-of-core matrix filling by repeating the integral calculations as needed. The important thing for an out-of-core filling of the matrix is to divide the large problem into a set of small problems that can fit into the in-core memory. If the data decomposition is properly implemented, as discussed in Chapter 2, the result will be a load-balanced code.

Since the data decomposition was already addressed in Chapter 2, now the discussion proceeds to the out-of-core matrix filling algorithm. The pseudocode for an efficient parallel matrix filling scheme is given in Figure 3.11.

As shown in Figure 3.11, at the beginning of the matrix filling algorithm, the matrix is partitioned into different slabs. As discussed in Section 2.4 (see Chapter 2), the number of slabs I_{slab} and the width of each slab K_i are determined by the application and the particular system architecture. Let each process go through a loop of out-of-core operations, from 1 to I_{slab}. Each process calculates K_i for the ith out-of-core slab and sets the global upper (*nend* in Fig. 3.11) and the lower bound (*nstart* in Fig. 3.11). For example, for the first slab ($i = 1$), the global lower bound is 1 and the upper bound is K_1. For the second slab, the global lower bound is K_1+1 and the upper bound is K_1+K_2. Each process fills a portion of the matrix in the same way as it occurs for an in-core matrix fill algorithm. However, each process is not concerned with the columns that fall outside the fill bound. After every process has completed the desired element computations, a synchronous write is executed to write the portion of the matrix into a file. Then, each process continues to the next slab associated with the loop.

Parallel Out-of-Core Matrix Filling:	
Partition A into different slabs ($[A_1] \mid [A_2] \mid [A_3] \mid ... \mid [A_{I\text{slab}}]$)	
Calculate_slab_bound (*nstart, nend*)	
	!calculate the global lower bound
	!and upper bound for the current slab
For each Slab:	
Do q = 1, *NumTriangles*	!loop over source triangles
flag(1: *NumTriangles*)=0	!set initial value of flags of integral for all the
	!field (testing) triangles to be zero
Do p = 1, *NumTriangles*	!loop over field (testing) triangles
Do e = 1, 3	!loop over edges of the source triangle
nn = edge_num(q,e)	!global index of eth edge on qth source triangle
If (nn.NE.-1 .and. nn is on this process .and. nn >= *nstart* .and. nn <= *nend*) then	
	!nnth edge is valid and on this process
	!and within current out-of-core slab
n = local_ index(nn)	!get the local index of the global index nn
Do f = 1, 3	!loop over edges of the field triangle
mm = edge_num(p,f)	!global index of fth edge of pth field triangle
If (mm.NE.-1.and. mm is on this process) then	
	!mmth edge is valid and on this process
If (flag(p) == 0) then	
compute_interactions(q,p)	!calculate integral on triangle pair (q,p)
flag(p)=1	!set flag of integral for pth testing triangle to be 1
Endif	
m = local_Index(mm)	!get the local index of the global index mm
$Z(m,n) = Z(m,n)$ + delta_ $Z(m,n)$	!add into the local impedance matrix
Endif	
Enddo	!end loop over edges of the field triangle
Endif	
Enddo	!end loop over edges of the source triangle
Enddo	!end loop over field (testing) triangles
Enddo	!end loop over source triangles
Write_file	!write this slab into hard disk
! Finish Filling Impedance Matrix	

Figure 3.11. A parallel out-of-core matrix filling algorithm.

This scheme avoids calculation of most of the redundant integrals related to the potential functions, which are used in calculation of the elements of the impedance matrix using a different process, by using a flag for each geometric element. This is similar to the in-core case (see the discussion on scheme C for a parallel in-core matrix filling algorithm given in Fig. 3.9). However, for the out-of-core case, a little additional overhead may be incurred by repeating the integral on the same process for different slabs.

By comparing with the in-core matrix filling algorithm, it can be found that for each slab, the algorithm is exactly the same. Most of the overhead for filling the out-of-core matrix, excluding that from the in-core matrix calculation for an individual slab, comes from two parts: (1) calculation of the redundant integrals performed on each process, for different elements of the impedance matrix, which belongs to different slabs; and (2) writing the matrix elements to the hard disk.

Once the entire dense complex MoM matrix is computed, the matrix solvers introduced in Chapter 2 can be utilized to solve this set of equations using either an in-core or an out-of-core solution algorithm. Numerical results of Figures 3.9 and 3.11 generated with the parallel schemes along with their levels of performance will be discussed in the next section.

3.9   NUMERICAL RESULTS FROM A PARALLEL IN-CORE MOM SOLVER

3.9.1   Numerical Results Compared with Other Methods

Before the performance of the various parallel codes is investigated, two examples are presented to validate the accuracy of the codes.

3.9.1.1 A PEC Cube. The first example involves a PEC cube with an edge length a, and $k_0 a = 2$ ($k_0 = 2\pi/\lambda$). The cube is meshed into 384 triangular patches. A plane wave E_z travelling along the $+y$ direction is used as the incident field. The model is included in Figure 3.12 (a). As shown in Figure 3.12 (a) and (b), the RCS calculated by the MoM code matches the results generated from an explicit FDTD algorithm. This validates the results of the parallel MoM codes.

3.9.1.2 A Combined Cube-and-Sphere PEC Model. The second example uses a combined PEC cube-and-sphere model. The edge length of the cube is a, and $k_0 a = \pi$. The diameter of the sphere $D = a$, and the center of the sphere is collocated at one of the vertices of the cube. The plane wave E_x is incident from the $-z$ direction to the scatterer. The normalized scattered far fields in the xoy plane and in the zox plane are shown in Figure 3.13 (a) and (b), respectively. The model and the incident wave are illustrated in Figure 3.13 (b). As a comparison, the results computed using Ansoft HFSS (a commercial electromagnetic software using the FEM method) analyzing the same problem is also plotted in the same figure.

3.9.2   Different Metrics Used to Assess the Degree of Parallel Efficiency

How to measure the performance of a parallelized code will be discussed in this section. Quantitatively defining the performance of a parallel program is more complex than just optimizing its execution time. This is because of the large number of variables that can influence the time taken to execute a parallel code.

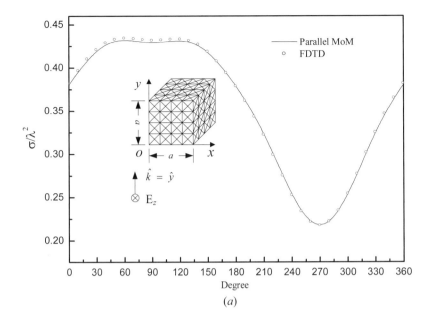

(a)

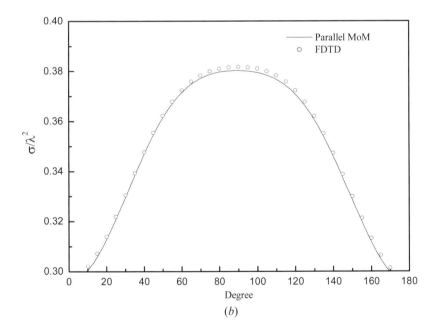

(b)

Figure 3.12. RCS of a cube: (a) in the *xoy* plane (0° starts from *x* axis); (b) in the *zox* plane (0° starts from *z* axis).

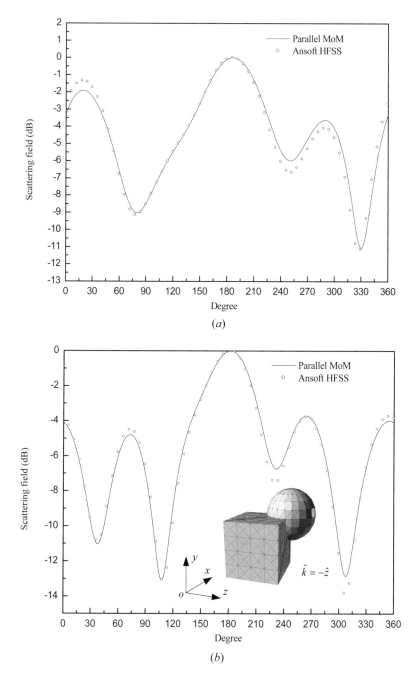

Figure 3.13. Normalized scattered field: (*a*) in the *xoy* plane (0° starts from *x* axis); (*b*) in the *zox* plane (0° starts from *z* axis).

Some of these variables are

- The number of processors (cores) over which the problem is being solved
- The size of the data being worked on and allocated to each processor
- Interprocessor communications can also limit the efficiency
- Available memory on each processor

It is important to understand how the execution time depends on the programming variables such as the size of the problem, number of processors (cores), and so on. Three metrics that are commonly used to evaluate the performance of a parallel code are

- Execution time
- Speedup
- Efficiency

The most common performance parameter is the execution time of a parallel program, or what is typically referred to as the *wall-clock time* (*wall time*). The execution time is defined as the time elapsed from when the first process starts executing a problem to the moment when the last process completes the execution. The total execution time T_e for a parallel code is composed of the following $T_e = T_{comp} + T_{comm} + T_{idle}$, where the execution time is divided between computing, communicating, or the processor just sitting idle.

- *Computation time, T_{comp}.* Computation time is the time a single processor (core) spends doing its part of the computation. It depends on the problem size and is specific to the processor (core). Ideally, this is just T_{serial}/P, but it may be different depending on the parallel algorithm that is being used. T_{serial} is the time that a serial program needs to finish the problem. P is the number of processors (here we assume that one processor or one core is used for one process).
- *Communication time, T_{comm}.* Communication time is the part of the execution time spent on the communication between processors (cores).
- *Idle time, T_{idle}.* When a processor (core) is not computing or communicating, it is just sitting idle. Good parallel algorithms try to minimize a processor's (core's) idle time with proper load balancing and efficient coordination between the various processors when the processor is carrying out computation as opposed to communication.

The next item, speedup, is the ratio of the execution time of a particular application on one processor (core) to the execution time of the same application on p processors (cores):

- *Speedup(p) = $T(1)/T(p)$.* This shows how much faster the problem can be solved using p processors (cores) than 1 processor (core). $T(p,n)$ is the time to solve a problem of size n on p processors (cores). Usually, n is omitted and $T(p,n)$ is written as $T(p)$. $T(1)$ is the serial or sequential time, using the best serial algorithm, which is not necessarily related to the parallel algorithm, with p set to 1.

Efficiency, in turn, is defined to be the average utilization of the p processors (cores) allocated to execute the parallel program:

- *Efficiency*(p) = Speedup$(p)/p$.

Speedup and efficiency are the most common ways to report the performance of a parallel algorithm, and these measures can provide insight on whether a parallel code is efficient when executing the code on a particular computer platform.

Scalability is also a parameter that can be used to evaluate the performance of a parallel algorithm. A scalable parallel algorithm has the property that if reasonable efficiency is attained when executing an application over a small number of processors, then a scaled efficiency can be achieved when a large number of processors are used to execute the same application. The size of the problem being solved can also be increased, provided the combined memories of the processors can handle the larger problem. Since the total amount of memory grows linearly with the number of processors, this also implies that the amount of data processed also grows linearly with the number of processors.

Besides these factors, robustness of the code is a fundamental requirement of a parallel code. The parallel algorithm must exhibit the same numerical stability as the sequential algorithm.

When testing the performance of a parallel in-core and a parallel out-of-core integral equation solver, one can use all the criteria described above to assess the quality of the code.

3.9.3   Efficiency and Portability of a Parallel MoM In-Core Solver

A PEC sphere with 1.0 meter (m) radius is taken as the model for the solution of a scattering problem. The excitation for this simulation is a plane wave that is polarized along the z axis and is propagating along the negative x-axis direction. Two problems are considered, one with 11,604 unknowns that is solved at an excitation frequency of 480 MHz, and the other with 19,101 unknowns with an excitation frequency of 600 MHz.

Such small problem sizes are poor examples for testing the parallel implementation of a code because each process has only a small-sized job to perform. These simulations are specifically chosen to illustrate that the parallel code indeed provides excellent parallel speedup for the filling of the impedance matrix even for a very small-sized problem [6].

The speedup, obtained by using the computer platform CEM-1 for the analysis of the sphere at an excitation frequency of 480 MHz and CEM-3 at an excitation frequency of 600 MHz, is plotted in Figures 3.14 and 3.15, respectively. As shown in these figures, the parallel speedup of the matrix filling is very high and nearly overlays the ideal linear speedup.

Test results on computing platforms CEM-2 and CEM-4 for the PEC sphere, at an excitation frequency of 600 MHz, are plotted in Figures 3.16 and 3.17, respectively. The performance obtained on CEM-2 is similar to that obtained on CEM-1 and CEM-3, as indicated in Figure 3.16. The observed speedup for matrix filling is very close to that of the ideal linear speedup. However, the performance on CEM-4 is different from that on CEM-1, CEM-2, and CEM-3.

In Figure 3.17, when the number of CPUs is less than 10, the speedup in matrix filling is almost equal to the theoretical value. The matrix filling efficiency diverges from the theoretical value when the number of CPUs is more than 10. This is due to the fact that as the number of CPUs increases, the computational load on each CPU decreases. When each CPU is very lightly loaded, any small imbalance in the loading forms a larger percentage of the total load. The data points plot the worst-case CPU time for each project executed. Some of the processes continue to achieve higher speedup in matrix filling, but the data points plot the time for the worst case, because the performance of a parallel code is only as good as the worst-case scenario.

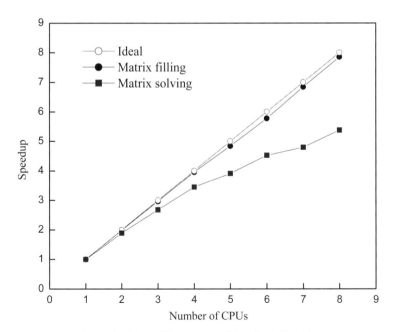

Figure 3.14. Parallel speedup achieved on CEM-1.

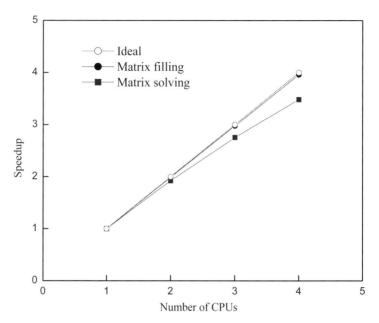

Figure 3.15. Parallel speedup achieved on CEM-3.

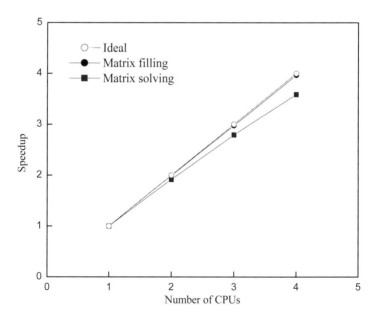

Figure 3.16. Parallel speedup achieved on CEM-2

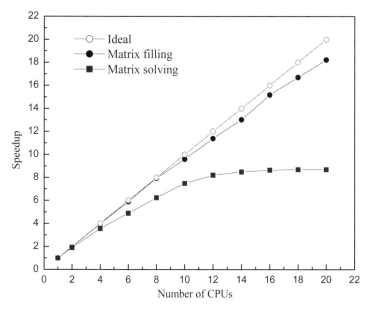

Figure 3.17. Parallel speedup achieved on CEM-4.

The procedure for solving a matrix equation is referred to as *matrix solving*. The performance in the solution of the matrix exhibits a type of degradation similar to that explained previously. However, there is an additional factor, which is the inherent additional communication required between the various processors encountered in a ScaLAPACK-based matrix solver. The amount of degradation can be observed from Figures 3.14 to 3.17. The plot of the speedup for the matrix solving diverges from the plot of the ideal speedup for each of them. For a fixed problem size, as the number of CPUs increases, the contribution from interprocessor communication, for passing messages between CPUs, to the total time becomes more significant during the matrix solving process.

Figure 3.18 shows the total parallel efficiency, including matrix filling and matrix solving, for both the larger and smaller example projects run on the same computer platform. The plot shows that the parallel efficiency for the solution of the larger problem with 19,101 unknowns is higher than for the solution of the smaller problem with 11,604 unknowns, although the efficiency for both of them is very good. This is quite a fortunate situation and is to be expected, since parallel algorithms are typically employed for the purpose of solving large problems using multiple CPUs. In this case, each CPU is assigned a larger load per core, which provides a more efficient performance than with a light load per core.

The portability of these codes is evident from the variety of platforms on which they have been successfully tested.

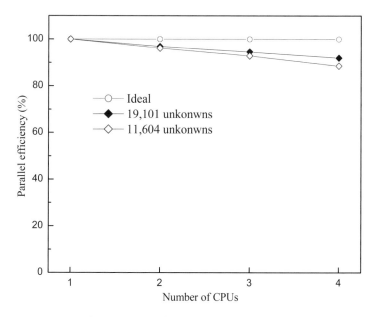

Figure 3.18. Parallel efficiency achieved on CEM-2.

3.10   NUMERICAL RESULTS FROM A PARALLEL OUT-OF-CORE MOM SOLVER

Results from the parallel out-of-core solver are presented next. Before discussing the expected performance and capabilities of the parallel out-of-core solver, in Section 3.10.1 we provide a comparison demonstrating that the parallel out-of-core solver can be as efficient as the parallel in-core solver. This may be a surprising result to some, but not when one considers the heavy demand on computational resources for solving a large dense matrix with complex elements, which often arises in the application of the MoM procedure to the solution of an integral equation, as compared to reading/writing of the matrix elements from/to the hard disk.

3.10.1   Parallel Out-of-Core Solver Can Be as Efficient as a Parallel In-Core Solver

Table 3.1 summarizes the solution times obtained from executing the parallel in-core and the parallel out-of-core codes on the same computer platform for two different problem sizes. The 1-m-radius sphere is again used as the geometric model with the excitation frequency increased to 1.0 GHz for the first two cases and 1.2 GHz for the third.

TABLE 3.1. Performance of a Parallel In-Core and Out-of-Core Solver on CEM-4

	Case 1-1	Case 1-2	Case 1-3
Parallel code	In-core	Out-of-core	Out-of-core
Unknowns	48,582	48,582	70,278
Num of CPUs	18	18	18
Memory used (GB)	37.8	36	36
Filling time (minutes)	4.2	4.6	10 (estimate: 8.8)
Solving time (minutes)	67.8	67.4	195 (estimate: 205.2)
Total time (minutes)	72	72	208 (estimate: 218)

First, consider cases 1-1 and 1-2. Both cases solve the same problem on the same computer platform (CEM-4), which has sufficient RAM to solve this problem in-core. Case 1-1 uses the parallel in-core code and each process can access up to 4 GB RAM, which is more than needed for this size of the problem. Case 1-2 uses the parallel out-of-core code and each process is assigned 2 GB RAM. The total memory assigned is 36 GB, which is less than the 37.8 GB needed to store the matrix. During the matrix filling stage, elements of the matrix need to be written to the hard disk associated with each of the nine compute nodes, where one hard disk is shared between two processes. Table 3.1 shows that, within the accuracy of the measurement of the total execution time, the wall times required for executing the same problem in-core and out-of-core are essentially the same.

This might be surprising as one might expect that the out-of-core solver needs to read/write from/to the hard disk, which would involve considerable execution time even for a small problem. But actually, for a Linux operating system, when a process writes to the hard disk, the kernel does not necessarily write the data to hard disk immediately if there is enough memory (RAM) to cache the data. Instead, it makes a copy of the block in memory and defers the disk update until later (an exception to this is for files that are open for synchronous writing). Therefore the computation will proceed without any delay of reading data from the hard disk. The problem size for case 1-1 and case 1-2 uses half the total amount of RAM on the cluster, and thus there has been sufficient RAM available to cache the data for the out-of-core solver.

Next, consider case 1-3, which is a larger sized problem requiring 79 GB of storage and could not be run in-core on this platform using nine compute nodes and 72 GB RAM. Using the execution time for the parallel in-core problem of case 1-1 as a benchmark, an estimate of the time required for both matrix filling and matrix solving can be calculated and compared to the actual observed times. The ratio of CPU time for solving the same problem using different numbers of unknowns, N_1 and N_2, should fall between $(N_2/N_1)^2$ and $(N_2/N_1)^3$. This is because the matrix filling portion of the code scales with the computational complexity $O(N^2)$ and the matrix solving, which uses the LU decomposition, scales with the computational complexity $O(N^3)$.

The parallel in-core performance is used as the benchmark because it predicts the best case performance, as if sufficient RAM were available to solve the problem. For matrix filling, the estimate is based on the relationship $(N_2/N_1)^2$ and is estimated to be 8.8 minutes. The measured time agrees with this prediction and demonstrates that the matrix filling algorithm is as good as the predicted time for an in-core solver.

For the matrix solving, the estimate is based on the relationship $(N_2/N_1)^3$ and is calculated to be 205.2 minutes. The actual solution time of 195 minutes is better than the predicted for an in-core solver! This is a reasonable result and indicates that the CPU time required for the parallel out-of-core solver can be comparable to that for the parallel in-core solver for a relatively small sized problem.

Here, for solving a relatively small-sized problem, the amount of storage required is slightly larger than the amount of RAM available on the computer platform. Therefore, the problem cannot be run using the in-core solver, however, it can be executed using the out-of-core solver as efficiently as if enough RAM was available and the in-core solver was used.

3.10.2   Scalability and Portability of the Parallel Out-of-Core Solver

The effectiveness of the out-of-core solver depends partly on the amount of in-core memory available and on the performance of the I/O system. For problems requiring larger than 80 GB of RAM on CEM-4, the performance of the parallel out-of-core solver would be slightly worse than the parallel in-core solver, due to the read/write of the data from/to the hard disk.

The example chosen for demonstration is a full-size PEC airplane model. The length, width, and height of the model airplane are 16.78 m, 11.04 m, and 2 m, respectively. The airplane structure was generated from a scaled toy model and was meshed into 8540 triangular patches at 100 MHz, as shown in Figure 3.19. Using a plane wave, polarized along the direction of the height and incident from a straight nose on of the model aircraft, the scattering pattern at 100 MHz is calculated and compared with references [7,8]. The results are plotted in Figure 3.20.

Next, three different cases are considered. For each case, scattering from the full-scale airplane model described above is considered. However, the frequency of the incident wave is increased to yield problems with a different number of unknowns. Table 3.2 summarizes the various parameters for the three different cases and lists the amount of hard-disk storage needed, which is calculated from the number of unknowns. For each of these cases, the amount of RAM is insufficient for an in-core solution. Each process is assigned 2.0 GB RAM as the in-core buffer, providing a total of 36 GB RAM available for the out-of-core integral equation solver when 18 processes are used. The time listed in Table 3.2 is the total "wall" time for matrix filling and solution of the matrix equations.

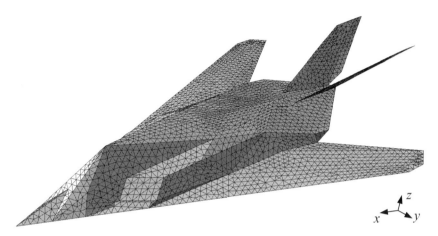

Figure 3.19. The triangulated mesh for the model airplane.

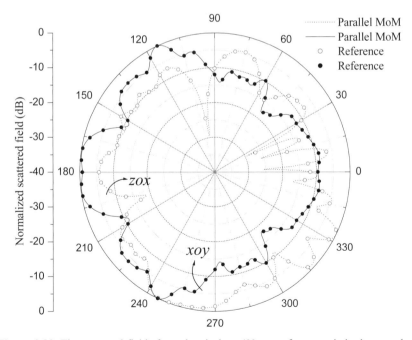

Figure 3.20. The scattered fields from the airplane (0° starts from *x* axis in the *xoy* plane and starts from *z* axis in the *zox* plane).

TABLE 3.2. Comparison of Wall Times Achieved on CEM-4

	Case 2-1	Case 2-2	Case 2-3
Frequency (MHz)	200	250	300
Unknowns	$N_1 = 79{,}986$	$N_2 = 123{,}744$	$N_3 = 186{,}798$
Dimension	$L = 11.19\,\lambda$ $W = 7.36\,\lambda$ $H = 1.333\,\lambda$	$L = 13.98\,\lambda$ $W = 9.2\,\lambda$ $H = 1.67\,\lambda$	$L = 16.78\,\lambda$ $W = 11.04\,\lambda$ $H = 2.0\,\lambda$
Memory (GB)	36	36	36
Storage (GB)	102.4	245	558.3
Filling time (minutes)	$T_{1f} = 11.5$	$T_{2f} = 28$ Estimate $1f$: 27.5	$T_{3f} = 64$ Estimate $1f$: 62.7 Estimate $2f$: 63.8
Solving time (minutes)	$T_{1s} = 290.5$	$T_{2s} = 1253.8$ Estimate $1s$: 1075.7	$T_{3s} = 4190$ Estimate $1s$: 3700.2 Estimate $2s$: 4312.9
Total time (minutes)	307	1290	4259

The simulations at different frequencies are executed on CEM-4 and the estimated times are listed in Table 3.2. In Table 3.2, T_{1f}, T_{2f}, and T_{3f} denote the measured matrix filling time at three different frequencies, and T_{1s}, T_{2s}, and T_{3s} denote the corresponding measured matrix solving time at each frequency. "Estimate $1f$" and "Estimate $2f$" denote the calculated matrix filling times based on T_{1f} and T_{2f} as the benchmark, respectively. "Estimate $1s$" and "Estimate $2s$" denote the calculated matrix solving times based on T_{1s} and T_{2s} as the benchmark, respectively.

For matrix filling, the time is estimated to be approximately 63 minutes for case 2-3 by using both the relationships $(N_3/N_1)^2$ and $(N_3/N_2)^2$ based on case 2-1 and case 2-2 as benchmarks, respectively. The measured time is 64 minutes. This indicates again that the matrix filling algorithm is as good as expected from a theoretical perspective.

The matrix solving times are compared using the $O(N^3)$ relationship. Table 3.2 shows that the actual solution time for case 2-3 falls between the times estimated using the two smaller problems as benchmarks, which were also executed using the parallel out-of-core code.

It is meaningful to compare the performance of the parallel out-of-core code with respect to solving the same problem using a parallel in-core code. The analysis, which follows, highlights how little the performance of the matrix solving procedure associated with a parallel out-of-core code deviates from the performance of a parallel in-core code.

Case 1-1 in Table 3.1 (in-core) is used as a benchmark to predict the matrix solving time for case 2-1 in Table 3.2 (out-of-core). The estimated time is $68 \times (79,986/48,582)^3 = 303.5$ minutes, and the measured time is 290.5 minutes. Also as had been illustrated for case 1-3 (sphere model), it is seen that the parallel out-of-core solver can be comparable in time to a parallel in-core solver for relatively small problems.

Consider the wall time for case 2-2, using the same in-core benchmark. The estimated matrix solving time is $68 \times (123,744/48,582)^3 = 1123.7$ minutes and the measured time is 1253.8 minutes. The performance of the out-of-core code for this medium-sized problem has degraded from the predicted in-core performance by about 10%.

Finally, consider the wall time for case 2-3, using the in-core benchmark. The estimated time is $68 \times (186,798/48,582)^3 = 3865.4$ minutes and the measured time is 4190 minutes. The performance of the out-of-core matrix solver for this large-sized problem has degraded from the predicted performance of the in-core code by about 8%. Here, a medium-sized problem requires a storage space several times larger than the amount of RAM used for the out-of-core solver on the computer platform. A large-sized problem is one that requires storage space 10 or more times greater than the amount of RAM available for the out-of-core solver on the computer platform. Figure 3.21 shows the current distribution over the surface of the aircraft at 300 MHz.

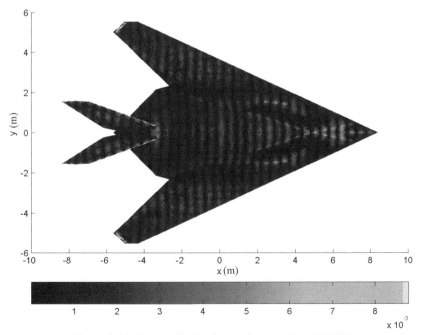

Figure 3.21. Current distribution on the aircraft at 300 MHz.

Simulations performed using the platforms CEM-6 and CEM-7, with the computational times listed in Tables 3.3 and 3.4, yield similar conclusions. In Table 3.3 (Table 3.4), T_{1f} and T_{1s}, denote the measured matrix filling time and matrix solving time for case 3-1 (case 4-1), respectively; "Estimate $1f$" denotes the calculated matrix filling time using T_{1f} as the benchmark; "Estimate $1s$" denotes the calculated matrix solving time using T_{1s} as the benchmark, respectively. Other symbols have corresponding similar meanings for each case.

TABLE 3.3. Time Comparison on CEM-6

	Case 3-1	Case 3-2
Frequency (MHz)	200	250
Unknowns	$N_1 = 79,986$	$N_2 = 123,744$
Filling time (minutes)	$T_{1f} = 28$	$T_{2f} = 62$
		Estimate $1f$: 67
Solving time (minutes)	$T_{1s} = 851$	$T_{2s} = 3324$
		Estimate $1s$: 3151
Total time (minutes)	886	3392

TABLE 3.4. Time Comparison on CEM-7

	Case 4-1	Case 4-2	Case 4-3
Frequency (MHz)	200	200	225
Unknowns	$N_1 = 79,986$	$N_2 = 79,986$	$N_3 = 104,592$
Hard disk (GB)	102	102	175
Processes	8	16	16
Filling time (minutes)	$T_{1f} = 36$	$T_{2f} = 20$	$T_{3f} = 29$
		Estimate $1f$: 18	Estimate $1f$: 30.8
			Estimate $2f$: 34.2
Solving time (minutes)	$T_{1s} = 680$	$T_{2s} = 349$	$T_{3s} = 790.5$
		Estimate $1s$: 340	Estimate $1s$: 760.2
			Estimate $2s$: 780.3
Total time (minutes)	716	371	820

When the number of unknowns (in Table 3.3) is increased from 79,986 to 123,744 on CEM-6, the performance of the parallel out-of-core code is degraded by 5.5%. Using one blade (8 processes) of CEM-7 as a benchmark, the parallel efficiency for two blades (16 processes) is as high as 97%. This indicates the excellent scalability achieved by the parallel out-of-core MoM solver.

When the number of unknowns (in Table 3.4) increases from 79,986 to 104,592 on CEM-7, the performance of the parallel out-of-core solver degrades by only 1.3%.

The purpose of the next example is to illustrate that even on a very simple PC cluster, the parallel out-of-core code works well and can go far beyond the 2 GB RAM limitation of an IA-32 system.

Two cases are executed on two nodes of CEM-5, which is constructed by connecting ordinary PCs using 10/100-Mbps (megabits per second) network interface cards. Table 3.5 shows the results for the model airplane simulated at 150 MHz using 49,368 unknowns. In Table 3.5, T_{1f} and T_{1s}, and "Estimate 1f" and "Estimate 1s" have meanings similar to those explained for Tables 3.3 and 3.4.

Executing the problem on one node, which is one PC with one CPU, and on two nodes, a 74.4% parallel efficiency can be obtained. If this was simulated using the parallel in-core code, it would require about 40 GB RAM and couldn't be executed on this simple PC cluster.

The scattered fields from the aircraft in the xoy plane at a distance of 1000 λ are shown in Figure 3.22 at five different frequencies.

The examples presented in this section are for different sized problems and have been executed on multiple computer platforms, using different number of computing nodes. These examples demonstrate the scalability and portability of the parallel code.

TABLE 3.5. Time Comparison on CEM-5

	Case 5-1	Case 5-2
Frequency (MHz)	150	150
Unknowns	49,368	49,368
Num of CPUs	1	2
Filling time (minutes)	$T_{1f} = 133$	$T_{2f} = 65$
		Estimate 1f: 66.5
Solving time (minutes)	$T_{1s} = 1367.3$	$T_{2s} = 944$
		Estimate 1s: 683.7
Total time (minutes)	1502	1009

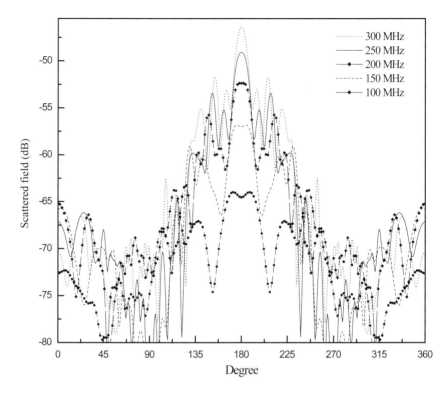

Figure 3.22. The plot of the scattered fields at five different frequencies (0° starts from x axis in the xoy plane).

Since the cost of using a hard disk is much lower than that of RAM of the same capacity, the solution of very large problems can be achieved with comparable computational efficiency, using storage on inexpensive hard disks rather than storing the data on expensive RAM. Analysis of large problems, using a parallel out-of-core code, incur only about a 10% penalty in matrix solving time over the parallel in-core solver. A parallel out-of-core MoM code with RWG basis functions offers a high-performance, low-cost choice for tackling real-world electromagnetic computational challenges.

Note that results presented in this chapter have not been optimized yet [9]. Further research on how to improve the performance of the parallel solver will be discussed in detail in Chapter 5.

3.11   CONCLUSION

In this chapter, efficient schemes for parallel implementation of MoM codes using the RWG basis functions are presented. The simulations of the developed parallel codes on several popular computer platforms have demonstrated its excellent performance, including parallel speedup, parallel efficiency, portability and scalability.

Although only the EFIE has been applied for PEC structures in this chapter to demonstrate the performance of the various parallel schemes, similar conclusions actually hold true for any complex, composite structures, which are numerically solved using the method of moments with RWG basis functions. However, they have not been presented here.

REFERENCES

[1]   R. F. Harrington, *Field Computation by Moment Methods*, IEEE Press Series on Electromagnetic Waves, Piscataway, NJ, 1993.

[2]   S. M. Rao, D. R. Wilton, and A. W. Glisson, "Electromagnetic Scattering by Surfaces of Arbitrary Shape," *IEEE Transactions on Antennas and Propagation*, Vol. 30, No. 5, pp. 409–418, 1982.

[3]   D. R .Wilton, S. M. Rao, A. W. Glisson, D. H. Schaubert, O. M. Al-Bundak, and C. M. Butler, "Potential Integrals for Uniform and Linear Source Distributions on Polygonal and Polyhedral Domains," *IEEE Transactions on Antennas and Propagation*, Vol. 32, No. 3, pp. 276–281, 1984.

[4]   T. F. Eibert and V. Hansen, "On the Calculation of Potential Integrals for Linear Source Distributions on Triangular Domains," *IEEE Transactions on Antennas and Propagation*, Vol. 43, No. 12, pp. 1499–1502, 1995.

[5]   V. I. Krylov, *Approximate Calculation of Integrals*, Macmillan, New York, 1962.

[6]   Y. Zhang, T. K. Sarkar, A. De, N. Yilmazer, S. Burintramart, and M. Taylor, "A Cross-Platform Parallel MoM Code with ScaLAPACK Solver," *IEEE Antennas and Propagation Society International Symposium,* pp. 2797–2800, Honolulu, HI, June 2007.

[7]   Y. Zhang, X. W. Zhao, M. Chen, and C. H. Liang, "An Efficient MPI Virtual Topology Based Parallel, Iterative MoM-PO Hybrid Method on PC Clusters," *Journal of Electromagnetic Waves and Applications*, Vol. 20, No. 5, pp. 661–676, 2006.

[8]   Y. Zhang, *Parallel Computation in Electromagnetics*, Xidian University Press, Xi'an, China, 2006.

[9]   Y. Zhang, M. C. Taylor, T. K. Sarkar, A. De, M Yuan, H. Moon, and C Liang, "Parallel In-Core and Out-of-Core Solution of Electrically Large Problems Using the RWG Basis Functions," *IEEE Antennas and Propagation Magazine*, Vol. 50, No. 5, pp. 84–94, Oct. 2008.

4

A PARALLEL MOM CODE USING HIGHER-ORDER BASIS FUNCTIONS AND SCALAPACK-BASED IN-CORE AND OUT-OF-CORE SOLVERS

4.0 SUMMARY

In this chapter, the solution of surface integral equations (SIEs) is introduced for dealing with frequency-domain analysis of radiation and scattering problems involving antennas and microwave structures composed of conductors and/or dielectrics. Higher-order polynomials over wires and quadrilateral plates are used as basis functions over larger subdomain patches so that this procedure can be approximately related to the use of entire domain basis functions. Polynomial expansions for the basis functions over larger subdomains result in a good approximation of the current distributions over large surfaces using approximately 10 unknowns per wavelength squared of surface area. Therefore, the use of polynomial basis functions over larger patches sharply reduces the number of unknowns required to approximate the unknown solution, compared with the number of piecewise RWG basis functions outlined in Chapter 3, required to carry out a similar approximation for the same unknown current.

With these advantages in mind, an efficient parallel matrix filling scheme is designed for the parallel in-core and the parallel out-of-core MoM codes using higher-order basis functions. Numerical results for problems simulated on various computer clusters indicate that the implementation of the higher-order basis functions in parallel integral equation solver codes creates a new powerful tool for solving electrically large, complex, real-world computational electromagnetic problems.

4.1 FORMULATION OF THE INTEGRAL EQUATION FOR ANALYSIS OF DIELECTRIC STRUCTURES

The Poggio–Miller–Chang–Harrington–Wu (PMCHW) [1,2] formulation is widely used for analysis of three-dimensional structures composed of both conductors and dielectrics. The PMCHW formulation will be derived in this section. An example will be presented to illustrate how to solve the general scattering problem as sketched in Figure 4.1.

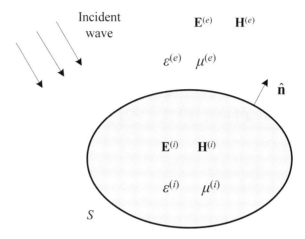

Figure 4.1. Electromagnetic scattering from a dielectric body.

The equivalent principle will be applied in two steps to obtain equivalent models for the exterior fields and the interior fields.

First, the application of the equivalent theorem leads to a model that can be used for computation of the external scattered fields. Assume that the interior fields produced by the problem are equal to zero. Once the fields internal to the structure are zero, the material constituting the internal region can be replaced by the material of the exterior region. We then place equivalent electric and magnetic currents, $\mathbf{J}^{(e)}$ and $\mathbf{M}^{(e)}$, on the surface of the body to preserve the continuity of the fields across the dielectric boundary as illustrated in Figure 4.2 (a). The respective boundary conditions are

$$-\hat{\mathbf{n}} \times [\mathbf{E}_{inc} + \mathbf{E}^{(e)}(\mathbf{J}^{(e)}, \mathbf{M}^{(e)})]|_{S_+} = \mathbf{M}^{(e)}, \tag{4.1}$$

$$\hat{\mathbf{n}} \times [\mathbf{H}_{inc} + \mathbf{H}^{(e)}(\mathbf{J}^{(e)}, \mathbf{M}^{(e)})]|_{S_+} = \mathbf{J}^{(e)}, \tag{4.2}$$

where S_+ implies the fields are evaluated on the exterior side of the surface S. $\hat{\mathbf{n}}$ is the unit normal vector directed outward from the interior to the exterior region. $\mathbf{E}_{inc}$ and $\mathbf{H}_{inc}$ are the electric and magnetic fields incident on the structure from the external region, respectively. $\mathbf{E}^{(e)}$ and $\mathbf{H}^{(e)}$ denote the scattered electric and magnetic fields in the exterior region, respectively.

Next, the equivalent principle is applied again to obtain the appropriate surface currents so as to produce the appropriate fields interior to the structure.

The exterior fields are equated to be zero as one is interested in the correct interior fields only. The exterior medium is now replaced by the same material that exists in the interior of the structure as the external fields are identically zero for this equivalent problem. Equivalent electric and magnetic currents, $\mathbf{J}^{(i)}$ and $\mathbf{M}^{(i)}$, are now placed on the surface of the structure to compensate for the field discontinuity across the boundary as shown in Figure 4.2 (b). In this case, the boundary conditions are

$$\hat{\mathbf{n}} \times \mathbf{E}^{(i)}(\mathbf{J}^{(i)}, \mathbf{M}^{(i)})\big|_{S_-} = \mathbf{M}^{(i)}, \tag{4.3}$$

$$-\hat{\mathbf{n}} \times \mathbf{H}^{(i)}(\mathbf{J}^{(i)}, \mathbf{M}^{(i)})\big|_{S_-} = \mathbf{J}^{(i)}, \tag{4.4}$$

where S_- implies that the fields are evaluated at the interior side of the surface S. $\mathbf{E}^{(i)}$ and $\mathbf{H}^{(i)}$ denote the scattered electric and the magnetic fields in the interior region, respectively.

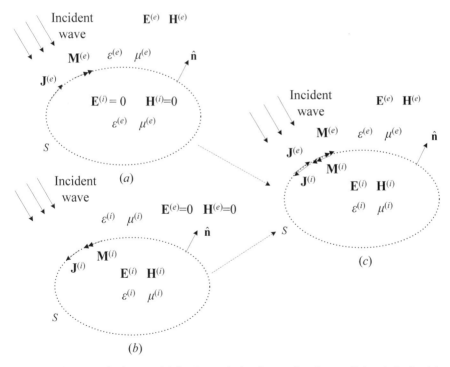

Figure 4.2. An equivalent model for the analysis of scattering from a dielectric body: (a) the equivalent model for the exterior region; (b) the equivalent model for the interior region; (c) the equivalent model for the entire region.

The equivalent model of Figure 4.2 (a) preserves the exterior fields of the original problem, and the equivalent model of Figure 4.2 (b) retains the same interior fields as that of the original problem. Therefore, the exterior part from Figure 4.2 (a) and the interior part from Figure 4.2 (b) can be used to construct a new model as in Figure 4.2 (c). For the original problem, the tangential components of the electric and the magnetic fields are continuous. The boundary conditions dictate that for the equivalent problem, the following conditions must hold:

$$\hat{\mathbf{n}} \times [\mathbf{E}_{\mathrm{inc}} + \mathbf{E}^{(e)}(\mathbf{J}^{(e)}, \mathbf{M}^{(e)})]\big|_{S_+} = \hat{\mathbf{n}} \times \mathbf{E}^{(i)}(\mathbf{J}^{(i)}, \mathbf{M}^{(i)})\big|_{S_-} , \tag{4.5}$$

$$\hat{\mathbf{n}} \times [\mathbf{H}_{\mathrm{inc}} + \mathbf{H}^{(e)}(\mathbf{J}^{(e)}, \mathbf{M}^{(e)})]\big|_{S_+} = \hat{\mathbf{n}} \times \mathbf{H}^{(i)}(\mathbf{J}^{(i)}, \mathbf{M}^{(i)})\big|_{S_-} . \tag{4.6}$$

Equations (4.5) and (4.6) are generally called the "PMCHW formulations". Substituting Eqs. (4.1)–(4.4) into Eqs. (4.5) and (4.6), we can reduce the four unknown currents to two as follows:

$$\mathbf{M}^{(e)} = -\mathbf{M}^{(i)} = \mathbf{M} , \tag{4.7}$$

$$\mathbf{J}^{(e)} = -\mathbf{J}^{(i)} = \mathbf{J} . \tag{4.8}$$

As a result, there are two unknown currents, $\mathbf{J}$ and $\mathbf{M}$, to be solved for using the two corresponding equations, as shown in Eqs. (4.5) and (4.6).

4.2   A GENERAL FORMULATION FOR THE ANALYSIS OF COMPOSITE METALLIC AND DIELECTRIC STRUCTURES

The main objective of a frequency-domain analysis of a composite, lossy, metallic, and dielectric structure is to evaluate the distribution of the electric and the magnetic fields inside and/or outside the material object as a function of frequency due to some impressed sources or incident fields. This objective can be achieved in several different ways. Here, a method is provided for the solution for the equivalent currents over the composite metallic and dielectric boundary surfaces using the surface integral equations (SIEs).

Presence of inhomogeneous dielectric structures can always be categorized by a combination of various piecewise homogeneous dielectric bodies [3–7]. Therefore, any composite metallic and dielectric structure can be represented as an electromagnetic system consisting of a finite number of finite-sized, linear, homogeneous, and isotropic regions, situated in an unbounded linear, homogeneous, and isotropic environment, as shown in Figure 4.3 (a). Most often, this environment is a vacuum, but it can also be another medium.

Some of the regions in Figure 4.3 can be perfect electric conductors (PEC). In the interior of any such region, the electromagnetic field is always zero. These regions can be referred to as *zero-field regions* and are collectively denoted by the region 0 as shown in Figure 4.3. Metallic wires and plates can be considered as special cases of the zero-field regions. Plates forming an open surface can be regarded as a degenerate case of a zero-field region, where one dimension (thickness) of the region is zero.

In all the remaining regions, there exist electromagnetic fields. These regions will collectively be referred to as *non-zero-field regions*. The total number of such regions is denoted by n. The medium representing region i is characterized by the complex permittivity $\varepsilon^{(i)}$ and permeability $\mu^{(i)}$, $i = 1,\ldots, n$, which include all the relevant losses. In any of these regions, there may exist impressed electric and magnetic fields, $\mathbf{E}_{inc}^{(i)}$ and $\mathbf{H}_{inc}^{(i)}$, $i = 1,\ldots, n$, whose angular frequency is ω.

Consider an arbitrary region with a nonzero impressed electromagnetic field, e.g., region i as shown in Figure 4.3 (a). According to the surface equivalence theorem, the effect of all the sources outside the region i can be replaced by equivalent currents placed at the boundary surface of region i, in which case the fields outside region i become zero, as shown in Figure 4.3 (b). The region outside region i is denoted as region $0 = i$ (region 0 with respect to region i, in the equivalent problem for region i). Since the field outside region i is zero, it can be homogenized with respect to region i; i.e., it can be filled by the same material as region i. Thus, a multiple-region problem (consisting of n regions) can be decomposed into n single-region problems.

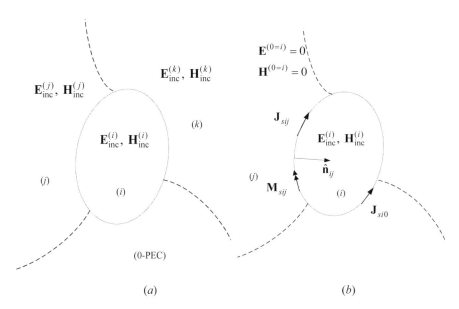

Figure 4.3. Decomposition of a multiple-region problem into single-region problems: (a) the original problem; (b) the equivalent problem for region i.

The densities of equivalent currents at the boundary surface between regions i and j are

$$\mathbf{J}_{sij} = \hat{\mathbf{n}}_{ij} \times \mathbf{H}^{(i)}, \qquad\qquad \mathbf{M}_{sij} = -\hat{\mathbf{n}}_{ij} \times \mathbf{E}^{(i)}, \qquad (4.9a,b)$$

where $\hat{\mathbf{n}}_{ij}$ is the unit normal vector directed from region j to region i, and $\mathbf{E}^{(i)}$ and $\mathbf{H}^{(i)}$ are the electric and the magnetic fields at the boundary surface, just inside region i. If the equivalent currents for region j, $\mathbf{J}_{sji}$ and $\mathbf{M}_{sji}$, are considered, they are evaluated according to Eq. (4.9), but with the indices i and j interchanged:

$$\mathbf{J}_{sji} = \hat{\mathbf{n}}_{ji} \times \mathbf{H}^{(j)}, \qquad\qquad \mathbf{M}_{sji} = -\hat{\mathbf{n}}_{ji} \times \mathbf{E}^{(j)}. \qquad (4.10a,b)$$

Since there are no currents across regions i and j, the fields $\mathbf{H}^{(i)}$ and $\mathbf{H}^{(j)}$, as well the fields $\mathbf{E}^{(i)}$ and $\mathbf{E}^{(j)}$ satisfy the following boundary conditions:

$$\hat{\mathbf{n}}_{ij} \times (\mathbf{H}^{(i)} - \mathbf{H}^{(j)}) = 0, \qquad\qquad \hat{\mathbf{n}}_{ij} \times (\mathbf{E}^{(i)} - \mathbf{E}^{(j)}) = 0. \qquad (4.11a,b)$$

After expressing the field vectors in terms of the equivalent currents, and using Eqs. (4.9) and (4.10), along with $\hat{\mathbf{n}}_{ij} = -\hat{\mathbf{n}}_{ji}$, one observes that the equivalent currents are related by

$$\mathbf{J}_{sij} = -\mathbf{J}_{sji}, \qquad\qquad \mathbf{M}_{sij} = -\mathbf{M}_{sji}. \qquad (4.12a,b)$$

Thus, n single-region problems are mutually coupled through the conditions of Eqs. (4.11) and (4.12). It has been observed that satisfaction of these conditions guarantees the uniqueness of the solution for the sources and the fields alike. In that case, the main objective of the frequency-domain analysis is to evaluate the distribution of the equivalent electric and the magnetic currents at the boundary surfaces of the single-region problems, which satisfy the conditions expressed in Eqs. (4.11) and (4.12).

The field integral equations can be derived from the boundary conditions given by Eq. (4.11). Starting with Eq. (4.11b), the total electric field in region i can be expressed as

$$\mathbf{E}^{(i)} = \sum_{\substack{k=0 \\ k \neq i}}^{n} \mathbf{E}^{(i)}(\mathbf{J}_{sik}, \mathbf{M}_{sik}) + \mathbf{E}_{\text{inc}}^{(i)}, \qquad (4.13)$$

where $\mathbf{E}^{(i)}(\mathbf{J}_{sik}, \mathbf{M}_{sik})$ represents the scattered field inside region i, which is produced by the currents placed on the boundary surface between regions i and k, and the corresponding incident field is $\mathbf{E}_{\text{inc}}^{(i)}$. The scattered fields, just inside region i, due to the currents placed on the boundary surface of regions i and k, are given by

$$\mathbf{E}^{(i)}(\mathbf{J}_{sik},\mathbf{M}_{sik}) = -L^{(i)}(\mathbf{J}_{sik}) + K^{(i)}(\mathbf{M}_{sik}), \tag{4.14a}$$

$$\mathbf{H}^{(i)}(\mathbf{J}_{sik},\mathbf{M}_{sik}) = -K^{(i)}(\mathbf{J}_{sik}) - \frac{\varepsilon^{(i)}}{\mu^{(i)}} L^{(i)}(\mathbf{M}_{sik}), \tag{4.14b}$$

where $L^{(i)}$ and $K^{(i)}$ are the operators defined by

$$L^{(i)}(\mathbf{X}_{sik}) = j\,\omega\,\mu^{(i)} \int_{S_{ik}} \left(\begin{array}{l} \mathbf{X}_{sik}(\mathbf{r}_{ik})\, g^{(i)}(\mathbf{r},\mathbf{r}') + \\[2mm] \dfrac{1}{\omega^2 \varepsilon^{(i)} \mu^{(i)}} \nabla_{sik} \bullet \mathbf{X}_{sik}(\mathbf{r}_{ik})\, \nabla g^{(i)}(\mathbf{r},\mathbf{r}') \end{array} \right) dS_{ik}, \tag{4.15a}$$

$$K^{(i)}(\mathbf{X}_{sik}) = \int_{S_{ik}} \mathbf{X}_{sik}(\mathbf{r}_{ik}) \times \nabla g^{(i)}(\mathbf{r},\mathbf{r}')\, dS_{ik}. \tag{4.15b}$$

In Eq. (4.15), $\mathbf{X}_{sik}$ can be either the electric or the magnetic current, and

$$g^{(i)}(\mathbf{r},\mathbf{r}') = \frac{e^{-\gamma^{(i)}R}}{4\pi R}, \qquad R = |\mathbf{r}-\mathbf{r}_{ik}|, \qquad \gamma^{(i)} = j\omega\sqrt{\varepsilon^{(i)}\mu^{(i)}}, \tag{4.15c,d,e}$$

where $g^{(i)}(\mathbf{r},\mathbf{r}')$ is Green's function for the homogenous medium i; $\mathbf{r}_{ik}$, which is denoted as $\mathbf{r}'$ in Green's function, is the vector position of the source point and $\mathbf{r}$ is the vector position of the field point. The divergence operator ∇_{sik} acts on $\mathbf{r}_{ik}$ (on surface S_{ik}) and the gradient operator ∇ acts on $\mathbf{r}$. After substituting Eqs. (4.13) and (4.14a) that hold for regions i and j into Eq. (4.11b), the integral equation can be expressed in the following form:

$$\hat{\mathbf{n}}_{ij} \times \left\{ \begin{array}{l} \displaystyle\sum_{\substack{k=0 \\ k \neq i}}^{n} \left[L^{(i)}\left(\mathbf{J}_{sik}\right) - K^{(i)}\left(\mathbf{M}_{sik}\right) \right] \\[4mm] - \displaystyle\sum_{\substack{k=0 \\ k \neq j}}^{n} \left[L^{(j)}\left(\mathbf{J}_{sjk}\right) - K^{(j)}\left(\mathbf{M}_{sjk}\right) \right] \end{array} \right\} = \hat{\mathbf{n}}_{ij} \times \left(\mathbf{E}_{inc}^{(i)} - \mathbf{E}_{inc}^{(j)} \right). \tag{4.16a}$$

In a similar manner, another integral equation using the boundary condition of Eq. (4.11a) can be obtained, which is dual to Eq. (4.16a) and is given by

$$
\hat{\mathbf{n}}_{ij} \times \left\{ \begin{array}{l} \displaystyle\sum_{\substack{k=0 \\ k \neq i}}^{n} \left(\frac{\varepsilon^{(i)}}{\mu^{(i)}} L^{(i)} \left(\mathbf{M}_{sik} \right) + K^{(i)} \left(\mathbf{J}_{sik} \right) \right) \\[2em] - \displaystyle\sum_{\substack{k=0 \\ k \neq j}}^{n} \left(\frac{\varepsilon^{(j)}}{\mu^{(j)}} L^{(j)} \left(\mathbf{M}_{sjk} \right) + K^{(j)} \left(\mathbf{J}_{sjk} \right) \right) \end{array} \right\} = \hat{\mathbf{n}}_{ij} \times \left(\mathbf{H}_{inc}^{(i)} - \mathbf{H}_{inc}^{(j)} \right). \quad (4.16b)
$$

These two sets of equations represent a general form of the PMCHW formulation. For the case when one of the two regions sharing a common boundary surface is a PEC, the magnetic currents are equal to zero at the boundary surface and therefore, the first of the equations above degenerates into the electric field integral equation (EFIE).

Note that the EFIE provides a solution not only for closed metallic bodies but also for open metallic surfaces and metallic wires. Particularly, for the case dealing with wires, the EFIE is based on the extended boundary conditions and in addition invokes the thin-wire approximation. So for the analysis of electromagnetic radiation and scattering from arbitrary structures composed of wires and plates, a set of PMCHW and EFIE equations for the unknown electric and magnetic currents can be obtained. This set of equations is solved using the MoM methodology. In order to obtain an efficient method for the analysis of radiation and scattering from any composite structures, special care should be devoted to the choice of the basis functions, which needs to be characterized in two distinct different steps: geometric modeling and approximation of the relevant currents.

4.3   GEOMETRIC MODELING OF THE STRUCTURES

Flexible geometric modeling can be achieved by using truncated cones for wires and bilinear patches to characterize other surfaces [3–5]. This is illustrated next.

4.3.1   Right-Truncated Cone to Model Wire Structures

Note that any structure composed of wires can be considered to be composed of thin PEC wires that meet the requirements of a thin-wire approximation. So in this thin-wire model, the circumferential variation of the currents on the wires is neglected, and in addition the length of the wire should be at least 10 times larger than its radius. A right-truncated cone is determined by the position vectors and the radii of its startpoints and endpoints, characterized by $\mathbf{r}_1$, a_1 and, $\mathbf{r}_2$, a_2, respectively. This is shown in Figure 4.4, where $\mathbf{r}_1$ represents the position vector of the beginning of the cone, s is a local coordinate along the cone reference generatrix, and s_1 and s_2 are the s coordinates of the beginning and the end of the cone generatrix.

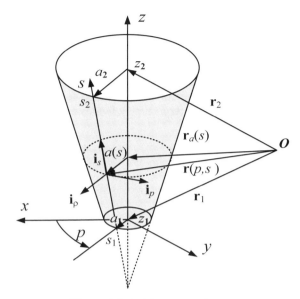

Figure 4.4. A right-truncated cone.

The reason for adopting the s-coordinate system rather than the z-coordinate system is that it will be assumed that the surface current density vector has only the s component and is not expressed in terms of z if the s and z axes are normal to each other. In order to define the parametric equations of the surface of a cone, a local cylindrical coordinate system where the z-coordinate axis coincides with the cone axis will be adopted. In that case the parametric equation of the cone surface can be written as

$$\mathbf{r}(p,s) = \mathbf{r}_a(s) + a(s)\,\mathbf{i}_\rho(p)\,, \ -1 \le s \le 1,\ -\pi \le p \le \pi , \tag{4.17a}$$

$$\mathbf{r}_a(s) = \mathbf{r}_1 + (s - s_1)\,\frac{\mathbf{r}_2 - \mathbf{r}_1}{s_2 - s_1},\ s_1 \le s \le s_2 , \tag{4.17b}$$

$$a(s) = a_1 + (s - s_1)\,\frac{a_2 - a_1}{s_2 - s_1},\ s_1 \le s \le s_2 , \tag{4.17c}$$

where s is the local coordinate along a cone generatrix, p is the circumferential angle measured from the x axis, and $\mathbf{i}_\rho(p)$ is the radial unit vector perpendicular to the cone axis.

Under some specific circumstances, the truncated cone degenerates into a right cylinder ($a_1 = a_2$), an ordinary cone ($a_2 = 0$), a flat disk ($a_2 = 0$, $\mathbf{r}_1 = \mathbf{r}_2$), or a frill ($\mathbf{r}_1 = \mathbf{r}_2$), if the appropriate parameters are chosen. The right-truncated cone and its degenerate forms can be used for the modeling of cylindrical wires with stepped variation of the radius as well as using flat and conical wire ends.

4.3.2   Bilinear Surface for Modeling Arbitrarily Shaped Surfaces

Metallic and dielectric surfaces will be modeled by bilinear surfaces. A bilinear surface is, in general, a nonplanar quadrilateral, which is defined uniquely by its four arbitrarily spaced vertices, as shown in Figure 4.5. Hence, it can be used for efficient modeling of both flat and curved surfaces. The parametric equation of such an isoparametric element can be written in the form

$$\mathbf{r}(p,s) = \mathbf{r}_{11}\frac{(1-p)(1-s)}{4} + \mathbf{r}_{12}\frac{(1-p)(1+s)}{4}$$
$$+ \mathbf{r}_{21}\frac{(1+p)(1-s)}{4} + \mathbf{r}_{22}\frac{(1+p)(1+s)}{4}, \qquad (4.18a)$$

$$-1 \le p \le 1, \quad -1 \le s \le 1, \qquad (4.18b)$$

where $\mathbf{r}_{11}$, $\mathbf{r}_{12}$, $\mathbf{r}_{21}$, and $\mathbf{r}_{22}$ are the position vectors of the four vertices, and p and s are the local coordinates. After some elementary transformations this equation can be written as

$$\mathbf{r}(p,s) = \mathbf{r}_c + \mathbf{r}_p\, p + \mathbf{r}_s\, s + \mathbf{r}_{ps}\, ps, \qquad (4.19a)$$

$$-1 \le p \le 1, \qquad -1 \le s \le 1, \qquad (4.19b)$$

where

$$\mathbf{r}_c = \frac{(\mathbf{r}_{11} + \mathbf{r}_{12} + \mathbf{r}_{21} + \mathbf{r}_{22})}{4}, \qquad \mathbf{r}_p = \frac{(-\mathbf{r}_{11} - \mathbf{r}_{12} + \mathbf{r}_{21} + \mathbf{r}_{22})}{4},$$
$$\mathbf{r}_s = \frac{(-\mathbf{r}_{11} + \mathbf{r}_{12} - \mathbf{r}_{21} + \mathbf{r}_{22})}{4}, \qquad \mathbf{r}_{ps} = \frac{(\mathbf{r}_{11} - \mathbf{r}_{12} - \mathbf{r}_{21} + \mathbf{r}_{22})}{4}.$$

Depending on the values of the vectors $\mathbf{r}_p$, $\mathbf{r}_s$, and $\mathbf{r}_{ps}$, a bilinear surface can take different degenerate forms: flat quadrilateral ($\mathbf{r}_p$, $\mathbf{r}_s$, and $\mathbf{r}_{ps}$ are coplanar), rhomboid ($\mathbf{r}_{ps} = 0$), or a rectangle ($\mathbf{r}_{ps} = 0$ and $\mathbf{r}_p \cdot \mathbf{r}_s = 0$). The maximum allowable length for the edges of the bilinear surfaces is at most two wavelengths. The current distribution over such large patches can still be successfully approximated by entire domain expansions consisting of polynomials. If the length of an edge for any patch is longer than two wavelengths, that patch is subdivided into a set of subpatches in such a manner that only edges longer than two wavelengths are subdivided into a minimum number of edges shorter than two wavelengths.

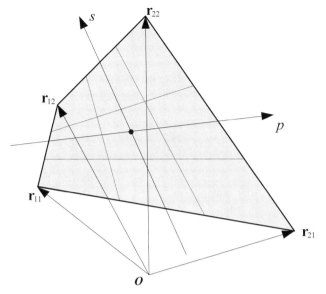

Figure 4.5. Example of a bilinear surface.

4.4   HIGHER-ORDER BASIS FUNCTIONS

Efficient functional approximation for the unknown currents can be achieved by using entire-domain expansions consisting of combinations of polynomials that can automatically satisfy the continuity equation. This guarantees the continuity of the current at arbitrary multiple metallic and/or dielectric junctions.

4.4.1   Current Expansion along a Thin PEC Wire

Consider an arbitrary shaped perfectly conducting wire structure, where one assumes that the surface current $J_s(s)$ is a function of the coordinate s only, i.e.

$$\mathbf{J}(p,s) = J_s(s)\,\hat{\mathbf{i}}_s(p,s)\,,\quad \hat{\mathbf{i}}_s(p,s) = \frac{\partial \mathbf{r}(p,s)/\partial s}{\left|\partial \mathbf{r}(p,s)/\partial s\right|}\,, \tag{4.20a,b}$$

where $\partial r(p,s)/\partial s$ represents the unitary vector along the generatrix of the truncated cone and $\hat{\mathbf{i}}_s(p,s)$ is the corresponding unit vector. The total current intensity per unit length along the cone can be defined as

$$I(s) = 2\,\pi\,a(s)\,J_s(s)\,. \tag{4.21}$$

Thus the corresponding charge per unit length can be expressed as

$$Q(s) = \frac{j}{\omega} \frac{dI(s)}{ds},$$ (4.22)

from which the continuity of the charge and the current for truncated cone can be obtained as

$$\rho(s) = \frac{Q(s)}{2\pi a(s)} = \frac{j}{2\pi a(s)\omega} \frac{dI(s)}{ds}.$$ (4.23)

Current distributions along the wires are approximated by polynomials that automatically satisfy the equation of continuity at the ends of the wires and junctions. The approximation for the current takes the form of a polynomial, and can be written as

$$I(s) = \sum_{i=0}^{N_s} a_i s^i, \; -1 \le s \le 1,$$ (4.24)

where a_i are the unknown coefficients and N_s is the order of the approximation. Coefficients a_0 and a_1 can be expressed in terms of the other unknown coefficients a_i $(i = 2,..., N_s)$ so that it satisfies the condition for the current at the ends of the wires, i.e., $I_1 = I(-1)$ and $I_2 = I(1)$. Using these boundary conditions, the expansion for the current in Eq. (4.24) can be written as

$$I(s) = I_1 N(s) + I_2 N(-s) + \sum_{i=2}^{N_s} a_i S_i(s),$$ (4.25)

where the node basis function, $N(s)$, and the segment basis functions, $S_i(s)$ $(i = 2,..., N_s)$, are expressed as

$$N(s) = \frac{1-s}{2}, \qquad S_i(s) = \begin{cases} s^i - 1 & i \text{ is even} \\ s^i - s & i \text{ is odd} \end{cases}.$$ (4.26a,b)

Note that all the basis functions in this expansion are equal to zero at the beginning and the end of a wire, except the (node) basis function corresponding to the unknown values of the current I_1 (I_2), which is equal to one at the beginning (end) of the wire. The continuity equation at the beginning (end) of a free wire end, which is not connected to any other structure, is satisfied by omitting the first (second) term in the expansion shown above. The continuity equation at a junction of the ith and the jth wires (the end of the ith wire coincides with the beginning of the jth wire) can be automatically satisfied if independent basis functions corresponding to I_{2i} and I_{1j} are replaced by a unique basis function in the form of a doublet as

$$D_{ij}(s_i, s_j) = \begin{cases} \dfrac{1+s_i}{2} & \text{along } i\text{th wire} \\ \\ \dfrac{1-s_j}{2} & \text{along } j\text{th wire} \end{cases} \tag{4.27}$$

Doublets are basis functions defined along two interconnecting wires. Similarly, basis functions corresponding to the unknowns a_i ($i = 2,.., N_s$) defined on a single segment are termed *singletons*. Note that such triangle doublets (usually used in subdomain piecewise linear approximation) represent a special case of an entire-domain approximation, obtained for $N_s = 1$. For a general case, the entire-domain approximation of currents along a complex thin-wire structure can be represented as a combination of overlapping doublets and singletons.

As an example, Figure 4.6 illustrates the singletons and the doublets along a generalized wire structure connected to other wires. In Figure 4.6, $s(1)$ and $s(-1)$ are the nodes of each wire, where different subscripts indicate that the nodes belong to different wires. For example, $s_3(-1)$ is the node located at $s = -1$ of the third wire. The doublets drawn in this figure are given by Eq. (4.27), and singletons along the wires are characterized by Eq. (4.26b).

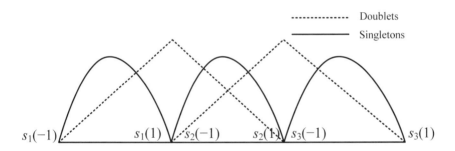

Figure 4.6. Sketch of the singletons and doublets along a wire.

The equation of continuity at the junction of two or more wires is satisfied by grouping the corresponding node basis functions into doublets. These doublets actually represent the usual triangle (rooftop) basis functions. One doublet automatically satisfies the continuity equation at a junction of two wires, and two doublets automatically satisfy the continuity equation at a junction of three wires, and so on.

4.4.2 Current Expansion over a Bilinear Surface

The surface current over a bilinear surface is decomposed into its p and s components. However, the p component current can be treated as the s component current defined over the same bilinear surface with interchanged p

and s coordinates. Thus, for the general case, a distribution of the surface currents can be represented by a sum of s components defined over bilinear surfaces that overlap or are interconnected. The initial approximations for the s components of the electric and the magnetic currents over a bilinear surface are expressed as

$$\mathbf{J}_s(p,s) = \sum_{i=0}^{N_p} \sum_{j=0}^{N_s} a_{ij} \mathbf{F}_{ij}(p,s), \qquad (4.28a)$$

$$\mathbf{M}_s(p,s) = \sum_{i=0}^{N_p} \sum_{j=0}^{N_s} b_{ij} \mathbf{F}_{ij}(p,s), \qquad (4.28b)$$

where N_p and N_s are the degrees of approximations along the coordinates; a_{ij} and b_{ij} are the unknown coefficients; $\mathbf{F}_{ij}$ are the basis functions. They are mathematically characterized by

$$\mathbf{F}_{ij}(p,s) = \frac{\boldsymbol{\alpha}_s}{\left| \boldsymbol{\alpha}_p \times \boldsymbol{\alpha}_s \right|} f_i(p) h_j(s), \qquad (4.29a)$$

$$\boldsymbol{\alpha}_p = \frac{\partial \mathbf{r}(p,s)}{\partial p}, \boldsymbol{\alpha}_s = \frac{\partial \mathbf{r}(p,s)}{\partial s}, \hat{\mathbf{i}}_p = \frac{\boldsymbol{\alpha}_p}{\left| \boldsymbol{\alpha}_p \right|}, \hat{\mathbf{i}}_s = \frac{\boldsymbol{\alpha}_s}{\left| \boldsymbol{\alpha}_s \right|}, \qquad (4.29b)$$

where $\boldsymbol{\alpha}_p$ and $\boldsymbol{\alpha}_s$ are the unitary vectors; $\hat{\mathbf{i}}_p$ and $\hat{\mathbf{i}}_s$ are the corresponding unit vectors and $\mathbf{r}(p,s)$ is given by Eq. (4.19). The relevant expressions in Eq. (4.29a) can be rearranged to automatically satisfy the continuity equation, assuming that there are no line charges present at the ends of the elements and at junctions. Let us assume the following:

$$f_i(p) = p^i , \ h_j(s) = s^j . \qquad (4.29c,d)$$

In that case

$$\mathbf{F}_{ij}(p,s) = \frac{\boldsymbol{\alpha}_s}{\left| \boldsymbol{\alpha}_p \times \boldsymbol{\alpha}_s \right|} p^i s^j , \qquad (4.29e)$$

and the expansion for the electric current can be written in the form

$$\mathbf{J}_s(p,s) = \frac{\boldsymbol{\alpha}_s}{\left| \boldsymbol{\alpha}_p \times \boldsymbol{\alpha}_s \right|} \sum_{i=0}^{N_p} \left(\sum_{j=0}^{N_s} a_{ij} s^j \right) p^i . \qquad (4.30)$$

Next, introduce an alternate set of unknowns through c_{i1} and c_{i2} $(i = 0,..., N_p)$, which are defined by

$$c_{i1} = \sum_{j=0}^{N_s} a_{ij}(-1)^j, \quad c_{i2} = \sum_{j=0}^{N_s} a_{ij}. \qquad (4.31a,b)$$

The coefficients for any i, a_{i0} and a_{i1} can be expressed in terms of the other unknown coefficients, a_{ij} ($j = 2,3, \ldots,N_s$), c_{i1}, and c_{i2}. In this case the expansion for the electric currents is given by Eq. (4.30) and can be written as

$$\mathbf{J}_s(p,s) = \sum_{i=0}^{N_p} \left\{ c_{i1}\,\mathbf{E}_{i1}(p,s) + c_{i2}\,\mathbf{E}_{i2}(p,s) + \sum_{j=2}^{N_s} a_{ij}\,\mathbf{P}_{ij}(p,s) \right\}, \qquad (4.32)$$

where the edge basis functions, $\mathbf{E}_i(p,s)$ ($i = 0,\ldots,N_p$), and the patch basis functions, $\mathbf{P}_{ij}(p,s)$ ($i = 0,\ldots,N_p; j = 2,\ldots,N_s$), are expressed as

$$\mathbf{E}_{ik}(p,s) = \frac{\boldsymbol{\alpha}_s}{\left| \boldsymbol{\alpha}_p \times \boldsymbol{\alpha}_s \right|} \begin{cases} p^i N(s), & k = 1 \\ p^i N(-s), & k = 2 \end{cases}, \qquad (4.33a)$$

$$\mathbf{P}_{ij}(p,s) = \frac{\boldsymbol{\alpha}_s}{\left| \boldsymbol{\alpha}_p \times \boldsymbol{\alpha}_s \right|}\, p^i S_j(s), \qquad (4.33b)$$

where $N(s)$ and $S(s)$ are defined in Eq. (4.26). When $k = 1$, the first edge located along $s = -1$ of a bilinear patch is connected to an edge of another bilinear patch, and when $k = 2$, the second edge located along $s = 1$ of a bilinear patch is connected to an edge of another bilinear patch.

The basis functions $\mathbf{E}_{i2}$ and $\mathbf{P}_{ij}$ are equal to zero along the first edge defined by $s = -1$, and the basis functions $\mathbf{E}_{i1}$ and $\mathbf{P}_{ij}$ are equal to zero at the second edge defined by $s = 1$. This means that the equation of continuity at the first (second) edge is automatically satisfied by omitting the basis functions $\mathbf{E}_{i2}(\mathbf{E}_{i1})$ and that the basis functions $\mathbf{E}_{i1}(\mathbf{E}_{i2})$ take part in satisfying the continuity equation along the interconnected first (second) edge.

If a surface patch is not connected to the other patches, then the current distribution over that patch is approximated only by the basis functions $\mathbf{P}_{ij}$. In what follows, such a current expansion will be termed as a patch expansion and the basis functions $\mathbf{P}_{ij}$ will be termed the *patch basis functions*. It is important to note that the basis functions $\mathbf{E}_{i1}$ and $\mathbf{E}_{i2}$ are used to satisfy the equation of continuity along the interconnected edges, and therefore they will be referred to as the *edge basis functions*. All edge basis functions that are used for the satisfaction of the equation of continuity at some junction will be grouped with the junction expansion functions. This implies that in the case of complex structures, the current distribution is approximated by patch and junction expansion functions.

To display the shape of the patch basis function and the edge basis function, the first several terms of Eq. (4.33) is shown in Figure 4.7. In Figure 4.7, i and j are the order of the basis function, and k is either 1 for the first edge or 2 for the second edge of a bilinear patch as mentioned before.

Figure 4.7 (a)–(d) depicts the different patch basis functions. Recall that in this section, only the s component current is considered. It is shown that the patch basis functions tend to zero at the edge, where $s = -1$ or $s = 1$. The purpose of the negative sign in front of the term $p^i S_j(s)$ is used for convenience.

Figure 4.7 (e)–(h) depicts the different edge basis functions. Edge basis functions are specified along different edges. If the second edge located along $s = 1$ of a patch as shown in Figure 4.7 (e) is interconnected to the first edge located along $s = -1$ of another patch as shown in Figure 4.7 (f), then the two edge basis functions can be grouped into one doublet. The doublet has to be of the same order of p. Similarly, if the second edge of a patch, as shown in Figure 4.7 (g), is interconnected to the first edge of another patch, as shown in Figure 4.7 (h), then these two edge basis functions can be grouped into a doublet. Again, note that the doublet has to be of the same order of p for both cases. In this way, the current at the junctions of the edges between different patches can be expressed in terms of the edge basis functions.

Note that the edge basis functions $\mathbf{E}_{i1}$ and $\mathbf{E}_{i2}$ have the same form with respect to the appropriate edge. (If the s coordinate is replaced by its negative value, then the basis function $\mathbf{E}_{i1}$ is transformed into the basis function $\mathbf{E}_{i2}$, and vice versa.) Therefore, any junction can be considered to be composed of isoparametric elements interconnected along the side $s = -1$. In that case, the corresponding junction expansion consists of the edge basis functions of the form $\mathbf{E}_{i1}$ multiplied by the unknown coefficients of the form c_{i1}. In order to simplify the notation, index "1" will be omitted from c_{i1} and $\mathbf{E}_{i1}$. The normal component of the ith edge basis function going out of the junction can be expressed as

$$\left\{ \mathbf{E}_i \left(p, -1 \right) \right\}_{\text{normal}} = \frac{p^i}{\left| \mathbf{a}_p \left(p, -1 \right) \right|}. \tag{4.34}$$

All interconnected isoparametric elements can be described by the same parametric equation along the junction. In that case, they also have the same unitary vectors along the junctions; thus, the expressions $\left| \mathbf{a}_p \left(p, -1 \right) \right|$ are equal for all interconnected isoparametric elements. It is thus concluded that all edge basis functions defined over interconnected elements have the same normal component going out of the junction. Note that the equation of continuity along this edge must be satisfied independently for each order i (i.e., for different values of the order p).

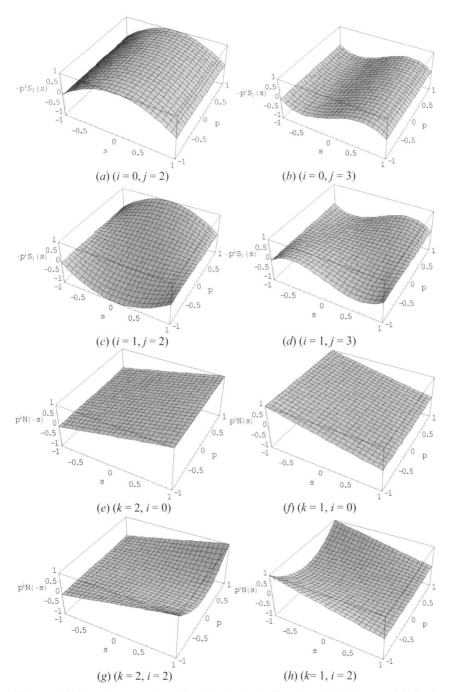

Figure 4.7. Several types of patch and edge basis functions: (*a*)–(*d*) are patch basis functions; (*e*)–(*h*) are edge basis functions.

Because of this property, it is possible to exactly satisfy the continuity equation at the junction by a proper combination of the edge basis functions without introducing fictitious line charges. Along the edges of free patches, the appropriate boundary condition is satisfied by omitting the corresponding edge basis functions (of all orders). The continuity equation for the ith order at the junction of two or more patches is satisfied by grouping the corresponding edge basis functions of the ith order into doublets of the ith order. These surface doublets represent generalized rooftop basis functions, which can be considered to be an extension of the doublets for the wires. One doublet of the ith order automatically satisfies the continuity equation for the ith order at a junction of two patches, two doublets of the ith order automatically satisfy the continuity equation for the ith order at a junction of three patches, and so on.

For the case of multiple dielectric junctions, the continuity equation must be satisfied for each region. In that case one doublet of the ith order automatically satisfies the continuity equation for the ith order at a junction of two patches in each region. However, all these doublets must enter the final solution with the same weighting coefficients. Hence, all these doublets are grouped into one multiplet basis function of the ith order. For the case of composite metallic and dielectric junctions [3–7] the equation of continuity for the ith order is satisfied by the proper combination of doublets and multiplets.

4.5 TESTING PROCEDURE

The integral equations are solved by using the Galerkin's method, which is a special implementation of the method of moments (MoM). Use of the higher-order basis functions results in a very efficient code, requiring only 3–4 unknowns per wavelength along electrically long wires, and 10–20 unknowns per wavelength squared of surface area for large metallic surfaces, enabling solution of real-life problems on personal computers.

4.5.1 Testing Procedure for Thin PEC Wires

As mentioned in Section 4.2, for the case where one of the two regions sharing a common boundary surface is a PEC, the magnetic currents are equal to zero at that boundary surface and the first of the PMCHW equations degenerates into the electric field integral equation (EFIE). For the case of PEC wires, only the EFIE is needed.

The area element for PEC wires dS can be expressed as

$$dS = a(s)\, ds\, dp \quad , \quad s_1 \le s \le s_2, \quad -\pi \le p \le \pi. \tag{4.35}$$

Note that the local p-coordinate system has been chosen to be circular. It is convenient to choose the origin of the p coordinates for any s in the following way. Consider a plane that contains the field point and a representative plane of symmetry of that circle. The intersection of this plane and the circle will always be chosen as the origin of the p coordinates corresponding to the chosen s coordinate. In this case $R(p) = R(-p)$ so that the integration over p can be performed from 0, rather than from $-\pi$, to π. According to Eq. (4.14), the fields finally become

$$\mathbf{E}(\mathbf{r}) = -j\omega\mathbf{A}(\mathbf{r}) - \nabla\Phi(\mathbf{r}), \tag{4.36a}$$

$$\mathbf{H}(\mathbf{r}) = \frac{1}{\mu}\nabla\times\mathbf{A}(\mathbf{r}), \tag{4.36b}$$

where the magnetic vector potential $\mathbf{A}(\mathbf{r})$ and the electric scalar potential $\Phi(\mathbf{r})$ are given by

$$\mathbf{A}(\mathbf{r}) = \frac{\mu}{2\pi}\int_0^\pi\int_{s_1}^{s_2} I(s)\left(\hat{\mathbf{i}}_s(p,s) + \hat{\mathbf{i}}_s(-p,s)\right) g(R)\, ds\, dp, \tag{4.37a}$$

$$\Phi(\mathbf{r}) = \frac{j}{\omega\varepsilon}\frac{1}{\pi}\int_0^\pi\int_{s_1}^{s_2} \frac{dI(s)}{ds} g(R)\, ds\, dp. \tag{4.37b}$$

In Eq. (4.37), Green's function is given by

$$g(R) = \frac{e^{-j\beta R}}{4\pi R}$$

with $\beta = \omega\sqrt{\mu\varepsilon}$. The distance between the source and the field point is $R = |\mathbf{r} - \mathbf{r}'(p,s)|$ in the expression for Green's function and $\mathbf{r}'(p,s)$ can be obtained using Eq. (4.17a). The vector $\hat{\mathbf{i}}_s(p,s)$ is given by Eq. (4.20b).

Often, for simplicity in the analysis, the exact kernel of Green's function is replaced by the reduced kernel. This simplifies the numerical computations significantly with negligible loss in accuracy if the wire structure is considered to be composed of electrically thin wires. The distance between the source and the field point in this case is defined by

$$R_a = \sqrt{\left\{|\mathbf{r} - \mathbf{r}_a(s)|\right\}^2 + a(s)^2}, \tag{4.38}$$

and since R_a is no longer a function of p, the expressions for the electric scalar potential and the magnetic vector potential reduce to the following form:

$$\mathbf{A}(s) = \mu \int_{s_1}^{s_2} I(s)\, \hat{\mathbf{i}}_s(s)\, g(R_a)\, ds\,, \tag{4.39a}$$

$$\Phi(s) = \frac{j}{\omega\varepsilon} \int_{s_1}^{s_2} \frac{d\,I(s)}{d\,s}\, g(R_a)\, ds\,. \tag{4.39b}$$

Thus the expressions for the electric and magnetic fields are given by

$$\mathbf{E}(\mathbf{r}) = -j\omega\mu \left\{ \int_{s_1}^{s_2} I(s)\, \hat{\mathbf{i}}_s(s)\, g(R_a)\, ds + \frac{1}{\beta^2} \int_{s_1}^{s_2} \frac{dI(s)}{ds} \nabla g(R_a)\, ds \right\}, \tag{4.40a}$$

$$\mathbf{H}(\mathbf{r}) = \int_{s_1}^{s_2} I(s)\, \nabla R_a \times \hat{\mathbf{i}}_s(s)\, \frac{dg(R_a)}{dR_a}\, ds\,, \tag{4.40b}$$

where

$$\nabla g(R_a) = \frac{d\,g(R_a)}{d\,R_a} \nabla R_a\,, \qquad \nabla R_a = \frac{\mathbf{r} - \mathbf{r}_a}{R_a}\,. \tag{4.41a,b}$$

If the usual boundary condition for the tangential components of the electric field is applied, i.e., $\left(\mathbf{E} + \mathbf{E}_{\mathrm{inc}}\right)_{\mathrm{tan}} = 0$, then the elements of the impedance matrix due to the singletons will be given by

$$Z_{i,j} = j\omega \int_{s_1}^{s_2} I_i(s)\, \hat{\mathbf{i}}_s(s) \cdot \mathbf{A}_{sj}(s)\, ds - \int_{s_1}^{s_2} \frac{d\,I_i(s)}{d\,s} \Phi_{sj}(s)\, ds\,, \tag{4.42}$$

where

$$\mathbf{A}_{sj}(s) = \mu \int_{s_1}^{s_2} I_j(s)\, \hat{\mathbf{i}}_s(s)\, g(R_a)\, ds\,, \tag{4.43a}$$

$$\Phi_{sj}(s) = \frac{j}{\omega\varepsilon} \int_{s_1}^{s_2} \frac{d\,I_j(s)}{d\,s}\, g(R_a)\, ds\,, \tag{4.43b}$$

and $I_i(s)$ is the ith testing function, $\mathbf{A}_{sj}(s)$ is the s component of the magnetic vector potential and $\Phi_{sj}(s)$ is the electric scalar potential along the generatrix of the truncated cone due to the jth basis function along the direction of the generatrix of the truncated cone.

If the extended boundary condition, $\left(\mathbf{E}_z + \mathbf{E}_{\mathrm{inc},z}\right) = 0$, is used, the same expression is valid, except that the s coordinate should be substituted by the z coordinate.

To evaluate the line integrals arising in the potentials and field vectors with the approximation of currents in the form of polynomials containing powers of functions, special care needs to be employed for the evaluation of the following integrals

$$I_{pi} = \int_{s_1}^{s_2} s^i g(R_a) ds ,\qquad\qquad (4.44a)$$

$$I_{qi} = \int_{s_1}^{s_2} s^i \frac{1}{R_a} \frac{d g(R_a)}{d R_a} ds ,\qquad\qquad (4.44b)$$

where $g(R_a) = e^{-j\beta R_a} / (4\pi R_a)$ is Green's function.

The integral in Eq. (4.44) can be separated into real and imaginary parts. The imaginary parts of the integrand,

$$-\frac{1}{4\pi} s^i \frac{\sin(\beta R_a)}{R_a} , \qquad -\frac{1}{4\pi} s^i \frac{\beta R_a \cos(\beta R_a) - \sin(\beta R_a)}{R_a^3} ,$$

can be efficiently numerically integrated using the Gauss–Legendre quadrature formulas as these functions are well behaved.

For the real part of the integrand $\mathrm{Re}(I_{pi}) = (1/(4\pi)) \int_{s_1}^{s_2} s^i \left[\cos(\beta R_a)/R_a \right] ds$,

and $\mathrm{Re}(I_{qi}) = -(1/(4\pi)) \int_{s_1}^{s_2} s^i \left[(\cos(\beta R_a) + \beta R_a \sin(\beta R_a))/R_a^3 \right] ds$, there is

an integrable singularity. In order to decrease the computational time and simultaneously increase the accuracy in the evaluation of the integrals, the quasisingularity of the kernel should be softened prior to the numerical integration of these integrals. First, $h(s) = s^i$ is expanded in a Taylor series about $s = s_0$, to yield

$$h(s) = s^i = s_0^i + i(s - s_0) s_0^{i-1} + \frac{1}{2} i(i-1)(s - s_0)^2 s_0^{i-2} + \cdots .\qquad (4.45)$$

Then, let $R_0 = \beta R_a$, and expand $\cos R_0$ in $\mathrm{Re}(I_{pi})$ and $\cos(R_0) + R_0 \sin(R_0)$ in the expression of $\mathrm{Re}(I_{qi})$ by a MacLaurin series about $R_0 = 0$. After some rearrangements of the terms, the singularity in the integrand of $\mathrm{Re}(I_{pi})$ and $\mathrm{Re}(I_{qi})$ can be analytically evaluated [7].

4.5.2   Testing Procedure for Bilinear Surfaces

To determine the unknown current coefficients over PEC and composite dielectric structures, the first of the coupled integral equations, Eq. (4.16a), is tested with the basis functions of the electric current, and the second of the coupled integral equations, Eq. (4.16b), is tested by using the basis functions of the magnetic current. The resulting matrix elements represent linear combinations of two types of impedance integrals, Z_{kl}^{L} and Z_{kl}^{K}, defined as

$$Z_{kl}^{L} = \left\langle L(\mathbf{F}_k), \mathbf{F}_l \right\rangle = \int_{S_l} \left(L(\mathbf{F}_k) \right) \cdot \mathbf{F}_l \, dS_l, \tag{4.46}$$

$$Z_{kl}^{K} = \left\langle K(\mathbf{F}_k), \mathbf{F}_l \right\rangle = \int_{S_l} \left(K(\mathbf{F}_k) \right) \cdot \mathbf{F}_l \, dS_l, \tag{4.47}$$

where L and K are linear operators given by Eq. (4.15). $\mathbf{F}_k$ is the kth basis function and the surface over which it is defined is represented by S_k; $\mathbf{F}_l$ is the lth testing function and the surface over which it is defined is denoted by S_l. Before applying the operators in the expression above, the notation of Eq. (4.15) is changed; order number of region is omitted, index of boundary surface ik is replaced by the index of the basis function k, and the position vector of the field point $\mathbf{r}$ is replaced by the position vector of the testing surface $\mathbf{r}_l$. After applying the operators L and K and after some suitable transformation, the integrals shown above, which are encountered in the evaluation of the elements of the impedance matrix, are obtained in a symmetric form as follows

$$Z_{kl}^{L} = j\omega\mu \int_{S_l} \int_{S_k} \left\{ \begin{array}{c} \mathbf{F}_k(\mathbf{r}_k) \cdot \mathbf{F}_l(\mathbf{r}_l) \; - \\ \dfrac{1}{\omega^2 \varepsilon\mu} \nabla \cdot \mathbf{F}_k(\mathbf{r}_k) \; \nabla \cdot \mathbf{F}_l(\mathbf{r}_l) \end{array} \right\} g(\mathbf{r},\mathbf{r}') \, dS_k \; dS_l, \tag{4.48}$$

$$Z_{kl}^{K} = \int_{S_l} \int_{S_k} (\mathbf{r}_k - \mathbf{r}_l) \cdot \left(\mathbf{F}_k(\mathbf{r}_k) \times \mathbf{F}_l(\mathbf{r}_l) \right) \frac{1}{R} \frac{dg(\mathbf{r},\mathbf{r}')}{dR} \, dS_k \; dS_l. \tag{4.49}$$

Since the patch basis functions, doublets, and multiplets are linear combinations of the initial polynomial basis functions, given by Eq. (4.28), the elements of the impedance matrix due to the initial basis functions are evaluated according to Eqs. (4.48) and (4.49) with the basis functions $\mathbf{F}_k$ and $\mathbf{F}_l$ replaced by the initial basis functions $\mathbf{F}_{i_k j_k}$ and $\mathbf{F}_{i_l j_l}$. According to Eq. (4.29a), these initial basis functions are written in the form

$$\mathbf{F}_{i_k j_k} = \frac{f_{i_k}(p_k)\, h_{j_k}(s_k)}{\left| \boldsymbol{\alpha}_{p_k} \times \boldsymbol{\alpha}_{s_k} \right|} \boldsymbol{\alpha}_{s_k}, \; \mathbf{F}_{i_l j_l} = \frac{f_{i_l}(p_l)\, h_{j_l}(s_l)}{\left| \boldsymbol{\alpha}_{p_l} \times \boldsymbol{\alpha}_{s_l} \right|} \boldsymbol{\alpha}_{s_l}, \tag{4.50a,b}$$

where p_k and s_k (p_l and s_l) are the local p and s coordinates of the kth (lth) element and $\boldsymbol{\alpha}_{p_k}$ and $\boldsymbol{\alpha}_{s_k}$ ($\boldsymbol{\alpha}_{p_l}$ and $\boldsymbol{\alpha}_{s_l}$) are the corresponding unitary vectors. Similarly, $\nabla \cdot \mathbf{F}_{i_k j_k}$ and $\nabla \cdot \mathbf{F}_{i_l j_l}$ are easily found, keeping in mind that

$$\nabla \cdot \mathbf{F}_{ij} = \frac{1}{\left| \boldsymbol{\alpha}_p \times \boldsymbol{\alpha}_s \right|} \, j \, p^i \, s^{j-1} \, . \tag{4.51}$$

Finally, the surface elements dS_k and dS_l, which correspond to the initial basis functions, can be written in the form

$$dS_k = \left| \boldsymbol{\alpha}_{p_k} \times \boldsymbol{\alpha}_{s_k} \right| dp_k \, ds_k \, , \quad dS_l = \left| \boldsymbol{\alpha}_{p_l} \times \boldsymbol{\alpha}_{s_l} \right| dp_l \, ds_l \, . \tag{4.52a,b}$$

The final expressions for the elements of the impedance matrix are

$$Z^L_{i_k j_k i_l j_l} = j\omega\mu \int_{-1}^{+1} \int_{-1}^{+1} \int_{-1}^{+1} \int_{-1}^{+1} \left\{ f_{i_k}(p_k) h_{j_k}(s_k) f_{i_l}(p_l) h_{j_l}(s_l) \boldsymbol{\alpha}_{s_k} \cdot \boldsymbol{\alpha}_{s_l} \right.$$
$$\left. - \frac{f_{i_k}(p_k) f_{i_l}(p_l)}{\omega^2 \varepsilon \mu} \frac{dh_{j_k}(s_k)}{ds_k} \frac{dh_{j_l}(s_l)}{ds_l} \right\} g(\mathbf{r},\mathbf{r}') \, dp_k \, ds_k \, dp_l \, ds_l \, , \tag{4.53}$$

$$Z^K_{i_k j_k i_l j_l} = \int_{-1}^{+1} \int_{-1}^{+1} \int_{-1}^{+1} \int_{-1}^{+1} \left[\begin{array}{c} f_{i_k}(p_k) h_{j_k}(s_k) f_{i_l}(p_l) h_{j_l}(s_l) \\ (\mathbf{r}_k - \mathbf{r}_l) \cdot \left(\boldsymbol{\alpha}_{s_k} \times \boldsymbol{\alpha}_{s_l} \right) \frac{1}{R} \frac{dg(\mathbf{r},\mathbf{r}')}{dR} \end{array} \right] dp_k \, ds_k \, dp_l \, ds_l \, . \tag{4.54}$$

Since the basis and the testing functions are adopted in the form of power series, the corresponding elements of the impedance matrix can be written as

$$Z^L_{i_k j_k i_l j_l} = \left(\mathbf{r}_{s_k} \cdot \mathbf{r}_{s_l} \right) I^L_{i_k, j_k, i_l, j_l} + \left(\mathbf{r}_{s_k} \cdot \mathbf{r}_{ps_l} \right) I^L_{i_k, j_k, i_l+1, j_l}$$
$$+ \left(\mathbf{r}_{ps_k} \cdot \mathbf{r}_{s_l} \right) I^L_{i_k+1, j_k, i_l, j_l} + \left(\mathbf{r}_{ps_k} \cdot \mathbf{r}_{ps_l} \right) I^L_{i_k+1, j_k, i_l+1, j_l}$$
$$- \frac{j_k \, j_l}{\omega^2 \varepsilon \mu} I^L_{i_k, j_k-1, i_l, j_l-1} \, , \tag{4.55}$$

$$I^L_{i_k, j_k, i_l, j_l} = j\omega\mu \int_{-1}^{+1} \int_{-1}^{+1} \int_{-1}^{+1} \int_{-1}^{+1} p_k^{i_k} s_k^{j_k} p_l^{i_l} s_l^{j_l} \, g(\mathbf{r},\mathbf{r}') \, dp_k \, ds_k \, dp_l \, ds_l \, , \tag{4.56}$$

$$
Z^K_{i_k j_k i_l j_l} = \left(\mathbf{t}_c \bullet \mathbf{t}_s \right) I^K_{i_k, j_k, i_l, j_l} + \left(\mathbf{t}_c \bullet \mathbf{t}_{ps} + \mathbf{r}_{p_l} \bullet \mathbf{t}_{s_l} + \mathbf{r}_{p_k} \bullet \mathbf{t}_{s_k} \right) I^K_{i_k+1, j_k, i_l+1, j_l}
$$

$$
+ \left(\mathbf{t}_{s_k} \bullet \mathbf{t}_c - \mathbf{r}_{p_l} \bullet \mathbf{t}_s \right) I^K_{i_k, j_k, i_l+1, j_l} - \left(\mathbf{t}_{s_l} \bullet \mathbf{t}_c - \mathbf{r}_{p_k} \bullet \mathbf{t}_s \right) I^K_{i_k+1, j_k, i_l, j_l}
$$

$$
- \mathbf{r}_{p_l} \bullet \mathbf{t}_{s_k} I^K_{i_k, j_k, i_l+2, j_l} - \mathbf{r}_{p_k} \bullet \mathbf{t}_{s_l} I^K_{i_k+2, j_k, i_l, j_l}
$$

$$
- \mathbf{r}_{p_l} \bullet \mathbf{t}_{ps} I^K_{i_k+1, j_k, i_l+2, j_l} + \mathbf{r}_{p_k} \bullet \mathbf{t}_{ps} I^K_{i_k+2, j_k, i_l+1, j_l}, \tag{4.57}
$$

$$
I^K_{i_k j_k i_l j_l} = \int_{-1}^{+1} \int_{-1}^{+1} \int_{-1}^{+1} \int_{-1}^{+1} p_k^{i_k} s_k^{j_k} p_l^{i_l} s_l^{j_l} \frac{1}{R} \frac{dg(\mathbf{r}, \mathbf{r}')}{dR} \, dp_k \, ds_k \, dp_l \, ds_l, \tag{4.58}
$$

where

$$
\mathbf{t}_c = \mathbf{r}_{c_k} - \mathbf{r}_{c_l}, \tag{4.59a}
$$

$$
\mathbf{t}_s = \mathbf{r}_{s_k} \times \mathbf{r}_{s_l}, \tag{4.59b}
$$

$$
\mathbf{t}_{s_k} = \mathbf{r}_{s_k} \times \mathbf{r}_{ps_l}, \tag{4.59c}
$$

$$
\mathbf{t}_{s_l} = \mathbf{r}_{s_l} \times \mathbf{r}_{ps_k}, \tag{4.59d}
$$

$$
\mathbf{t}_{ps} = \mathbf{r}_{ps_k} \times \mathbf{r}_{ps_l}. \tag{4.59e}
$$

In Eq. (4.59), the vectors $\mathbf{r}_{c_k}, \mathbf{r}_{p_k}, \mathbf{r}_{s_k}, \mathbf{r}_{ps_k}$ ($\mathbf{r}_{c_l}, \mathbf{r}_{p_l}, \mathbf{r}_{s_l}, \mathbf{r}_{ps_l}$) define the kth (lth) bilinear surface according to Eq. (4.19). Recall that these expressions are derived for the basis and testing functions directed along the s coordinate. If the basis and/or testing functions directed along the p coordinate are used, the same expressions are obtained for each type of elements of the impedance matrix, except that the vectors $\mathbf{r}_{p_k}$ and $\mathbf{r}_{s_k}$, and/or vectors $\mathbf{r}_{p_l}$ and $\mathbf{r}_{s_l}$, and the subscripts i and j, are interchanged.

Note that evaluation of the elements of the impedance matrix is now reduced to evaluating only two classes of integrals given by Eqs. (4.56) and (4.58). Special care is now devoted to the efficient evaluation of these integrals.

The evaluation of these integrals cannot be performed analytically. Pure numerical evaluation suffices if the kth and the lth surfaces are not close to each other. If they are either close, or coincide, then quasisingular and singular parts of these integrals are extracted and evaluated analytically, while the remainder is evaluated numerically. The quasisingular and singular parts of the surface integrals of the field vectors can be treated in a manner similar to that outlined in the literature [8], but more attention needs to be paid to the coordinate transformation between the Cartesian system and the local bilinear system. In all cases, the Gauss–Legendre formulas are used for numerical evaluations. The main part of the CPU time used for evaluation of these integrals is due to the evaluation of Green's function at the sampling points of the integrand.

In order to minimize the computation time, for each pair of integration points, one belonging to the kth surface and another one to the lth surface, Green's function is evaluated only once for all the integrals in Eqs. (4.56) and (4.58). Thus, a relatively short matrix filling time is realized. Since the order of the integration formula along one coordinate is directly proportional to the order of the current approximation along this coordinate, the fill time for one matrix element depends only slightly on the order of the approximation.

The electric and magnetic fields due to the s component of the electric surface–current–density vector and the assumed current expansion over any bilinear surfaces takes the form

$$\mathbf{E} = \sum_{j=1}^{N_p} \sum_{i=1}^{N_s} a_{ij} \, \mathbf{E}_{ij} \, , \quad \mathbf{H} = \sum_{j=1}^{N_p} \sum_{i=1}^{N_s} a_{ij} \, \mathbf{H}_{ij} \, , \tag{4.60a,b}$$

where

$$\mathbf{E}_{ij} = -j \, \omega \mu \int_{s_1}^{s_2} \int_{p_1}^{p_2} h_j(s) f_i(p) \, \frac{\partial \mathbf{r}'(p,s)}{\partial s} \, g(R) \, dp \, ds$$

$$- \frac{j}{\omega \varepsilon} \int_{s_1}^{s_2} \int_{p_1}^{p_2} \frac{dh_j(s)}{ds} f_i(p) \nabla R \, \frac{d\,g(R)}{d\,R} \, dp \, ds, \tag{4.61a}$$

$$\mathbf{H}_{ij} = \int_{s_1}^{s_2} \int_{p_1}^{p_2} \frac{d\,g(R)}{d\,R} \, \nabla R \, \times \, \frac{\partial \mathbf{r}'(p,s)}{\partial s} \, h_j(s) f_i(p) \, dp \, ds \, . \tag{4.61b}$$

Using these formulations, one can analyze the various interactions in structures consisting of both truncated conical wires and bilinear surfaces.

4.6   PARALLEL IN-CORE AND OUT-OF-CORE MATRIX FILLING SCHEMES

In applying the MoM, an integral equation is first discretized into a matrix equation by using a set of expansion and testing functions. To obtain the current distribution on the structure for a given excitation, this matrix equation needs to be solved. Hence, the parallelization of the solution procedure involves two steps. The first step is the matrix filling and the second step is the solution of the matrix equation. Both of these must be handled efficiently. Furthermore, efficient parallel matrix filling for MoM with higher-order basis functions introduces new challenges and is quite different from the procedure used in a MoM formulation using the more typical subdomain basis functions.

To parallelize the solution of the large dense matrix in a MoM problem, typically one needs to divide the matrix between processes in such a way that two important conditions are fulfilled: (1) each node should store approximately the same amount of data, and (2) the computational load should be equally distributed among the processes that run on different nodes. Fulfilling these conditions, which affect both the matrix filling and matrix solution, is typically determined by which kind of solver is chosen. In this chapter, the ScaLAPACK library package is selected to solve the matrix equation, and this choice determines the details of the matrix filling. In the next two sections, the parallel matrix filling scheme is presented, first for in-core execution, and then for the out-of-core solver.

4.6.1   Parallel In-Core Matrix Filling Scheme

The parallelization of the generation of the impedance matrix should be designed to produce a linear speedup with an increase in the number of processes. If the parallelization is carried out appropriately, this linear speedup will be evident in the execution of the code. Distributing the computation of the elements of the matrix enables a user to solve very large problems in a reasonable time. The impedance matrix has the dimension of $N \times N$, where N is the number of unknowns. The matrix is dense and its elements are complex. Each node performing the computation of the elements of the matrix also stores a portion of the complete matrix. The minimum number of nodes required for the solution of the problem is determined by dividing the number of unknowns for the problem by the number of unknowns that can be held in the memory of a single node. A node is the entity that has an Internet Protocol (IP) address (internal or external), and each node may have multiple processors, and in the case of multicore processors, each processor can have multiple instances of the code, one on each core. This type of configuration can efficiently utilize large amounts of RAM per node. An IP address is a numerical identification (logical address) that is assigned to devices participating in a computer network utilizing the Internet Protocol for communication between its nodes.

The impedance matrix is constructed by looping over the number of geometric elements (number of wires and plates) and performing the calculation of the elements of the impedance matrix. There is an outer and inner loop that cycles over the number of elements. The iterations of the outer loop are independent of each other, except for the accumulations of the final result that need to take place at the end of the inner loop.

A process is a single serial instance of the code: several processes run simultaneously in a parallel code. Each process entering the loop participates in the iterative process and performs the task that it is assigned to do and all the processes go through the iterations in parallel. Results prior to performing the accumulation are saved, and the actual accumulations are performed afterward. All of the calculations to generate the impedance matrix are distributed over the number of processes performing the generation of the complete impedance matrix. This parallel execution provides an increase in the computational speed for matrix filling.

The parallel filling of the matrix could be the same regardless of the choice of the basis functions. However, the matrix filling scheme will be most efficient if the characteristics of the basis functions are taken into account.

In the following paragraphs, a description of an efficient matrix filling scheme, using the higher-order basis functions over wires and quadrilateral plates, is provided. The same concept applies to the matrix filling scheme when using other types of basis functions, such as the RWG functions discussed in Chapter 3. However, the actual procedure to achieve maximum computational efficiency will be quite different when using these two types of different basis functions [9−11]. A detailed efficient matrix filling scheme for a MoM code using the RWG basis functions was presented in the previous chapter. Here, it is seen that different steps than described in the last chapter need to be taken to deal with the higher-order basis functions.

When using higher-order basis functions over quadrilateral plates, the surface can be defined using two coordinate directions, p and s. The goal is to find the current components along the p and s directions. However, there is an advantage in using the polynomial basis functions because the intermediate results obtained in evaluating the elements of the impedance matrix for lower-order can be used in the computation of the elements of the impedance matrix when using higher-order polynomials. This advantage improves the efficiency of the matrix filling for the higher-order basis functions when employed over wires and quadrilaterals and can be implemented quite easily in both the serial and parallel codes.

For parallel matrix filling, an additional improvement can be made to further increase the efficiency of the code. The objective is to eliminate redundant calculations for each process. For the most efficient code, this concept can be applied regardless of the choice of the basis functions. However, the specific details for implementing this are quite different for different basis functions. The pseudocode describing an efficient scheme to fill the impedance matrix for higher-order basis functions is shown in Figure 4.8. In this scheme, within each process, redundant calculations related to the evaluation of the potential functions are avoided by using a flag, set true or false, for each order of the polynomial on a geometric element. This flag is initially set to be false. After a process performs an integration, the flag status is changed to be true. The flag status is checked before each integration is performed for the elements of the impedance matrix, and if the flag is true, the redundant calculation is avoided and the calculated value is retrieved from the RAM.

When dealing with the RHS of the matrix equation in MoM, the parallel matrix filling algorithm is much easier to design, compared to the parallel filling algorithm for the impedance matrix.

Load balancing is critical to obtain an efficient operation of a parallel code. When performed correctly, the matrix filling is executed across all the available resources in an accurate and efficient manner. Little communication between nodes is necessary during the initial matrix filling, and parallel speedup can be carefully tracked to ensure proper implementation.

Parallel In-Core Matrix Filling Scheme:

```
Do  k = 1,nel       ! loop geometric elements for basis functions
Do  kp = 1,nep(k)   ! loop p-direction subdivisions of kth geometric element
Do  ks = 1,nes(k)   ! loop s-direction subdivisions of kth geometric element
Do  l = 1,nel       ! loop geometric elements for testing functions
Do  lp = 1,nep(l)   ! loop p-direction subdivisions of lth geometric element
Do  ls = 1,nes(l)   ! loop s-direction subdivisions of lth geometric element
    find_Zmn_index(k,kp,ks,l,lp,ls)      !find the index of the element, i.e., (m,n)
    flag(ls)=0            !flag is set false before initial order of integration
    … inner loops start here …
    If (m,n belongs to this process) then
        if (flag(ls)==0) then            !if flag is false, then perform integration
        Compute_integral
        flag(ls)=1                       !flag is set true after integration is performed
        endif
        Calculate the value of Z(m,n)
    Endif
    … inner loops end here …
Enddo
Enddo
Enddo
Enddo
Enddo
Enddo
```

Figure 4.8. A parallel in-core matrix filling scheme.

During the process of finding the solution of the matrix, balancing the computational load between all the processors is also important. However, it is less easy to track, since an increase in the number of unknowns or the number of nodes executing the solution can increase the amount of communication required between the nodes. This increase in communication will decrease the gains of parallel speedup. However, as a rule of thumb, more nodes typically means less wall time for solving large problems, even though the overall parallel efficiency may sometimes be increased by using fewer nodes to solve the same problem!

4.6.2   Parallel Out-of-Core Matrix Filling Scheme

The reason for developing an out-of-core matrix filling algorithm is to enable one to solve large matrix equations, where the impedance matrix may be too large to be stored in the main memory (RAM) of the system. Compared with the in-core matrix filling algorithm, where the matrix is filled once and kept in the RAM, the main idea of designing an out-of-core algorithm is to fill a portion of the matrix at a time and then write this portion to the hard disk rather than keeping it in the RAM. The pseudocode for the efficient parallel matrix filling scheme is displayed in Figure 4.9.

Parallel Out-of-Core Matrix Filling Scheme:

Partition A into different slabs ($[A_1] \mid [A_2] \mid [A_3] \mid ... \mid [A_{I\text{slab}}]$)
Calculate_slab_bound (*nstart*, *nend*)   ! calculate the global lower bound
                                                 ! and the upper bound for current slab

For each slab:

Do $k = 1$,nel	!loop geometric elements of surfaces and wires
Do $kp = 1$,nep(k)	!loop p-direction subdivisions of kth geometric element
Do $ks = 1$,nes(k)	!loop s-direction subdivisions of kth geometric element
Do $l = 1$,nel	!loop geometric elements of surfaces and wires
Do $lp = 1$,nep(l)	!loop p-direction subdivisions of lth geometric element
Do $ls = 1$,nes(l)	!loop s-direction subdivisions of lth geometric element

    find_Zmn_index(k,kp,ks,l,lp,ls)   !find the index of the element, e.g., (m,n)
    flag(ls)=0   !flag is set false before initial order of integration
    ... inner loops start here ...
    If (m,n belongs to this process .and. n is within this slab) then
       if (flag(ls)==0) then   !if flag is false, then perform integration
       Compute_integral
       flag(ls)=1   !flag is set true after integration is performed
       endif
       Calculate the value of $Z(m,n)$
    Endif
    ... inner loops end here ...
Enddo
Enddo
Enddo
Enddo
Enddo
Enddo
Write_file   !write this slab to hard disk

! Finish Filling Impedance Matrix

Figure 4.9. A parallel out-of-core matrix filling scheme.

As shown in Figure 4.9, at the beginning of the matrix filling algorithm, the matrix is partitioned into different slabs. As discussed in Section 2.4 (see Chapter 2), the number of slabs I_{slab} and the width of each slab K_i are determined by the nature of the application and the particular system architecture of the computing platform on which the job is executed.

Each process goes through a loop of slabs or blocks, from 1 to I_{slab}. Each process calculates the elements for the K_i, which represents the ith out-of-core slab and it sets the global upper (*nend* in Fig. 4.9) and lower bound (*nstart* in Fig. 4.9). For example, for the first slab ($i = 1$), the global lower bound is 1 and the upper bound is K_1. For the second slab, the global lower bound is K_1+1 and the upper bound is $K_1 + K_2$. Each process fills a portion of the matrix in the same way as the in-core fill algorithm. However, each process pays no attention to the columns that fall outside the fill bound. After every process has completed the desired action of filling the appropriate portion of the matrix, a synchronous write is called to write the portion of the matrix into a file. Then, each process enters the loop corresponding to the next slab. This procedure avoids the calculation of most of the redundant integrals related to the potential functions, which are used to calculate the elements of the impedance matrix. This is accomplished by using a flag for each order of the geometric element, as described previously for the in-core matrix filling scheme.

By comparing with the in-core matrix filling algorithm, it can be found that for each slab, the algorithm is exactly the same. Most of the overhead for filling the out-of-core matrix, excluding that from the in-core matrix calculation for an individual slab, comes from two parts: (1) calculation of the redundant integrals performed on each process, for different elements of the impedance matrix, which belongs to different slabs; and (2) writing the matrix elements to the hard disk. The overhead in this case is similar as discussed in Chapter 3.

After the matrix is filled, the solution of the matrix equation is essentially the same regardless of the type of basis functions used in the MoM formulation. In this chapter, the ScaLAPACK routine is used to solve the matrix equation in-core. As mentioned in Chapter 1, a custom ScaLAPACK in-core solver was developed for the Itanium 2 platform. For the out-of-core matrix solver, the methodology is based on the ScaLAPACK implementation as introduced in Chapter 2.

4.7   NUMERICAL RESULTS COMPUTED ON DIFFERENT PLATFORMS

Three different models have been chosen to illustrate the performance of the parallel in-core and the out-of-core codes for solutions of large EM problems using different computer platforms. To illustrate the solution of small-sized and medium-sized problems using the in-core solver, the electromagnetic scattering from a perfect electric conducting (PEC) sphere is presented. To illustrate the solution of a composite problem, a large array of Vivaldi antennas is used. For the solution of an electrically large problem, the electromagnetic scattering from an airplane is demonstrated at high frequencies.

In Section 4.7.1, results for these models using the parallel in-core solver are presented, and in Section 4.7.2, results using the parallel out-of-core solver are shown. Results presented in Section 4.7.2.2 provide a comparison between the parallel in-core and the parallel out-of-core integral equation solvers.

4.7.1   Performance Analysis for the Parallel In-Core Integral Equation Solver

4.7.1.1 Comparison of Numerical Results Obtained on Single-Core and Multicore Platforms. Two test results are presented here, using a model with a small number of unknowns. A PEC sphere of 1 m radius is used as the structure to be analyzed. As shown in Figure 4.10, a plane wave polarized along the z axis and propagating along the direction of the minus x axis is used as an excitation for this model. This same PEC sphere model is executed on both CEM-1, a single-core CPU platform, and CEM-2, a dual-core CPU platform, but the analysis is done at two different frequencies.

For the computations on CEM-1, the frequency of the incident plane wave is set to be 1.53 GHz, which makes the diameter of the sphere approximately 10λ. The sphere is modeled with 96 elements and the number of segments per quarter-circumference is 4. For this example, a seventh-order polynomial is used along each of the two directions. The number of unknowns required is the product of 7^2 times the two directions times the number of elements resulting in $7 \times 7 \times 2 \times 96 = 9408$ unknowns. For a total of 9408 unknowns and using double-precision (complex) arithmetic for the computation, about 2 GB of RAM is needed to execute this simulation. The CPU time required to analyze this problem is listed in Table 4.1. Here "total time" means the time for obtaining the coefficients of the current and then writing them to the hard disk. Table 4.1 shows that the matrix filling time and the matrix solution time needed by the two CPUs is about half of that needed by one CPU. However, as the number of CPUs increase, even if the total time continues to decrease, the parallel efficiency also decreases as indicated in Figure 4.11.

TABLE 4.1. Time Required on CEM-1 Single-Core Server for Analyzing the PEC Sphere at 1.53 GHz

	Number of CPUs Used in the Computation							
	1	2	3	4	5	6	7	8
Matrix filling (minutes)	4.9	4.4	3.3	3.0	2.7	2.5	2.4	2.2
Matrix solving (minutes)	95.9	49.2	34.4	27.0	22.1	19.0	16.9	15.4
Total time (minutes)	101.0	56.0	40.0	33.0	28.0	24.0	23.0	22.0

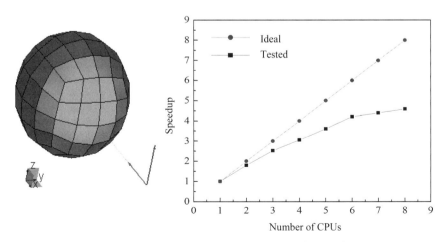

Figure 4.10. The PEC sphere model.   **Figure 4.11.** Parallel speedup on CEM-1.

There are two reasons why the efficiency drops with an increase of the number of processors to carry out the computation:

1. A linear speedup for matrix filling can be challenging. When the large impedance matrix is divided into small blocks, as required by the matrix solution method, the integral for the potential function for each element of the impedance matrix has to be repeated for different processes even though the parallel matrix filling scheme efficiently avoids the redundant calculation on each individual process. To remedy this, one possibility is to use communication between nodes to preclude repeating calculations of the integral on different nodes. This is accomplished by broadcasting the result from a process that first calculates the integral to other processes that need it. However, on a distributed-memory computer system, communication between the nodes can also be a significant time-consuming element that could ultimately reduce the efficiency when the number of nodes and/or the problem size is very large. The tradeoff between redundant calculations on different processes versus increased communication between the nodes has to be carefully considered for the selected environment. For the examples dealing with a small number of unknowns, it is seen that repeating the redundant calculations results in a better efficiency.

2. On a single-core platform, as the number of CPUs increases, the computational load for each core becomes less, which should provide a speedup. However, the contribution of the increased communication time between the different CPUs counteracts this speedup and, in the worst-case scenario, the CPU and the communication times could become comparable. Careful understanding of both the properties of the computer platform and the parameters of the problem are required to achieve maximum efficiency. This may mean using fewer nodes than available on a particular platform to solve a problem of small size!

A second test is performed using CEM-2, a dual-core platform, and again using the same PEC sphere model, but this time with an incident plane wave operating at 1.98 GHz. As before, the radius of the sphere is 1 m, but now the electrical diameter is about 13λ and the sphere is modeled with 1296 elements with 4 segments per quarter-circumference. There are 21,936 unknowns and about 8 GB of RAM is needed to perform the analysis. Figure 4.12 shows the meshing of the PEC sphere model at 1.98 GHz.

Table 4.2 lists the time for matrix filling, matrix solving, and total times taken to solve this example using one, two, three, and four cores. This table shows that the total time to compute the current coefficients decreases, as expected, with increasing number of cores. The parallel speedup for this example and that of the first example are plotted together in Figure 4.13. This plot shows that on these two platforms using different hardware and software, the same parallel speedup can be achieved on CEM-1, using eight single-core CPUs, and CEM-2, using two dual-core CPUs. This indicates that the portability of this parallel integral equation based EM code is quite good.

TABLE 4.2. Time Required on the CEM-2 Dual-Core Workstation for Analysis of the PEC Sphere at 1.98 GHz

	Number of Cores Used in the Computation			
	1	2	3	4
Matrix filling (minutes)	7.9	6.7	5.1	4.2
Matrix solving (minutes)	100.4	52.0	35.1	27.0
Total time (minutes)	109.0	61.0	43.0	36.0

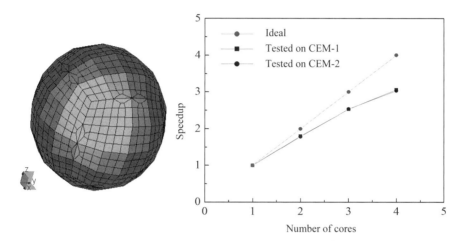

Figure 4.12. Meshed PEC sphere model.   **Figure 4.13.** Overlays of parallel speedups.

Figure 4.14 shows the results of the simulation. Figure 4.14 (*a*) plots the radar cross section (RCS), and Figure 4.14 (*b*) shows the current distribution on the sphere.

In both the previous examples, the speedup when using two CPU cores is much closer to the ideal value than when using three or more cores. This indicates that if each core is assigned a job using as much memory as it can access (this means that it takes the largest job that it can handle), better parallel efficiency can be achieved than when using more nodes, where each node accesses less memory. The parallel solver in this chapter is designed for analysis of large-scale EM simulation. In this case, each CPU core is assigned as much RAM as it can access.

These two examples illustrate that the same parallel speedup can be achieved executing two similar models on two different platforms with different hardware and software. This demonstrates that the parallel in-core solver has good portability.

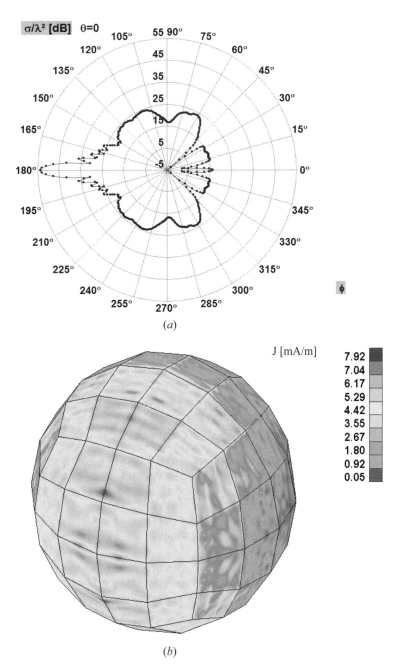

Figure 4.14. Numerical results for an electrically large PEC sphere: (*a*) bistatic RCS (0° starts from *x* axis in the *xoy* plane); (*b*) the current distribution.

4.7.1.2 Numerical Results Obtained on Single-Core Platforms. The solution of two large scale problems is presented next not only to illustrate the efficiency and portability of the code discussed in the previous section but also, in addition, to demonstrate the scalability of the solver. These examples will illustrate the performance on two single-core platforms, CEM-3 and CEM-4.

4.7.1.2.1 Radiation from a Vivaldi Antenna Array. The Vivaldi antenna has gained interest as a compact ultrawideband array for civilian and military applications [12,13]. The composite antenna structure has also gained the interest of CEM practitioners because it is a challenging structure to model and can be analyzed with high accuracy. The Vivaldi antenna array has become a benchmark project [14] with which to test the capability and speed of computational codes and high-performance computing platforms. For these reasons, the Vivaldi array model is presented in this chapter. The parallel solver can accurately and quickly provide the response of this complex antenna and the results of the computation on four different platforms are presented. The first two computer platforms, CEM-3 and CEM-4, both contain single-core CPUs. The former is IA-64, a true 64-bit Intel microprocessor called the Itanium 2 containing four different processors; the latter is EM64T, with two Intel Xeon processors per node, and 10 nodes in the blade cluster chassis. These two platforms have the resources to solve a large Vivaldi array. In Section 4.7.1.3, results are presented from analyzing smaller Vivaldi arrays on two small multicore platforms, CEM-6 and CEM-7.

A 60-element dual-polarized Vivaldi array is shown in Figure 4.15 (*a*). The composite structure including the feed of a single Vivaldi antenna element is shown in Figure 4.15 (*b*). This antenna element is fed by a stripline terminated with a radial stripline stub. The wire excitation is located at the other end of the stripline. A junction is located between the wire and the stripline. The stripline feeds the slots on both sides of the dielectric substrate through coupling. The slots are composed of a circular slotline cavity and extend to an exponentially tapered slot. The dimensions of a single element are 76 mm × 25.6 mm × 6 mm. The dielectric constant of the substrate is $\varepsilon_r = 2.2$. The operating frequency of this Vivaldi array is from 1 to 5 GHz, where it actually behaves as a traveling-wave antenna. However, the structure is analyzed at the highest frequency of interest. A top view of the element is shown in Figure 4.15 (*c*). The light gray plates are metallic, while the dark gray ones are dielectric. The array is backed by an infinitely large PEC plate.

Modeling this structure requires 34,390 geometric elements, which are composed of wires, plates, and junctions. A discretized surface integral equation with 45,500 unknowns for the electric currents and 4320 unknowns for the magnetic currents is used to calculate the current distribution on the array. The number of unknowns required for the analysis of this array is 49,820, which will occupy approximately 40 GB of RAM when double-precision arithmetic is used for computation. The execution times on the two single-core platforms are listed in Table 4.3.

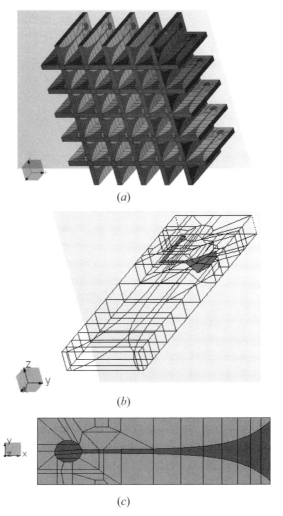

Figure 4.15. A Vivaldi antenna array with two views of a single element: (*a*) the structure of a 60-element dual-polarized Vivaldi array; (*b*) a single Vivaldi antenna element; (*c*) top view of a single antenna element.

TABLE 4.3. Time Required on CEM-3 and CEM-4

	Platform					
	CEM-3			CEM-4		
Serial in-core (hours)	75			—		
	Filling	Solving	Total	Filling	Solving	Total
Parallel in-core (minutes)	139.2	262.4	429	46	45.5	106

On CEM-3, the analysis time using a single processor and using a serial code takes approximately 75 hours (270,225 seconds) of CPU time. Using a parallel integral equation solver and all four Itanium 2 CPUs, the total solution time is reduced to 7 hours (429 minutes). This represents a significant reduction in the total time by fully utilizing the entire computing platform. The same problem solved on CEM-4 using 20 CPUs is also listed in Table 4.3. Because of the large number of elements used in this challenging computational model, the CPU times required for matrix filling and matrix solution are both approximately 46 minutes. The total time needed to compute the current distribution is 106 minutes.

The radiation pattern for this example is shown in Figure 4.16, where the results from the parallel in-core solver routines exactly overlay with the results from using a serial out-of-core solver developed for a single processor system [14,15]. Figure 4.16 (a) shows the azimuth pattern, and Figure 4.16 (b) shows the elevation pattern. This check on the numerical accuracy of the parallel code by using a single processor and the serial code is time-consuming, but necessary to guarantee the accuracy of the parallel solver.

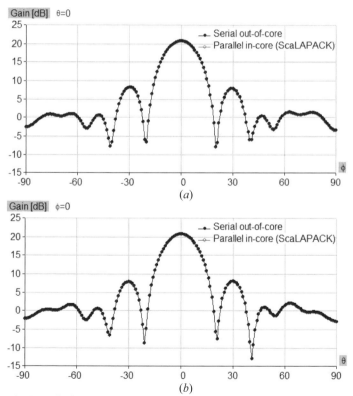

Figure 4.16. Radiation pattern of the 60-element Vivaldi array: (a) azimuth pattern (0° starts from x axis in the xoy plane); (b) elevation pattern (0° starts from x axis in the xoz plane).

4.7.1.2.2 Scattering from a Full-Size Airplane. The scattering from a full-size airplane is simulated here to further demonstrate the high efficiency achieved using the parallel implementation of the in-core integral equation solver. The aircraft structure is 11.6 m long, 7.0 m in width, and 2.92 m in height. It is modeled as a PEC surface with a plane of symmetry along the *yoz* plane. The plane wave is incident from the minus *y*-axis direction and is polarized along the *z* axis. Two different frequencies are simulated to scale the problem size and observe the response.

First, the airplane model is simulated at 1.25 GHz. The wavelength at 1.25 GHz is 0.24 m, and hence the structure is about 48.3λ long. The number of unknowns required to analyze this problem is 60,653. Figure 4.17 (*a*) shows the meshed airplane structure. Figure 4.17 (*b*) shows the RCS along the *yoz* plane with the incident wave polarized along the θ direction. This plot shows the results obtained using the parallel in-core solver and the serial out-of-core solver [14,15], to validate the accuracy of the solution. The two results at the frequency of 1.25 GHz agree with each other very well. The execution time obtained on CEM-3 and CEM-4 for this model is listed in Table 4.4.

TABLE 4.4. Comparison of Execution Times on CEM-3 and CEM-4 for Analysis of the 48.3λ-Long Airplane

		Platform				
		CEM-3			CEM-4	
Serial in-core (hours)		119			—	
	Filling	Solving	Total	Filling	Solving	Total
Parallel in-core (minutes)	76.2	469.3	559	18.5	80.9	119

As indicated in Table 4.4, using only a single processor on CEM-3 requires approximately 119 hours to analyze scattering from this 48.3λ-long structure. Using all four CPUs and the parallel solver, the same problem can be solved in 9 hours 19 minutes. Executing the same model on the CEM-4 platform using 20 CPUs requires just less than 2 hours of wall time!

Figure 4.17 (*b*) plots the RCS for the same example run at 2.20 GHz. This simulation requires 70,995 unknowns to accurately calculate the currents. At 2.20 GHz, the airplane is about 85.1λ long, 51.3λ wide, and 21.4λ high. On CEM-3, this simulation took about 854 minutes to obtain the currents. Executing the same problem on the CEM-4 using 20 CPUs requires about 190 minutes, as shown in Table 4.5.

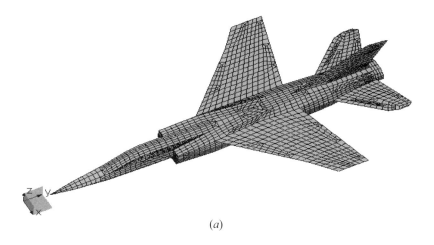

(*a*)

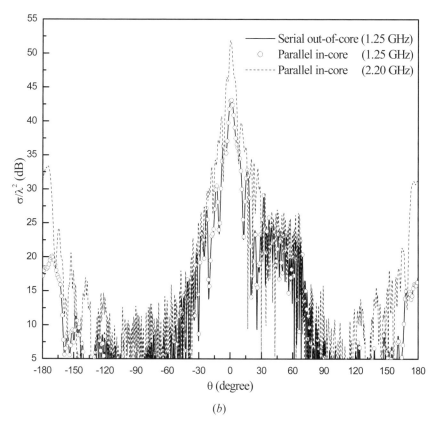

(*b*)

Figure 4.17. Model and simulation results for a full-size airplane: (*a*) model of the airplane; (*b*) bistatic RCS evaluated at different frequencies (0° starts from *y* axis in the *yoz* plane).

TABLE 4.5. Comparison of Execution Times on CEM-3 and CEM-4 for Analysis of
the 85.1λ-Long Airplane

	Platform	
	CEM-3	CEM-4
Parallel in-core (minutes)	854	190

The ratio of the total time needed to solve the 70,995 unknowns (N_2) problem as opposed to the solution for the 60,653 unknowns (N_1) should theoretically be between $(N_2/N_1)^2 = 1.370$ and $(N_2/N_1)^3 = 1.604$. This is because the computation associated with matrix filling scales with the complexity of $O(N^2)$ and the computation associated with the matrix solution, using the LU decomposition methodology, scales with the complexity $O(N^3)$, where N is the number of unknowns.

The ratio of the total computed time between the solution of these two problems on the CEM-3 system is 854 minutes/559 minutes ≈ 1.528. The ratio on CEM-4 is 190 minutes/119 minutes ≈ 1.597. On two different computer platforms using different hardware and software, the maximum theoretical ratio of 1.604 is not exceeded in the solution of these two problems using different numbers of unknowns. These examples and their comparisons support the conclusion that the parallel integral equation–based EM solver demonstrates not only portability but also good scalability.

4.7.1.3 Numerical Results Obtained on Multicore Platforms. In this section, two smaller Vivaldi antenna array models will be used for analysis on two small multicore CPU platforms. CEM-6 and CEM-7 are built with the AMD dual-core and Intel quad-core CPUs, respectively. The computing power in just two of these nodes consisting of these powerful processors is evident in their ability to solve problems in a short time. However, because the RAM on each of these platforms is smaller than on either the CEM-3 or CEM-4, the larger Vivaldi array could not be solved using the in-core method on these two platforms.

The models simulated here are a 24-element array and a 40-element array consisting of Vivaldi antennas. The results are listed in Table 4.6. The size of the matrix required to solve in case 6-2 is about twice the size of case 6-1. As discussed in the previous section, the ratio between the number of unknowns is proportional to the total time to execute different-sized models on the same platform. In this comparison, the number of cores used is also doubled so that the RAM/core is kept constant to solve the larger problem. When the number of unknowns is increased from 22,308 to 31,496, the total execution time should increase between $(N_2/N_1)^2 = 1.99$ and $(N_2/N_1)^3 = 2.814$ if the same number of cores is used. However, in this comparison, twice as many cores are used to solve the larger array. Ideally the time would be halved, if the increased communication time and latency is ignored. This gives a maximum ideal ratio of 1.41, as shown in the last column of Table 4.6.

TABLE 4.6. Comparison of Execution Times on CEM-6 and CEM-7

	Case 6-1			Case 6-2			Total Time Ratio
Cores	4			8			
Unknowns (RAM)	22,308 (7.96 GB)			31,496 (15.87 GB)			Ideal: 1.41
	Filling	Solving	Total	Filling	Solving	Total	Actual:
CEM-6 (minutes)	14.5	28.9	44	23.3	41.9	65.6	1.49
CEM-7 (minutes)	12.3	19	32	19.1	29.3	49	1.53

The last column in Table 4.6 lists this calculated ratio, along with the actual ratio of 1.49 for CEM-6 and 1.53 for CEM-7. These ratios indicate that 94%–91% efficiency in scalability can be expected on these platforms as the problem size scales along with the number of cores. This again demonstrates that good portability and scalability of the code developed in this chapter apply to platforms operating with either Intel or AMD multi-core CPUs.

4.7.2   Performance Analysis for the Parallel Out-of-Core Integral Equation Solver

The numerical results for two examples are presented in this section to demonstrate the efficiency, scalability, and portability of the parallel out-of-core integral equation solver. These results will illustrate the excellent performance of this integral equation solver.

First, a 60-element Vivaldi antenna array is presented to demonstrate that a challenging benchmark problem dealing with a composite dielectric body problem can be solved on a two-node IA-32 based PC cluster in less than one day. To quantify the performance of this code, a single-processor workstation is used as a benchmark. This same problem is also executed on a high-performance cluster, which has four dual-core processors operating with EM64T processors. These platforms have insufficient RAM to solve this large problem in-core.

4.7.2.1 Vivaldi Antenna Array — a Large Problem Solved on Small Computer Platforms. Consider the 60-element dual-polarized Vivaldi array as described in Section 4.7.1.2.1. The array is simulated on three different platforms. The first platform, a Dell 370 Precision PC computer with a 3.8 GHz CPU, provides the benchmark performance. This 60-element Vivaldi array was analyzed in approximately 86 hours using the serial out-of-core solver [14]. The difference, in this section, is that the parallel out-of-core solver is executed utilizing two threads on the CPU and therefore it reduced the total time to half, as shown in Table 4.7. A "thread" (http://en.wikipedia.org/wiki/Thread) in computer science terminology is short for a thread of execution. Threads are a way for a program to split itself into two or more simultaneously (or pseudo-simultaneously) running tasks. Threads and processes differ from one operating system to another but, in general, a thread is contained inside a process and different threads in the same process share some resources while different processes do not. On a single processor, multithreading generally occurs by time-division multiplexing in very much the same way as the parallel execution of multiple tasks (computer multitasking): the processor switches between different threads. This context switching can happen so fast as to give the illusion of simultaneity to an end user. On a multiprocessor or multi-core system, threading can be achieved via multiprocessing, wherein different threads and processes can literally run simultaneously on different processors or cores.

TABLE 4.7. A 60-Element Vivaldi Antenna Array Simulated on Different Platforms

Computer	CPUs	Cores	Total Time (minutes)	Estimated Time (minutes)
Dell 370 Precision	1 CPU (2 threads)	1	2466	—
CEM-5 (1 node)	1 CPU (single-core)	1	2216	3124
CEM-5 (2 nodes)	2 CPUs (single-core)	2	1414	1562
CEM-6	4 CPUs (dual-core)	8	326	418

The second platform used is CEM-5, which is a six-node IA-32 PC cluster, and can be used with one node or two nodes. CEM-6, the IBM LS41 blade server with four dual-core processors, is the third platform, which will also be used for comparison in these performance analyses.

Table 4.7 presents the total time versus the estimated time needed on each computer platform. The time needed for the PC cluster is less than that predicted by using two threads on the Dell 370 used as the benchmark and taking into account the difference in CPU speeds. Even though the PCs are connected by 10/100-Mbps network adapters, in this cluster a 78% parallel efficiency still can be achieved. This example shows that problems requiring larger than 2 GB of RAM can still be solved on IA-32 systems by using the parallel out-of-core integral equation solver.

The measured time on CEM-6 is 326 minutes and is less than the estimated time of 418 minutes, using the same benchmark and taking into account the differences in quantity and speed of the cores. This is reasonable since the RAID 0 hard disk configuration of CEM-6 offers two spindles for disk read/write.

4.7.2.2 Solution for a Full-Size Airplane — Parallel Out-of-Core Solver Can Be as Efficient as the Parallel In-Core. Next, the full scale airplane model at 1.25 GHz is simulated to demonstrate that the performance of the parallel out-of-core solver can be as efficient as that of the parallel in-core solver. In this example, approximately 59 GB RAM is needed for the in-core solver using double-precision arithmetic for computation of the 60,653 unknowns problem. Even though CEM-4 has sufficient RAM to run this same problem in-core, the out-of-core solver is also run for comparison and each process is assigned 2 GB RAM. This gives a total of 36 GB RAM available for the out-of-core solver when 18 processes on nine nodes are used. For this platform, one process is run on each single-core CPU.

The induced current over the surface of the airplane is shown in Figure 4.18. Figure 4.19 indicates that the field inside the aircraft structure is zero. The total times listed in Table 4.8 indicate that the parallel out-of-core solver is about 6 minutes slower than parallel in-core solver. The reason for this comes from both matrix filling and matrix solving. First, the matrix filling will be addressed.

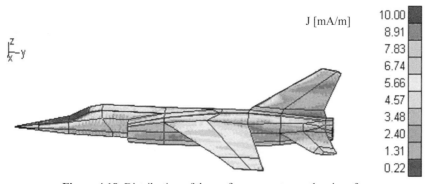

Figure 4.18. Distribution of the surface current over the aircraft.

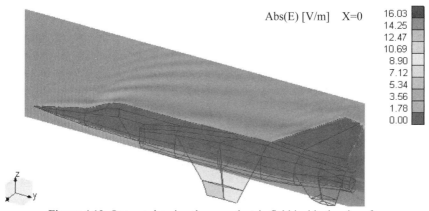

Figure 4.19. One cut showing the zero electric field inside the aircraft.

TABLE 4.8. Comparison of Performance between the Integral Equation Solvers

	Parallel Out-of-Core	Parallel In-Core
Platform	CEM-4 (9 nodes)	CEM-4 (9 nodes)
Memory (GB)	36	59
Time (minutes)	148	142

The parallel out-of-core solver generates blocks of matrix at two different levels. For any given slab of the matrix, the parallel matrix filling is the same as the parallel in-core solver. However, for the parallel out-of-core solver, the elements of the impedance matrix generated from one source on the surface may belong to a different slab and thus cause the integral to be repeated. This will slightly degrade the efficiency of the parallel matrix filling.

Regarding the matrix solution procedure, the parallel out-of-core solver needs to seek data on the hard disk and read it into the RAM. Then, it performs the LU decomposition and writes the intermediate results back onto the hard disk. The reading and writing steps of the parallel out-of-core solution process clearly consume time when compared to the in-core solution technique, in which case all data are operated on directly in the RAM with no hard-disk access required. However, minimal degradation in performance is observed, because each CPU has 2 GB of additional RAM available at each node. The additional RAM is used as cache for the read/write operation. Keeping the data in RAM reduces the I/O access time.

4.7.2.3 Solution for a Full-Size Airplane — Scalability and Portability of the Parallel Out-of-Core Solver. The scattering from the full-size airplane is simulated at four different frequencies on two different computer platforms to demonstrate the excellent scalability and portability of the code for very large problems. The electromagnetic responses at four different frequencies from 1.25 to 2.8 GHz are provided in this section. Computations using two different platforms, CEM-4, a single-core CPU cluster; and CEM-7, a quad-core CPU cluster, are demonstrated. The time needed for the different models simulated on CEM-4 are listed in Table 4.9. Table 4.9 also lists the electrical dimensions of the airplane for each frequency of operation.

In these examples, the amount of storage needed on the hard disk is shown in Table 4.9. It is calculated from the number of unknowns used in the analysis at each different frequency. Also, for these cases, the amount of RAM is insufficient to carry out an in-core solution. As the in-core buffer, 2 GB RAM is assigned to each process, resulting in a total available RAM of 36 GB for the out-of-core integral equation solver when 18 processes (one process per core) are used. The time listed in Table 4.9 is the total time taken for matrix filling and for solution of the matrix equation.

TABLE 4.9. Results for a Full-Size Airplane on CEM-4

	Case 9-1	Case 9-2	Case 9-3	Case 9-4
Frequency (GHz)	1.25	2.2	2.5	2.8
Unknowns	$N_1=60{,}653$	$N_2=70{,}995$	$N_3=93{,}091$	$N_4=110{,}994$
Length	48.3λ	85.1λ	96.7λ	108.3λ
Width	29.2λ	51.3λ	58.3λ	65.3λ
Height	12.2λ	21.4λ	24.3λ	27.3λ
Memory (GB)	36	36	36	36
Storage (GB)	58.9	80.6	138.7	197.1
Time (minutes)	$T_1=148$	$T_2=236$	$T_3=527$	$T_4=885$

$\dfrac{T_2}{T_1}$	$\left(\dfrac{N_2}{N_1}\right)^3$	$\dfrac{T_3}{T_1}$	$\left(\dfrac{N_3}{N_1}\right)^3$	$\dfrac{T_4}{T_1}$	$\left(\dfrac{N_4}{N_1}\right)^3$
1.595	1.604	3.561	3.615	5.980	6.128

$\dfrac{T_3}{T_2}$	$\left(\dfrac{N_3}{N_2}\right)^3$	$\dfrac{T_4}{T_2}$	$\left(\dfrac{N_4}{N_2}\right)^3$
2.233	2.254	3.750	3.821

$\dfrac{T_4}{T_3}$	$\left(\dfrac{N_4}{N_3}\right)^3$
1.679	1.695

As discussed in Section 4.7.1.2, the ratio of the total time taken for the solution of the problem using different number of unknowns, N_2 and N_1, should be between $(N_2/N_1)^2$ and $(N_2/N_1)^3$. The ratio of the time taken to solve different-sized problems using different numbers of unknowns is listed in Table 4.9 and this scaling holds true for all the cases tested. For example, the ratio of the wall-clock times between cases 9-1 and 9-4 in Table 4.9 illustrates this relationship. Using the ratio T_4/T_1, which is proportional to $(N_4/N_1)^3$, the maximum total time taken to solve case 9-4 using case 9-1 as a benchmark can be predicted. The total time for case 9-4 is predicted to be $T_4 = T_1 (N_4/N_1)^3 = 148$ minutes$\times(110{,}994 / 60{,}653)^3 = 907$ minutes. This is the maximum amount of time that is estimated, since the solution procedure scales as $O(N^3)$. It is expected that the actual time will be less than this estimate. Since the actual time $T_4 = 885$ minutes, it is less than the predicted, and this falls well within the estimated wall-clock time, thus demonstrating the scalable behavior on a single-core cluster.

The four cases presented in Table 4.10 provide the benchmark results using 16 processes on CEM-7, a quad-core platform. From Table 4.10, the scalability can also be demonstrated for the CEM-7 platform using reasoning similar to that previously used for the computations carried out on the platform CEM-4. In Table 4.10, the ratio between the predicted theoretical time and the actual time taken to solve the problem is presented. For example, the theoretical ratio between the computational times used in cases 10-1 and 10-4 is $(N_4/N_1)^3 = 6.128$. This ratio is the upper bound since the calculation uses the $O(N^3)$ relation. It is expected that the ratio of the measured time would be less than this, and in fact, that ratio is $T_4/T_1 = 5.102$, which is less than the one predicted. Since the actual time ratios are less than the predicted for all cases, this illustrates good scalability behavior of the parallel out-of-core computations on quad-core CPUs.

TABLE 4.10. Results for a Full-Size Airplane on CEM-7

	Case 10-1	Case 10-2		Case 10-3		Case 10-4	
Frequency (GHz)	1.25	2.2		2.5		2.8	
Unknowns	N_1=60,653	N_2=70,995		N_3=93,091		N_4=110,994	
Memory (GB)	30.4	30.4		30.4		30.4	
Length	48.3λ	85.1λ		96.7λ		108.3λ	
Width	29.2λ	51.3λ		58.3λ		65.3λ	
Height	12.2λ	21.4λ		24.3λ		27.3λ	
Storage (GB)	58.9	80.6		138.7		197.1	
Time (minutes)	T_1=225	T_2=330		T_3=680		T_4=1148	
		$\dfrac{T_2}{T_1}$	$\left(\dfrac{N_2}{N_1}\right)^3$	$\dfrac{T_3}{T_1}$	$\left(\dfrac{N_3}{N_1}\right)^3$	$\dfrac{T_4}{T_1}$	$\left(\dfrac{N_4}{N_1}\right)^3$
		1.467	1.604	3.022	3.615	5.102	6.128
				$\dfrac{T_3}{T_2}$	$\left(\dfrac{N_3}{N_2}\right)^3$	$\dfrac{T_4}{T_2}$	$\left(\dfrac{N_4}{N_2}\right)^3$
				2.061	2.254	3.479	3.821
						$\dfrac{T_4}{T_3}$	$\left(\dfrac{N_4}{N_3}\right)^3$
						1.688	1.695

4.7.2.4 Solution for a Full-Size Airplane — a Very Large Problem Solved on Nine Nodes of CEM-4. The full capability of this parallel out-of-core integral equation solver is highlighted by the solution of a full-scale airplane at 4.1 GHz. The corresponding wavelength at 4.1 GHz is 0.073 m, and hence the electrical dimensions of the structure are 158.5λ in length, 95.7λ in width, and 39.9λ in height. At this frequency, 232,705 number of unknowns are necessary to approximate the unknown current distribution on the surface and this dictates that approximately 866.4 GB RAM is needed to solve this problem using an in-core solver using double-precision (complex) arithmetic for the computations. Since a cost-effective solution is sought, the hard disk available on the CEM-4 cluster is used. For the simulation at 4.1 GHz, all the available RAM and the hard disk associated with all the compute nodes in the CEM-4 cluster are used. This is the worst-case performance of the parallel out-of-core solver because essentially all the computer resources have been used.

The maximum total time, can be predicted, for the solution of the RCS at 4.1 GHz using the time taken for the solution at 1.25 GHz as a benchmark. Instead of the parallel out-of-core solution time, the parallel in-core time will be used as the benchmark to assess the performance. The reason for this is that it is of interest to compare the performance of the out-of-core versus the in-core solver, as if sufficient RAM were available to solve such a large problem in-core. Using the parallel in-core performance at 1.25 GHz, as given in Table 4.8, the total time for the solution at 4.1 GHz is predicted to be 142 minutes$\times(232,705/60,653)^3 = 133.7$ hours. This is the maximum amount of total clock time that would be predicted if 867 GB RAM was available to solve this problem in-core. The actual time for the parallel out-of-core solver is 166.9 hours. This shows that the parallel out-of-core performance is degraded by about 25% from the performance predicted if sufficient RAM were available and were able to use the parallel in-core solver. It is significant to note that this is the worst-case scenario for the parallel out-of-core solver and therefore, the maximum degradation from the in-core performance that can be expected. Considering the cost of purchasing 867 GB RAM versus 867 GB hard disk and accepting less than 25% degraded performance, it is concluded that the parallel out-of-core integral equation solver offers a high-performance, low-cost solution for tackling real-world electromagnetic computational problems.

Figure 4.20 plots the RCS in dB along the *yoz* plane as a function of the elevation angle θ for all five frequencies simulated in this chapter. This plot shows that the RCS normalized to the square of the wavelength increase with the frequency.

It is worthwhile to point out that the out-of-core solver used here is currently under development. At the time of this writing [15], the parameters of the code have not been completely optimized. Parameters, such as the amount of in-core buffer storage available in the working array associated with the out-of-core matrix, are currently being investigated to further optimize the performance. For example, on CEM-7, if the amount of in-core buffer on each process is changed from 1.9 to 1.7 GB, the time needed for the simulation of the airplane at 2.8 GHz was reduced from 1148 to 1047 minutes, using 16 cores of the CPU. The time needed for the simulation of the airplane at 1.25 GHz was reduced from 583 to 332 minutes, using 8 cores of the CPU. This indicates that if the appropriate parameters are properly selected, the efficiency of the parallel out-of-core solver can be further improved, which would further reduce the time differences between the in-core and the out-of-core integral equation solvers. Similarly, for CEM-4, the performance can be further improved by using RAID 0 on two hard drives and adjusting the size of the in-core buffer. Optimization of the performance of the parallel out-of-core integral equation solver on different hardware configurations, including Intel and AMD multicore CPUs, will be discussed in Chapter 5.

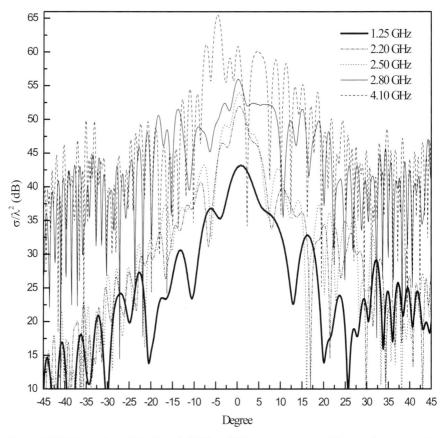

Figure 4.20. Evaluation of the bistatic RCS at different frequencies (0° starts from y axis in the yoz plane).

4.8   CONCLUSION

In this chapter, a general-purpose parallelized integral equation solver based on the MoM methodology is presented for the parallel in-core solution of electromagnetic scattering and radiation problems. The schemes for executing on multiple single-core CPUs versus multicore single CPUs are similar, although not the same. This chapter presents the key concepts for designing parallel codes to run efficiently on multicore processor platforms. These concepts apply to codes that can be executed efficiently on various computer platform configurations, from multiple single-core CPU servers, to a single dual-core CPU desktop to a cluster with hundreds of quad-core nodes. Several examples have been presented for different computer platforms to illustrate the goals of portability, scalability, and efficiency. This solver can be used to accurately analyze composite structures containing both metallic and dielectric structures. With this methodology, large problems that are difficult to solve on a single processor, can be solved within a reasonable time frame. This chapter provides the foundations of the formulation and the results for parallel in-core integral equation solvers.

These concepts are then applied to the parallel out-of-core integral equation solvers. Examples containing both metallic and dielectric structures have been presented to illustrate the accuracy, portability, scalability, and efficiency of the parallel out-of-core code. With this methodology, one can analyze problems as large as a 159 wavelength full-scale airplane using modest computational resources. The problems shown here were previously impossible to solve on a single processor in reasonable time. This research then will provide a powerful tool for electromagnetic analysis of electrically large targets simulated in their natural environments.

REFERENCES

[1]   R. F. Harrington, "Boundary Integral Formulations for Homogenous Material Bodies," *Journal of Electromagnetic Waves and Applications*, Vol. 3, No. 1, pp. 1–15, 1989.

[2]   S. M. Rao, C. C. Cha, R. L. Cravey, and D. L. Wilkes, "Electromagnetic Scattering from Arbitrary Shaped Conducting Bodies Coated with Lossy Materials of Arbitrary Thickness," *IEEE Transactions on Antennas and Propagation*, Vol. 39, No. 5, pp. 627–631, May 1991.

[3]   B. M. Kolundzija, J. S. Ognjanovic, T. K. Sarkar, and R. F. Harrington, *WIPL: Electromagnetic Modeling of Composite Wire and Plate Structures*, Artech House, Boston, 1995.

[4]   B. M. Kolundzija, J. S. Ognjanovic, and T. K. Sarkar, *WIPL-D: Electromagnetic Modeling of Composite Metallic and Dielectric Structures, Software and User's Manual*, Artech House, Boston, 2000.

[5]   B. M. Kolundzija, J. S. Ognjanovic, and T.K. Sarkar, "Analysis of Composite Metallic and Dielectric Structures — WIPL-D Code," *Proceedings of 17th Applied Computational Electromagnetics Conference*, pp. 246–253, Monterey, CA, March 2001.

[6]   B. M. Kolundzija, "Electromagnetic Modeling of Composite Metallic and Dielectric Structures," *IEEE Transactions on Microwave Theory Techniques*, Vol. 47, pp. 1021–1032, July 1999.

[7]   B. D. Popovic and B. M. Kolundzija, *Analysis of Metallic Antennas and Scatterers*, IEE Electromagnetic Wave Series, No. 38, London, 1994.

[8]   D. R. Wilton, S. M. Rao, A. W. Glisson, D. H. Schaubert, O. M. Al-Bundak, and C. M. Butler, "Potential Integrals for Uniform and Linear Source Distributions on Polygonal and Polyhedral Domains," *IEEE Transactions on Antennas and Propagation*, Vol. 32, No. 3, pp. 276–281, March 1984.

[9]   Y. Zhang, T. K. Sarkar, A. De, N. Yilmazer, S. Burintramart, and M. C. Taylor, "A Cross-Platform Parallel MoM Code with ScaLAPACK Solver," *IEEE Antennas and Propagation Society International Symposium,* pp. 2797–2800, Honolulu, HI, June 2007.

[10] Y. Zhang, T. K. Sarkar, H. Moon, A. De, and M. C. Taylor, "Solution of Large Complex Problems in Computational Electromagnetics using Higher Order Basis in MoM with Parallel Solvers," *IEEE Antennas and Propagation Society International Symposium*, pp. 5620–5623, Honolulu, HI, June 2007.

[11] Y. Zhang, T. K. Sarkar, P. Ghosh, M. C. Taylor, and A. De, "Highly Efficient Parallel Schemes Using Out-of-Core Solver for MoM" (invited paper). *IEEE Applied Electromagnetics Conference*, Kolkata, India, Dec. 2007.

[12] R. N. Simons and R. Q. Lee, "Characterization of Miniature Millimeter-Wave Vivaldi Antenna for Local Multipoint Distribution Service," *49th ARFTG/MTT Conference Digest*, pp. 95–99, 1997.

[13] K. Chang, T. Y. Yun, M. Li, and C. T. Rodenbeck, "Novel Low-Cost Beam Steering Techniques," *IEEE Transactions on Antennas and Propagation*, Vol. 50, No. 5, pp. 618–627, May 2002.

[14] M. Yuan, T. K. Sarkar, and B. Kolundzija, "Solution of Large Complex Problems in Computational Electromagnetics Using Higher-Order Basis in MoM with Out-of-Core Solvers," *IEEE Antennas and Propagation Magazine*, Vol. 48, No. 2, pp. 55–62, April 2006.

[15] Y. Zhang, M. C. Taylor, T. K. Sarkar, H. Moon, and M. Yuan, "Solving Large Complex Problems Using a Higher-Order Basis: Parallel In-Core and Out-of-Core Integral-Equation Solvers," *IEEE Antennas and Propagation Magazine*, Vol. 50, No. 4, pp. 13–30, Aug. 2008.

5

TUNING THE PERFORMANCE OF A PARALLEL INTEGRAL EQUATION SOLVER

5.0   SUMMARY

Parallel implementations of MoM-based integral equation solvers were discussed in the previous chapters. Several popular computer platforms were used to execute the parallel codes. Now the question that needs to be addressed is whether it is possible to get the best parallel efficiency on all the computing platforms using the same parameters, for example, using the same amount of in-core buffer for an out-of-core solver. The answer is obviously "no" since the performance of a parallel code is intimately related to the configuration of the computer hardware, such as the size of cache and RAM, the revolutions per minute (rpm) of hard disk, and the speed of frontside bus (FSB). To optimize the performance of a parallel integral equation solver, some numerical benchmarks need to be developed and executed on different computer platforms to quantify their effectiveness in solving a given problem in the shortest amount of time.

After reviewing the list of variables that affect the performance of a parallel out-of-core solver, two parameters, i.e., the block size and the shape of the process grid, are investigated since they both affect the performance of both parallel in-core and out-of-core solvers. The discussion is then focused on the third parameter, the size of the in-core buffer for each process, which is unique to an out-of-core solver and is denoted by IASIZE (*in*-core *a*rray *size*). At the end of this chapter, examples executed on the high-performance clusters are presented to illustrate that using the proper relevant parameters can ensure excellent performance of a parallel code.

5.1   ANATOMY OF A PARALLEL OUT-OF-CORE INTEGRAL EQUATION SOLVER

Before addressing the optimization of the performance of a parallel out-of-core solver, several important topics are discussed to further support the claim that the performance of the parallel out-of-core solver is comparable to that of an in-core one. In addition, significant emphasis should be placed on tuning the performance of a parallel code as the reward is quite large. The discussion in this section starts with a review of the various components of a parallel out-of-core solver, which are affected by the RAM, CPU, network, and hard disk. The comparison of the CPU time between a parallel out-of-core solver and a parallel in-core solver is also performed. The influence of the ratio of the used space on a hard disk to its total capacity on the overall performance of the parallel code is then examined.

5.1.1   Various Components of a Parallel Out-of-Core Solver that Can Be Observed through Ganglia and Tuned

The basic mechanism of an out-of-core solver is (1) read a portion of the impedance matrix into the main memory — the RAM, and (2) perform an LU decomposition of that block, then (3) write the intermediate results back to the hard disk when the computation for that portion is finished. The procedure is repeated with the next block of the matrix until the whole matrix has been LU factored.

Ganglia is a scalable distributed monitoring software for high-performance computing systems such as clusters. It is based on a hierarchical design targeted at federations of clusters. Ganglia uses the eXtensible Markup Language (XML) to represent data, eXternal Data Representation (XDR) for compact binary data transfers, and an open-source package called RRDTool for data storage (in round robin databases) and graphical visualization. It uses carefully engineered data structures and algorithms to achieve very low per-node overheads and high concurrency. It provides a complete, real-time monitoring and execution environment [1].

Figure 5.1 shows a portion of a Ganglia screenshot from the platform CEM-4 during the simulation of a problem with 148,273 unknowns. CEM-4 has one head node and nine compute nodes. As shown in this figure, the head node is named as CEM4.local. The compute node names are compute-1-2.local, compute-1-3.local, …, compute-1-10.local. The load on the head node is empty because only compute nodes are involved in the solution of the problem.

An edited version of the screenshot related to the 10th node of CEM-4 is shown in Figure 5.2, presenting the memory occupancy on the top, the CPU usage in the middle, and the network throughput at the bottom. The beginning of the central tracing (for percentage of CPU usage) in this figure is enlarged as an inset to show different slabs in the matrix filling.

When each process is assigned 3.2 GB of RAM as an in-core buffer memory, to hold a portion of a slab of the entire matrix, a total of 6.4 GB of RAM is then used on each node, which consists of two CPUs (one process employing one CPU). The remaining RAM on the system is used as cache for file input/output (I/O) operations.

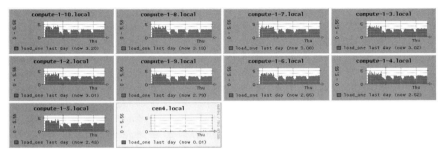

Figure 5.1. Computing load on different nodes of CEM-4 as shown by Ganglia.

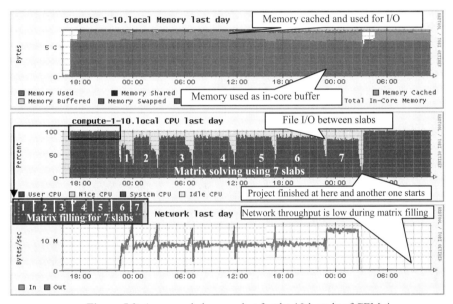

Figure 5.2. An expanded screenshot for the 10th node of CEM-4.

While filling each slab of the out-of-core matrix, the CPU is 100% utilized and the network throughput is close to 0 bytes per second. This is due to the fact that the parallel code has been designed in such a way so that no communication is required between different processes during the filling of the large impedance matrix.

After the LU decomposition is completed on the first slab, the intermediate results are written back to the hard disk. When the data is being written, the CPU usage becomes less due to the I/O operation.

For the solution of a project with 148,273 unknowns, one needs a storage space of approximately 352 GB on the hard disk. To solve the entire problem, we obtain *ceiling* [352 GB/(18 processes×3.2 GB)] = 7, and so seven cycles of I/O is carried out for each process so that the whole impedance matrix is decomposed into seven different slabs. Here the *ceiling* function returns the smallest integer greater than or equal to the number. The slabs are then written one at a time to the hard disk. The filling time for the last slab is typically less than that for the other slabs, because the rightmost slab located at the fringes has fewer columns than the slabs to its left. As shown in the inset of Figure 5.2, the filling of the seventh slab (fringe slab) takes less time.

When all the slabs are filled, the first slab is read back into the RAM and the LU factorization begins. During the LU decomposition, communication is required between processes and the data are synchronized; therefore, network throughput increases and the CPU usage percentage degrades as illustrated in Figure 5.2.

The bottom part (tracing) of Figure 5.2 represents the network throughput. The central part of Figure 5.2 represents the percentage usage of the CPU, where the "user CPU" is the percentage of CPU used by the parallel code. The "system CPU" is the percentage of CPU used by the operating system. "Idle CPU" is the percentage of CPU unused. "Nice CPU" is the percentage of CPU used by processes running with an altered scheduling priority. All user processes start with the same priority of execution. This priority of the processes can be adjusted with the *nice* and *renice* commands, which can alter their degree of accessibility to the main CPU.

The factored first slab is then written back to the hard disk, and the second slab is read into the RAM. The procedure of writing and reading are marked as I/O in Figure 5.2. The CPU usage is very low during this I/O phase because it needs to wait for the data to be ready for processing in the RAM.

Then the LU decomposition of the second slab begins, followed by writing the results back to the hard disk. The same procedures are applied to the remaining slabs until all of them are factored. Note that the time required for the LU decomposition of a slab increases gradually when the slab moves toward the right-hand side of the matrix (but before it reaches the fringe slab), as shown in Figure 5.2. This is because all of the data to the left of the active slab needs to be read in one panel at a time and used for computation. The total time for the computations associated with the fringe slab is determined by its time for LU factorization and pertinent computations related to all the slabs to the left of it. The computation associated with the fringe slab could be faster or slower depending on the width of the fringe slab.

After the LU decomposition is finished, the code begins the solution procedure by using substitution. For the example shown in Figure 5.2, the LU decomposition took 102,099.90 seconds, while the solution procedure took 941.94 seconds, which is too small to be depicted clearly in Figure 5.2. Once the final solution is obtained, a parallel postprocessing of the data (i.e., computation of the near and far fields and the like using the computed currents) is initiated.

As shown in Figure 5.2, approximately 100% and 90% of the CPU resources is used during the matrix filling and LU decomposition procedure, respectively. Therefore, most of the total CPU time required for a typical simulation is spent on computations, which involves constructing the matrix equation and then using the LU factorization to solve it. Before tuning the performance of an out-of-core solver, it is useful to study the difference between the CPU times for a parallel in-core and an out-of-core solver.

5.1.2   CPU Times of Parallel In-Core and Out-of-Core Integral Equation Solvers

Speedup and efficiency have been discussed as metrics for evaluating the performance of a parallel code in the previous chapters. Here, the focus is on another metric — the execution time, which involves the wall time and the CPU time.

Before checking the performance of a parallel out-of-core solver, we first compare its CPU time (for filling the matrix and solving the matrix equation) with that of a parallel in-core solver. IASIZE, the memory allocated to each process for an out-of-core solver to hold a portion of a slab of data, is taken as 1.7 GB on CEM-2 and 3.2 GB on CEM-4 in the following tests. The default process grid used is 1×(number of processes) in this chapter unless specified otherwise.

The first set of tests is implemented on CEM-2. The RCS of a PEC cube of length 2.0 m is simulated at different frequencies to generate different numbers of unknowns. The required number of unknowns and its corresponding storage (which refers to RAM when using the in-core solver and the amount of space used on the hard disk when using the out-of-core solver) are listed in Table 5.1.

Table 5.1 indicates that the CPU time required for an out-of-core solver is slightly longer than that required for an in-core solver. The difference between these times increases as the number of unknowns increases. An out-of-core code typically incurs an overhead of 8.6% CPU time over that of an in-core code when solving a problem requiring nearly all the RAM on this computer (recall that CEM-2 has 8 GB RAM). For comparison purposes, the measured CPU time for solving the matrix equations are also plotted in Figure 5.3.

TABLE 5.1. Comparison of CPU Times between the In-Core and the Out-of-Core Solvers on CEM-2

Frequency (GHz)	Number of Unknowns	Storage (GB)	Parallel In-Core (seconds)		Parallel Out-of-Core (seconds)	
			Filling	Solving	Filling	Solving
1.28	6,912	0.76	21.32	61.13	26.92	69.99
1.48	8,712	1.21	28.80	107.04	35.59	121.46
1.52	9,408	1.42	34.67	142.46	47.35	155.73
1.58	16,104	4.15	120.96	646.45	146.17	685.53
1.68	17,136	4.70	146.43	768.90	177.19	808.23
1.98	21,936	7.70	243.93	1611.82	316.87	1698.82

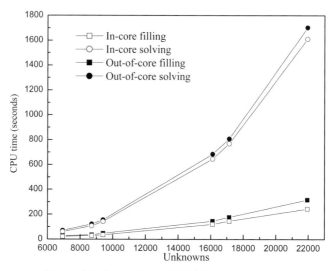

Figure 5.3. Comparison of CPU times on CEM-2.

Another set of tests is executed using a composite PEC and dielectric structure, which in this case is an array of Vivaldi antennas. The Vivaldi antenna array, consisting of different numbers of antenna elements, is used as a model, with each element identical to that described in Section 4.7.1.2 of Chapter 4.

The storage required and the measured CPU time for solving each problem are listed in Table 5.2 along with the wall time. The wall time for the out-of-core solver, which denotes the total time for each simulation, is equal to or slightly longer than that of the in-core solver. This is due to the fact that CEM-4 has 80 GB RAM and therefore can provide sufficient cache memory for the reading/writing operation for the out-of-core solver.

The CPU time for solving the matrix equations is given in Table 5.2 and is also plotted in Figure 5.4. It shows that with a smaller number of unknowns (NUN) CPU times required by an in-core and an out-of-core solver are similar.

TABLE 5.2. Comparison of Execution Times on CEM-4

Number of Antenna Elements	Number of Unknowns	Storage (GB)	Parallel In-Core		Wall Time (minutes)	Parallel Out-of-Core		Wall Time (minutes)
			CPU Time (seconds)			CPU Time (seconds)		
			Filling	Solving		Filling	Solving	
12	10,088	1.63	129.8	33.9	4	129.4	31. 7	4
24	20,052	6.43	467.9	212.8	14	491.5	205.0	14
40	33,096	17.53	1308.8	912.1	42	1274.9	874.4	43
60	49,820	39.71	2930.6	2990.3	111	2989.2	2908.1	116

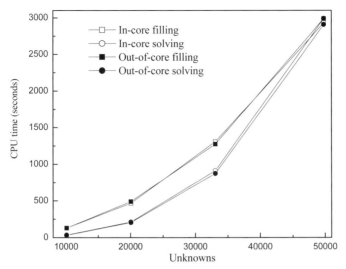

Figure 5.4. Comparison of CPU times on CEM-4.

To investigate the performance of the parallel out-of-core solver when solving a problem with a larger number of unknowns on CEM-4, the RCS of the full-size aircraft model (described in Section 4.7.1.2 of Chapter 4) at different operating frequencies are simulated using nine nodes. The CPU times between the in-core and the out-of-core solver are compared in the third set of tests.

The storage and the CPU time required to solve the various sized problems are listed in Table 5.3. The total time, namely the wall time, is also included in the table. Figure 5.5 plots the CPU time given in Table 5.3, for the solution of the matrix equations using the parallel in-core and the parallel out-of-core solvers.

TABLE 5.3. Comparison of CPU Times between an In-Core and an Out-of-Core Solver

Frequency (GHz)	Number of Unknowns	Storage (GB)	Parallel In-Core		Wall Time (minutes)	Parallel Out-of-Core		Wall Time (minutes)
			CPU Time (seconds)			CPU Time (seconds)		
			Filling	Solving		Filling	Solving	
1.25	24,993	9.99	230.2	414.4	14	242.2	403.8	14
1.50	34,502	19.05	426.7	1031.1	31	448.5	1014.1	31
1.75	44,595	31.82	657.2	2174.6	56	663.1	2087.5	59
2.00	57,732	53.33	1050.6	4615.3	110	1073.1	4458.6	120
2.10	63,277	64.06	1263.2	6037.9	141	1459.0	6289.9	169

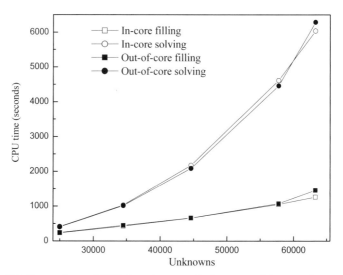

Figure 5.5. Comparison of CPU times between the in-core and the out-of-core solvers.

As shown in Figure 5.5, the CPU time required for the out-of-core solver is essentially the same as that of the in-core solver when the RAM on CEM-4 is not depleted for cases when solving a problem with a relatively small number of unknowns. The wall times of the in-core and the out-of-core solvers, when using different number of unknowns, are also essentially the same for these cases, because CEM-4 has sufficient cache memory for the read/write operation of the out-of-core solver when the problem size is not very large. The solution of a problem having a larger number of unknowns will definitely require more RAM. For the case of CEM-4, the CPU time required by the out-of-core solver will be longer than that of the in-core solver because the advantage of having sufficient cache for the reading/writing operation of the data is no longer available. The wall time required for the out-of-core solver is slightly longer than the in-core solver because of the degradation in performance.

For simulations where there does not exist sufficient cache memory, the parallel out-of-core solver is also efficient in terms of usage of CPU time. As the problem size increases, at most a 6% increase in the CPU time is required for the out-of-core solver over the in-core solver. However, for the largest size problem that has been tested, the degradation in the wall time for the out-of-core solver is at most 20% greater than the time required by the in-core solver as seen from Table 5.3.

Since the wall time of the parallel out-of-core solver will further degrade as the number of unknowns is increased over the problems solved so far, it is necessary to study this problem further. Consequently, the following section will discuss how to optimize the wall time for an out-of-core solver, where the size of the impedance matrix is so large that it cannot be fit into the RAM and therefore the in-core solution methodology will not be employed.

5.1.3 Performance of a Code Varies with the Amount of Storage Used on the Hard Disk

Since the hard disk is an indispensable hardware device for an out-of-core solver, the issue that needs to be addressed is whether the code can perform well when the storage requirement of the parallel out-of-core MoM-based integral equation solver increases until it fills the entire hard disk. To address this question the RCS of the airplane model presented in Chapter 4 is analyzed at different frequencies.

Two compute nodes (compute-1-5 and compute-1-6) of CEM-4 are employed for the simulation at each frequency. Two processes are used on each node. Two hard disks are installed on each node, and the problem is solved using only the first hard disk on each node. The first hard disk has a total of 140,062,276 kilobytes (kB) of accessible space, as shown in Figure 5.6 (the sum of all the values inside the two dashed-line boxes). Other than the storage space occupied by the system, the disk is partitioned as "/state/partition1". The size of the storage space available for calculation in "/state/partition1" is 123,059,956 kB on each node, as shown in Figure 5.6 (marked in the box).

Screenshots from SSH (secure shell) before and during execution of this simulation at a frequency of 2.88 GHz are given in Figures 5.6 and 5.7. The desired numbers are marked in these figures to highlight the data of interest.

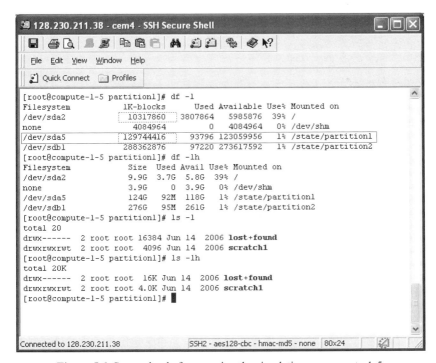

Figure 5.6. Screenshot before running the simulation on compute-1-5.

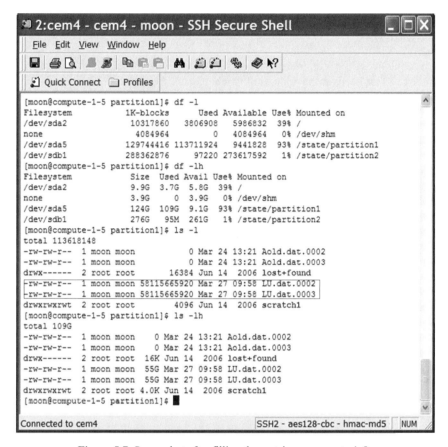

Figure 5.7. Screenshot after filling the matrix on compute-1-5.

At this frequency, this simulation requires 120,371 unknowns and needs a total of 231,826,842 kB of storage, with an estimated 57,956,710,564 bytes for each process if the load is well balanced.

Note that a perfect load balance cannot be achieved for any real application. The load on each process is determined by the process grid, the block size, and the dimension of the matrix. As shown in Figure 2.4 of Chapter 2, when the 9×9 matrix is distributed into six processes; the load for process 0 is different from that for process 1. It is also indicated in Section 2.3.2 that the unbalanced part of the load is different when using a different block size and process grid. As shown in Figure 5.7 (marked by the box) on compute-1-5, the files "LU.dat.0002" and "LU.dat.0003" are both 58,115,665,920 bytes, which is slightly different from the estimated one.

The results from all the tests are listed in Table 5.4. The estimated CPU times and wall times for different simulations are also included, using the corresponding times for executing the minimum number of unknowns as a benchmark. The ratios (in percentage) of the storage required by the parallel out-of-core solver to the disk space available for the calculation in "/state/partition1", and to the accessible hard disk space are also listed in Table 5.4.

TABLE 5.4. Results from Using Different Hard Disk Fill Sizes

Item	Values					
Frequency (GHz)	2.72	2.8	2.82	2.85	2.88	2.92
Unknowns	107,313	110,994	113,801	118,104	120,371	122,993
Required storage (GB)	184.26	197.11	207.21	223.18	231.83	242.04
% of available space in "/state/partition1"	74.90	80.13	84.23	90.72	94.24	98.39
% of the accessible hard disk on CEM-4	65.81	70.40	74.00	79.71	82.80	86.44
	CPU Time (seconds)					
Fill	16,256	17,953	19,250	21,129	22,211	23,405
Measured LU	133,361	148,356	159,883	178,120	190,386	201,783
Estimated LU	—	147,561	159,042	177,773	188,208	200,777
Substitution	161	174	184	189	200	211
	Wall Time (seconds)					
Fill	16,980	19,200	19,680	21,900	23,100	24,060
Measured LU	155,953	172,843	187,682	209,120	223,990	236,637
Estimated LU	—	172,558	185,984	207,889	220,091	234,789
Substitution	1,892	2,052	2,211	2,385	2,562	2,678

The CPU and the wall times used by the LU decomposition are plotted in Figure 5.8. These plots show that the measured time matches well with the predicted time. It also scales well with the increase of the percentage of the storage space used on the hard disk. The greatest difference occurs when the number of unknowns is 120,371 (94.24% available space in "/state/partition1" is used). Table 5.4 shows that for this case the measured wall time for LU decomposition is $(223,990 - 220,091) = 3899$ seconds longer than the estimated wall time. The measured CPU time is $(190,386 - 188,208) = 2178$ seconds longer than the estimated CPU time. This indicates that the total time for the computation other than that used by the CPU for calculation is only 1721 seconds and is slightly longer than the predicted value. Compared with the total wall time of 63 hours for the LU decomposition, a 0.5 hour (1721 seconds) difference is less than 1% of the total time. This demonstrates that the performance of this parallel code on this platform is quite good.

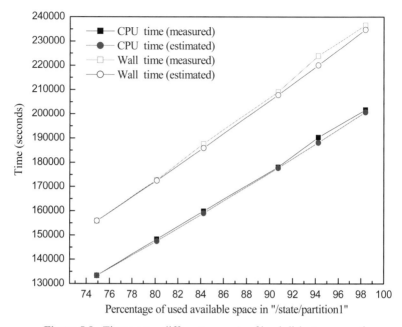

Figure 5.8. Time versus different amounts of hard-disk storage used.

For the benchmark problem, when using the largest number of unknowns, the difference between the predicted and the measured wall time for LU decomposition is 1848 seconds. This difference in the CPU time is about 1006 seconds. Therefore, the difference between the estimated and the measured times for the overhead during the computation other than the one used by the CPU is only 15 minutes. This supports the conclusion that using larger percentages of the hard disk will not significantly degrade the performance of the parallel out-of-core solver.

The next question related to the hard disk is whether one can obtain a better performance if more hard disks are used per compute node in the RAID 0 configuration than a single hard disk.

First, the significance of RAID is outlined. As discussed in the reference (http://en.wikipedia.org/wiki/Redundant_array_of_independent_disks/), RAID stands for redundant array of inexpensive (or independent) disks. It is a technology that employs the simultaneous use of two or more hard-disk drives to achieve greater levels of performance, reliability, and/or larger data volume sizes. When several physical disks are set up to use RAID technology, they are said to be *in* a *RAID* array. This array distributes data across several disks, but the array is seen by the computer user and operating system as one single disk. RAID can be set up to serve several different purposes. There are various combinations of these approaches giving different tradeoffs of protection against data loss, capacity, and speed. RAID levels 0, 1, and 5 are the most commonly found, and cover most requirements. The different terminologies of RAID levels are described next:

- RAID 0 (striped disks) distributes data across several disks in a way that gives improved speed and full capacity, but all data on all disks will be lost if any one disk fails.

- RAID 1 (mirrored disks) could be described as a backup solution, using two (possibly more) disks that each stores the same data so that data are not lost as long as one disk survives. Total capacity of the array is just the capacity of a single disk. The failure of one drive, in the event of a hardware or software malfunction, does not increase the chance of a failure nor decrease the reliability of the remaining drives (second, third, etc.).

- RAID 5 (striped disks with parity) combines three or more disks in a way that protects data against loss of any one disk; the storage capacity of the array is reduced by one disk.

- RAID 6 (less common) can recover from the loss of two disks.

- RAID 10 (or 1+0) uses both striping and mirroring.

To address the question posed in the previous paragraph, a project involving 92,729 unknowns dealing with analysis of the Vivaldi antenna array is executed using only one node (eight cores, as each node contains two quad-core processors) on the platform CEM-9. The CEM-9 system has three 15,000-rpm 300-GB SAS [serial attached SCSI (small computer system interface)] hard disks per node, so that one, two, or three hard disks can be chosen to be used with this out-of-core solver.

The SAS is a data transfer technology designed to move data to and from computer storage devices such as hard drives and tape drives. It is a point-to-point serial protocol that replaces the parallel SCSI bus technology that first appeared in the mid 1980s in corporate data centers, and uses the standard SCSI command set. The SCSI is a set of standards for physically connecting and transferring data between computers and peripheral devices. The SCSI standards define commands, protocols, and electrical and optical interfaces. SCSI is most commonly used for hard disks and tape drives, but it can connect a wide range of other devices, including scanners and CD drives. The SCSI standard defines command sets for specific peripheral device types; the presence of "unknown" as one of these types means that in theory it can be used as an interface to almost any device, but the standard is highly pragmatic and addressed toward commercial requirements (http://en.wikipedia.org/wiki/Hard_drive).

As Table 5.5 indicates, the choice of more SAS hard disks, using the RAID controller per node, is better, and there is no bottleneck associated with the corresponding I/O process. Use of a single hard disk is not the best hardware solution for the parallel out-of-core solver. Furthermore, the use of a RAID configuration using a RAID controller is very important.

TABLE 5.5. Different Numbers of Hard Disks Used in the Simulation

Number of Disks	RAID Configuration	Total (minutes)	LU (seconds)	Solve (seconds)
1	—	1222	46417	1973
2	RAID 0	1128	41702	1273
3	RAID 0	1125	41675	1080

RAID arrays that use striping improves performance by splitting up files into small pieces and distributing them to multiple hard disks. Most striping implementations allow the creator of the array control over two critical parameters that define how the data are broken into chunks and sent to the various disks. Each of these factors has an important impact on the performance of a striped array. The first key parameter is the *stripe width* of the array, that is, the number of parallel stripes that can be written to or read from simultaneously. This is, of course, equal to the number of disks in the array. So a four-disk striped array would have a stripe width of 4. Read/write performance of a striped array increases as stripe width increases, all else being equal. The reason is that adding more drives to the array increases the *parallelism* of the array, allowing access to more drives simultaneously. The second important parameter is the *stripe size* of the array, sometimes also referred to by terms such as *block size, chunk size, stripe length,* or *granularity*. This term refers to the size of the stripes written to each disk. RAID arrays that stripe in blocks typically allow the selection of block sizes in kB (kilobytes) ranging from 2 to 512 kB (or even higher) in powers of two (meaning 2 kB, 4 kB, 8 kB, etc.). In this test, the in-core buffer per core is 2.4 GB, the process grid is 2×4 and the stripe size for the RAID 0 configuration is 128 kB.

The analysis in this section indicates that the parallel out-of-core solver is comparable in performance to the parallel in-core solver. In the following sections, the discussion will focus on further improving the performance of the out-of-core solver.

5.2   BLOCK SIZE

The key to achieving high performance on modern cache-based processors is to use blocked algorithms, which permit reuse of data and use up all the available memory without paging or using the memory for performing system tasks such as executing MPI. These are important aspects since they relate to all types of computer architectures as the performance drops significantly with slight changes in the optimal problem size or block size. Commercial software vendors have performed many tests and may offer their own optimized parameters for parallel in-core solvers, to account for different machine configurations of memory hierarchy, cache and register sizes. For example, the row block size m_b and the column block size n_b are used for data distribution and for computational granularity. The block size determines the amount of data that is handled by each process. Choosing the wrong block size can have a significant negative effect on the performance. A small block size is better for load balancing. Picking a too small block size increases the amount of communication between processors. The best values depend on the ratio of the time used for computation as opposed to the communication between the various nodes of the computer cluster platform that is being tested. Acceptable block sizes typically range from 32 to 256.

The Intel Cluster Math Kernel Library (CMKL) implementation of ScaLAPACK is tolerant to the choice of the different block size by the users [2]. The block size for CMKL was suggested to be 104 by Dell for CEM-4. The block size for AMD Core Math Library (ACML) was suggested to be 112 by AMD for CEM-6 during this study. Throughout the rest of this book, the block size for CMKL is set to be 104 for Intel CPUs, and 112 for ACML on the AMD CPU, unless specified otherwise. These parameters are also applied to the parallel out-of-core solver, which is based on the ScaLAPACK routines.

One way to determine the optimum size for a given platform is to execute a batch of tests with different block sizes and observe for which value the optimum result is obtained. Several matrix equations with different number of unknowns are solved using the parallel in-core codes along with different block sizes to compare their degree of performance. Only (both row and column) block sizes of 104 and 112 are considered for the first example. Simulations are executed on CEM-6 and CEM-7 and the results are listed in Table 5.6. Results obtained on CEM-6 using the ACML library package shows that the block size of 112 is better than the block size of 104. Results obtained on CEM-7 shows that for the Intel CPU, the block size of 104 performs better than the block size of 112 as the problem size increases.

Another set of tests are performed using Cluster-2. A problem involving 44,600 unknowns is simulated using the parallel in-core MoM code and two nodes of Cluster-2. The problem with 63,282 unknowns is simulated using the parallel in-core MoM code and four nodes of Cluster-2. The problems are executed using different block sizes ranging from 64 to 256. The measured wall clock time when solving the two different sized problems using different block sizes are plotted in Figures 5.9 and 5.10. As seen in these figures, when matrix solution is considered, code performance using the block size of 64 is slightly better than that using the block size of 128. When computation of the matrix elements is taken into account, the block size of 128 is a better choice. Note that the smaller the block size, the higher the possibility for the need of performing redundant calculations of the elements of the impedance matrix in the MoM formulation using the higher-order or RWG basis functions.

TABLE 5.6. Execution Times When Using Different Block Sizes

		Case 1			Case 2			Case 3		
Cores		4			8			16		
Unknowns/ Memory (GB)		22,308 / 7.96			31,496 / 15.87			43,250 / 29.93		
System	Block Size	CPU Time (seconds)		Wall Time (seconds)	CPU Time (seconds)		Wall Time (seconds)	CPU Time (seconds)		Wall Time (seconds)
		Filling	Solving		Filling	Solving		Filling	Solving	
CEM-6	104	870	1736	2640	1397	2512	3900	—	—	—
CEM-6	112	862	1735	2640	1384	2505	3900	—	—	—
CEM-7	104	740	1140	1920	1147	1757	2940	1706	2425	4200
CEM-7	112	736	1140	1860	1163	1754	2940	1792	2457	4320

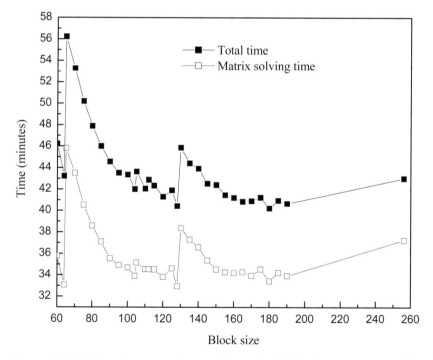

Figure 5.9. Wall time when executing a problem with 44,600 unknowns on two nodes of Cluster-2 using different block sizes.

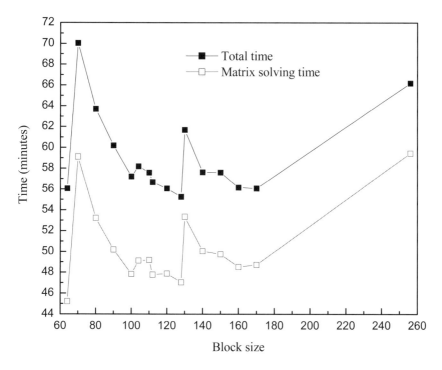

Figure 5.10. Wall time when executing a problem with 63,282 unknowns on four nodes of Cluster-2 using different block sizes.

5.3   SHAPE OF THE PROCESS GRID

Another parameter is the shape of the process grid $P_r \times P_c$, which a user can tune to optimize the performance of a code. The parameters can be set to maximize the performance for a given architecture. When performing an LU decomposition for a dense matrix, it is preferable that the large matrix be distributed on a square grid, or a nearly square grid. The $P_r \times 1$ grid incurs high communication costs in determining the pivot row, and the $1 \times P_c$ grid has a serial bottleneck. If at each step of the algorithm a block of columns and/or rows needs to be broadcast to the other processors, as in the right-looking variant of the LU factorizations, then it is possible to pipeline this duration of communication and overlap it with some other computations. As an example of this strategy, one can look at the paper by Desprez et al. [3] where the optimization of an LU factorization routine is carried out using an overlap between communication and computation.

The best shape of the process grid is determined by both the algorithm implemented in the library and the underlying physical network [4,5]. Usually P_r and P_c are approximately equal, with P_c slightly larger. This rule works well for a mesh or switched network with which each processor is connected. On a nonswitched, simple Ethernet network, performance is limited to flatter process grids like 1×4, 1×8, 2×4, and so on. This affects the scalability of the algorithm, but reduces the overhead caused by message collisions.

It is possible to predict the best shape of the process grid given the total number of processes. As examples, problems involving 110,994 unknowns (for the aircraft model described in Section 4.7.2.3) and 232,705 unknowns (for the aircraft model described in Section 4.7.2.4) are executed on the IBM eServer BladeCenter HS21 platform (Cluster-7) using 112 and 64 cores, respectively. Note that the simulations for case 1-1 to case 1-4 are essentially executed in an in-core mode, even though the parallel out-of-core code is used since sufficient RAM is available.

For cases where 112 CPU cores are used, the five process grids are chosen as 1×112, 2×56, 4×28, 8×14, and 8×14. For cases involving 64 cores, the process grids chosen are 1×64, 2×32, 4×16, 8×8, and 1×64, as listed in Table 5.7.

For case 1-1 to case 1-4, the matrix filling time does not change much, while the matrix solving time (here matrix solving time refers to the time used to solve a matrix equation, which includes the LU factorization time and substitution time) is reduced to be less than one-fifth of the original time (the time obtained for case 1-1) when the process grid changes from 1×112 to 8×14. Similarly, for cases 2-1 to 2-4 , the matrix filling time is similar, but the matrix solving time decreases to approximately 70% of the original time (the time obtained in case 2-1) after the process grid is changed from 1×64 to 8×8.

TABLE 5.7. Different Choices for the Shape of the Process Grid on Cluster-7

Case	NUN	Storage (GB)	Process Grid $P_r \times P_c$	Total (minutes)	Fill (minutes)	LU (seconds)	Substitution (seconds)
			Simulations Using 112 Cores				
Case 1-1	110,994	197.1	1×112	1183	26	68,542	869
Case 1-2	110,994	197.1	2×56	494	23	27,413	834
Case 1-3	110,994	197.1	4×28	267	20	13,915	880
Case 1-4	110,994	197.1	8×14	193	19	8,946	1469
Case 1-5	232,705	866.4	8×14	1237	88.6	65,446	3458
			Simulations Using 64 Cores				
Case 2-1	110,994	197.1	1×64	413	30	21,774	1219
Case 2-2	110,994	197.1	2×32	295	30	14,561	1336
Case 2-3	110,994	197.1	4×16	260	29	12,698	1155
Case 2-4	110,994	197.1	8×8	293	36	14,267	1182
Case 2-5	110,994	197.1	1×64	776	28	44,293	600

Using case 1-4 as the benchmark, the estimated time for executing case 1-5 should be $[(232,705/110,994)^3 \times 193] = 1778.59$ minutes, which is longer than the measured 1237 minutes. This indicates that the parallel out-of-core solver offers satisfactory scalability.

For case 2-5, the test is performed by allowing "mpiexec" to distribute the 64 processes to the 14 available nodes. The "mpiexec" spreads the processes onto all 14 nodes, resulting in some nodes having four processes and some having five. This simulation was not as efficient as the simulation where the processes were packed onto 8 nodes as in case 2-1.

The total wall times, matrix filling times and matrix solving times, required for case 1-1 to case 1-4 and for case 2-1 to case 2-4, are plotted in Figure 5.11. The horizontal numbers 1–4 denote case 1-1 to case 1-4, and 5–8 denote case 2-1 to case 2-4. Results show that for this platform, the performance of the LU decomposition gets better as the shape of the process grid approaches a square for a given number of processes. Also, note that the use of more processors does not necessarily provide a faster solution for a given problem. For example, by employing a proper process grid, the problem can be solved faster when using 64 cores for case 2-2 than using 112 cores for case 1-2. Therefore, using more CPU cores does not guarantee faster simulation unless the code is executed with a properly designed process grid.

The optimal shape of the process grid depends on the physical network and may vary from one computer platform to another. To illustrate this point, consider as an example, the same project involving 110,994 unknowns as in case 2-1 but executed on a different IBM BladeCenter HS21 platform (Cluster-8).

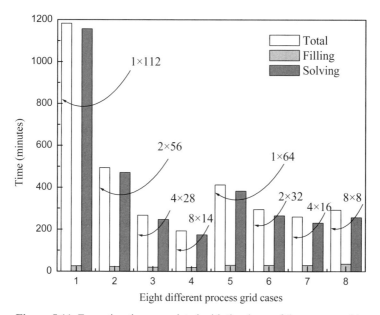

Figure 5.11. Execution time associated with the shape of the process grids.

The results listed in Table 5.8 indicate that the choice of the 2×32 process grid (in case 3-1) is better than the choice of the 4×16 process grid (in case 3-2). Recall that on Cluster-7, better results are obtained using a process grid of 4×16 rather than 2×32. This validates the statement that the choice of the optimum process grid changes with the hardware configuration. It is possible to develop models for the optimum shape of the process grid, which depends solely on the computing platform and on parameters of the problem on hand to accurately estimate the impact of modifying the shape of the grid on the total execution time [5].

TABLE 5.8. Different Choices for the Shape of the Process Grid on Cluster-8

Case	Unknowns	Storage	Grid $P_r \times P_c$	Total (minutes)	Fill (minutes)	LU (seconds)	Substitution (seconds)
Case 3-1	110,994	197.1	2×32	235	21	11987.27	846.29
Case 3-2	110,994	197.1	4×16	281	27.5	13647.75	1562.4

When a comparison is made between case 3-1 and 2-3, it is found that the simulation runs faster when using CPUs with a higher clock speed on Cluster-8 than when using a slower CPU, like Cluster-7. However, comparing case 3-2 with case 2-3 it is seen that a faster CPU does not guarantee better performance. Properly choosing the number of cores and the shape of the process grid is the key to attaining the best performance.

Another question related to the process grid is how many processes and what process grid should be used assuming at most N_P processes are available. Depending on N_P, it is not always possible to factor $N_P = P_r \times P_c$ to create the appropriate grid. For example, if N_P is a prime, the only possible grid sizes are $1 \times N_P$ and $N_P \times 1$. When such grids are bad choices for performance, it would be beneficial to let some processors (cores) remain idle, so the processes on the other processors (cores) can be shaped into a "squarer" grid. These problems can be analyzed as a function of the machine and the size of the problem to be solved [5].

5.4   SIZE OF THE IN-CORE BUFFER ALLOCATED TO EACH PROCESS

Block size and process grid are parameters that are relevant for tuning the performance of both the parallel in-core and the out-of-core solvers. In the following section, it is illustrated that the choice of the size of the in-core buffer allocated to each process, IASIZE, is quite relevant. The choice of this parameter, which is unique only to the out-of-core solvers, is discussed next.

5.4.1   Optimizing IASIZE for a Parallel MoM Code Using Higher-Order Basis Functions

The execution time for a typical parallel out-of-core simulation consists of the CPU time performing the computations, file I/O time, and the network communication time. All are dependent on the size of the in-core working buffer on each process, IASIZE, which can hold a partition of a slab. Therefore, the value of IASIZE is a critical factor for obtaining optimum performance.

For the parallel out-of-core solver, a larger value of IASIZE will lead to higher CPU usage, and this will result in faster computation. However, reading/writing a large file from/to the hard disk will need more time if the RAM is not properly allocated. In the following sections, we investigate how the performance of a cluster is affected by the choice of the IASIZE parameter.

5.4.1.1 Case A: Available 2 GB of RAM/Core

5.4.1.1.1 Overview of Wall Time with Different IASIZE. The first set of simulations is performed on one compute node of CEM-7. Since the simulations are executed on one blade, there is no necessity for communication between CPUs through the switch. With the attention solely focused on the IASIZE rather than on network communications, these simulations help to analyze the performance of the parallel out-of-core solver.

The full-scale aircraft model described in Chapter 4 is used for the analysis. The RCS of the aircraft at 2.05 GHz is calculated using 60,653 unknowns. This project requires 59 GB of storage and can not be analyzed using the in-core solver with 16 GB of RAM. Several modified Ganglia screenshots are shown in Figure 5.12 (*a*)–(*d*), with dashed line marked on each (the region between two dashed lines for each figure is of interest), indicating cases with 1.0 GB, 1.4 GB, 1.7 GB, and 2.0 GB of IASIZE for the parallel out-of-core solver.

As shown in Figure 5.12 (*a*), the "user" CPU is less than 90%. In the inset that magnifying the concerned region of Figure 5.12 (*a*), the gap between neighboring horizontal dashed line represents 10% CPU usage. In Figure 5.12 (*d*), the user CPU usage percent is higher than 90%. This indicates that the larger the IASIZE is for each process, the higher the CPU usage.

The number of slabs, which equals to the value obtained from *ceiling*[storage/(number of processes)/IASIZE], where the *ceiling* function returns the smallest integer greater than or equal to the number, decreases from 8 as marked in the inset of Figure 5.12 (*a*) to 5 as in Figure 5.12 (*c*). Even though the gap in the time between the computations of the neighboring slabs is almost the same for the cases shown in Figure 5.12 (*a*)–(*c*), it suddenly increases in Figure 5.12 (*d*) to be comparable with the computing time of a slab. The value of IASIZE chosen for the case shown in Figure 5.12 (*d*) will significantly degrade the performance.

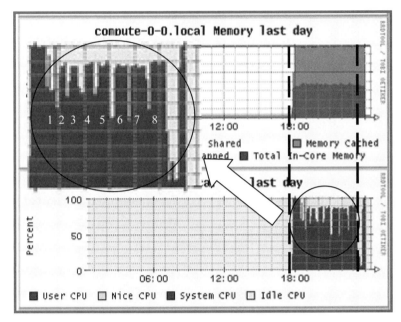

(a)

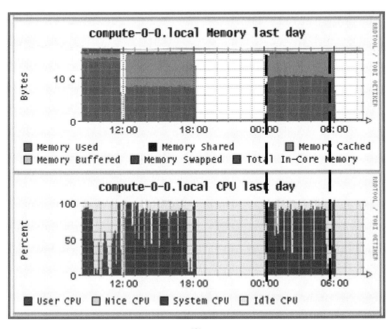

(b)

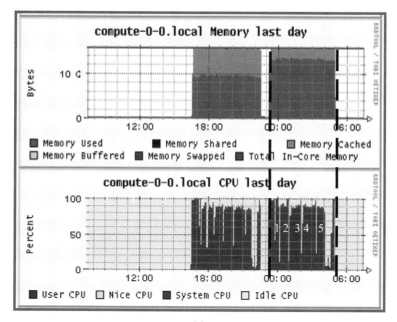

(c)

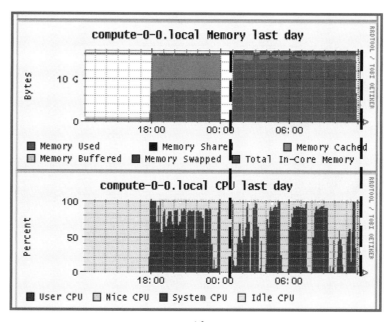

(d)

Figure 5.12. Edited screenshots from Ganglia when using different values of IASIZE: (a) 1.0 GB; (b) 1.4 GB; (c) 1.7 GB; (d) 2.0 GB.

The measured wall time as a function of the IASIZE is plotted in Figure 5.13. The performance of the parallel out-of-core solver for this particular case does not change much when the IASIZE is less then 1.7 GB. However, when the IASIZE is larger than this, the performance degrades dramatically.

To confirm the fact that the performance of the parallel out-of-core solver degrades rapidly when the IASIZE exceeds a certain value, this project was also executed on CEM-6 using one node. Figure 5.14 plots the wall times for different values of IASIZE. The results for CEM-7 are also included in this figure for comparison of the performance between these two platforms.

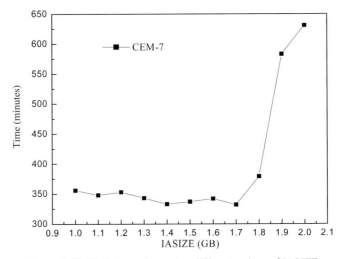

Figure 5.13. Wall time when using different values of IASIZE.

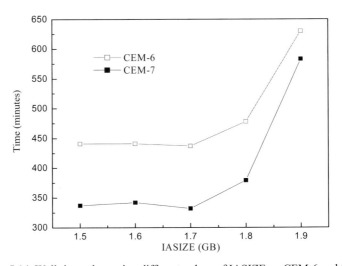

Figure 5.14. Wall time when using different values of IASIZE on CEM-6 and CEM-7.

As indicated in Figure 5.14, the value of the IASIZE exhibits similar influence on the performance of the codes on these two platforms. Recall that for the previous tests, which involve only one node, the communications through the switch can be neglected.

Next, the aircraft model is simulated on CEM-7 using two nodes for the calculation of the RCS at 2.8 GHz. This test involves 110,994 unknowns. The measured wall time using different IASIZE is plotted in Figure 5.15.

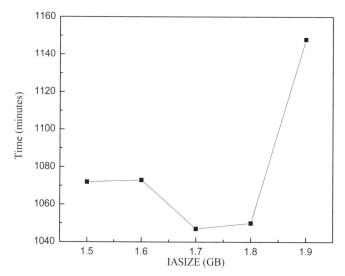

Figure 5.15. Wall time for 110,994 unknowns when using different values of IASIZE.

The code gives the best performance when the IASIZE is chosen as 1.7 GB. This indicates that the IASIZE affects the performance of the code when executing on two nodes in much the same way as it does on a single node.

5.4.1.1.2 Details on Matrix Filling and Matrix Solving. The previous analysis focused on the wall (total) time of the simulations. The total time for a simulation includes the time for filling the matrix, and the time for solving the matrix equation. To investigate the relationship between the matrix filling time and IASIZE, the wall time in the following simulations is separated into the wall time for matrix filling and the wall time for matrix solving.

The first example involves a wire antenna (only antenna 1 is present in Fig. 5.16) mounted on the airplane. The operating frequency is set to 1.6 GHz and this results in 79,431 unknowns for this problem. Two simulations are performed using one node with 8 processes and two nodes with 16 processes on CEM-7. The measured wall time is listed in Table 5.9.

To investigate the performance using a finer increment of the IASIZE, a problem of coupling between two monopole wire antennas mounted on the airplane (full model without symmetry) is chosen. This second example is shown in Figure 5.16, where both antenna 1 and antenna 2 are present simultaneously. The frequency is set to be 1.6 GHz and this results in the number of unknowns being 79,772. The length and radius of each wire antenna are $\lambda/4$ and $\lambda/200$, respectively. These simulations are also executed using one node with 8 processes and two nodes with 16 processes as done earlier, with the interval of IASIZE set to be 50 MB (megabytes) rather than 100 MB as in the first example. The observed execution time is listed in Table 5.10.

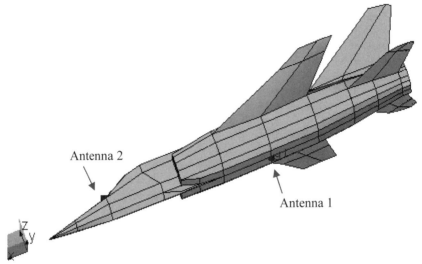

Figure 5.16. Coupling analysis for antennas mounted on a full-size aircraft.

TABLE 5.9. Wall Times When Using Different IASIZE on CEM-7

IASIZE (GB)	Number of CPUs	Time (minutes)		
		Filling	Solving	Wall
1.6	8	34	612	646
1.7	8	34	611	645
1.8	8	34	592	626
1.9	8	32	599	631
1.6	16	22	358	380
1.7	16	21	358	379
1.8	16	23	339	362
1.9	16	38	413	451

TABLE 5.10. Wall Times When Using Different IASIZE on CEM-7

IASIZE (GB)	Number of CPUs	Time (minutes)		
		Filling	Solving	Wall
1.65	8	35	630	665
1.70	8	34	630	664
1.75	8	34	612	646
1.80	8	34	613	647
1.85	8	34	616	650
1.65	16	22	376	398
1.70	16	21	377	398
1.75	16	23	357	380
1.80	16	24	358	382
1.85	16	43	462	505

The data obtained in Table 5.9 and 5.10 are plotted in Figures 5.17–5.19. The time for matrix filling is shown in Figure 5.17. The matrix filling time does not change significantly when executing the problem using one node with an increasing IASIZE. On two nodes the matrix filling time increases rapidly when the IASIZE is larger than 1.80 GB.

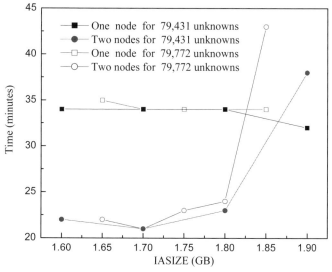

Figure 5.17. Wall time for matrix filling when using different values of IASIZE.

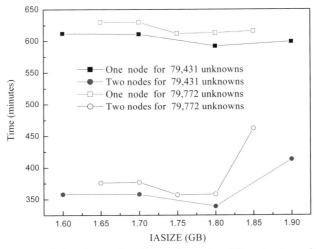

Figure 5.18. Wall time for matrix solving when using different values of IASIZE.

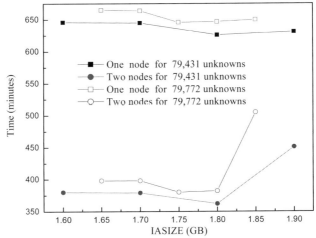

Figure 5.19. Wall time for the complete simulation process when using different values of IASIZE.

The matrix solving time, when using one node, also does not change significantly when the IASIZE is less than 1.80 GB. The matrix solving time, when using two nodes, increases rapidly when the IASIZE is greater than 1.80 GB, as shown in Figure 5.18.

Figure 5.19 plots the total time, which includes both the matrix filling and matrix solving time. The parallel out-of-core solver shows good performance when the IASIZE is less than 1.80 GB. The rapid degradation in the performance when the IASIZE is larger than a certain value is caused by use of the virtual memory of the computer, which should be avoided at all cost.

Physical memory is a finite resource on a computer system. Only a certain number of processes can fit into the physical memory at one time, although many more may actually be ready to be run or executed. When the computer needs to run programs that are bigger than its available physical memory, most modern operating systems use a technique called "swapping", in which chunks of memory are temporarily stored onto the hard disk while other data are moved into the physical memory space. Swapping is a process whereby a page of memory is copied to the preconfigured space, called "swap space", on the hard disk, to free up that page of the memory. Swap space is an area on the disk that temporarily holds the image of a process memory page. When the demand of the physical memory is sufficiently low, the images of the process memory are brought back into the physical memory from the swap area on the disk.

For example, the swap space in Linux is used when the amount of physical memory is full. Linux divides its physical RAM into chunks of memory called "pages". If the system needs more memory resources and the physical memory is full, inactive pages in memory are moved to the swap space.

Swapping and paging algorithms allow processes or portions of processes to move between physical memory and a mass storage device. This frees up the space in the physical memory. Having sufficient swap space enables the system to keep some physical memory free at all times. This type of memory management is often referred to as the *virtual memory* and allows the total number of processes to exceed the physical memory.

Swapping is necessary for two important reasons: (1) when the system requires more memory than the physically available one, the kernel swaps out less used pages and gives memory to the current application (process) that needs the memory immediately; and (2) a significant number of the pages used by an application during its startup phase may be used only for initialization and then never used again. The system can swap out those pages and free the memory for other applications or even for the disk cache.

To investigate the influence of the swap size on the performance of the parallel out-of-core solver, the aircraft is simulated on CEM-7 using different swap sizes and different values of IASIZE. The observed results are listed in Table 5.11.

TABLE 5.11. Quality of Performance as a Function of the Size of the Swap Space

Case	IASIZE (GB)	Swap Space Size (GB)	Time (minutes)
Case 1-1	1.7	1	332
Case 1-2	1.7	2	330
Case 1-3	1.7	8	332
Case 2-1	1.9	1	583
Case 2-2	1.9	2	507
Case 2-3	1.9	8	558

These results illustrate that when the in-core buffer is of proper size, the size of the swap space does not essentially influence the performance, as indicated by the lower curve in Figure 5.20. For the case where the size of the in-core buffer is too large to be considered, a relatively larger swap space may improve performance over a very small size of the swap space. Note that too large a swap space does not necessarily enhance the performance. For example, the best performance is seen in case 2-2, where the swap space is 2 GB. The performance is not as good for both case 2-1 with a 1 GB swap space and case 2-3 with an 8 GB swap space. An appropriately large swap space cannot compensate for the degradation in the performance caused by an improperly large size of the in-core buffer. Picking too large an in-core buffer memory size will cause nodes to swap, resulting in a significant drop in the performance.

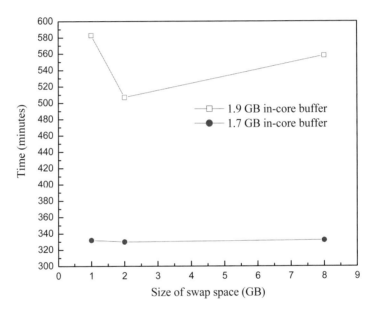

Figure 5.20. Wall time when using different size of the swap space on CEM-7.

While the size of the swap space can improve the performance in computers with a small amount of RAM, it should not be considered as a replacement of actual RAM. Swap space is located on hard drives, which have a longer access time than the physical memory. If a system has to rely too heavily on the virtual memory, a significant drop in performance will be noticed, since the read/write speed of a hard drive is much slower than that of the RAM. All the previous simulations lead to the same conclusion that the value of IASIZE has an influence on the performance of the code, in either a positive or in a negative way. Now the question arises as to how to find a reasonable IASIZE for the out-of-core solver to maintain a good degree of performance.

The answer will vary from case to case, but at least it can be seen from the examples above that there is an upper bound for the IASIZE. When the value of the IASIZE is chosen beyond that bound, the performance will degrade rapidly.

One way to find this upper bound value of IASIZE is to check the available RAM on the system. This task can be implemented by using the "top" command on a Linux system and the task manager when using a Windows operating system. Choosing IASIZE slightly less than the available amount of RAM will lead to good performance. As a general rule of thumb, approximately 90% of the free RAM is a good value for the size of the in-core buffer. Note that the amount of free RAM on a platform does not remain the same because the overhead of the system varies with time and/or other applications.

As an example, we use an EM64T Lenovo laptop (CEM-10) with a Windows Vista operating system as the computational platform to analyze antennas mounted on the aircraft involving 32,689 unknowns. This computer has 4 GB RAM and uses a dual-core CPU. Before starting the simulation, it may be observed that about 30% of the total physical RAM is used by the system, as shown in Figure 5.21. The optimum IASIZE obtained through benchmarks is around 1.3 GB as shown Figure 5.22, which is approximately 90% of the free RAM of the system for each process.

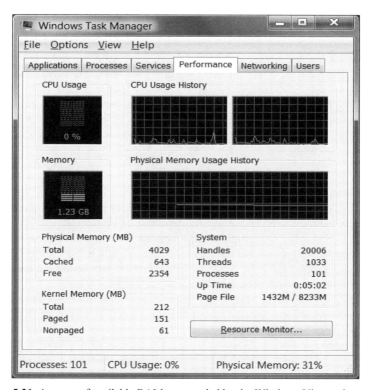

Figure 5.21. Amount of available RAM as recorded by the Windows Vista task manager.

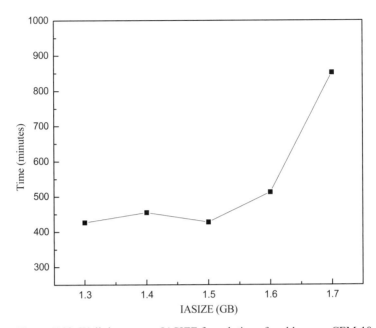

Figure 5.22. Wall time versus IASIZE for solution of problems on CEM-10.

Even if a good initial approximation of the IASIZE can be obtained by checking for the free RAM on the platform, the optimum value varies with operating systems and hardware. The optimum IASIZE is best determined by running a batch of simulations with different sizes for the in-core buffer. This section discussed IASIZE for platforms having 2 GB of RAM per core. In the next section, the optimum IASIZE for platforms with 4 GB of RAM per core is explored.

5.4.1.2 Case B: Available 4 GB of RAM/Core. In the first example, the code is used to calculate the isolation between two antennas mounted on the airplane model (no symmetry is employed for modeling) with 69,921 unknowns, as shown in Figure 5.16. The length of each antenna is 0.06 m, which is a quarter of a wavelength at 1.25 GHz, and the radius of each antenna is 0.0048 m.

Before running the code, the "top" command is used to find out the amount of free RAM on CEM-4. The free RAM of the system on each node (with two CPUs) is 7,441,568 kB, as shown in Figure 5.23 (marked by the box). Therefore each process (using one CPU) can only use less than 3.7 GB as the size of the in-core buffer to obtain good performance.

The measured time is listed in Table 5.12, and the total wall time is plotted in Figure 5.24. These problems were run twice to double-check the results. The results indicate that the code performs best with an IASIZE of 3.3 GB, which is approximately 90% of the amount of free RAM for each process.

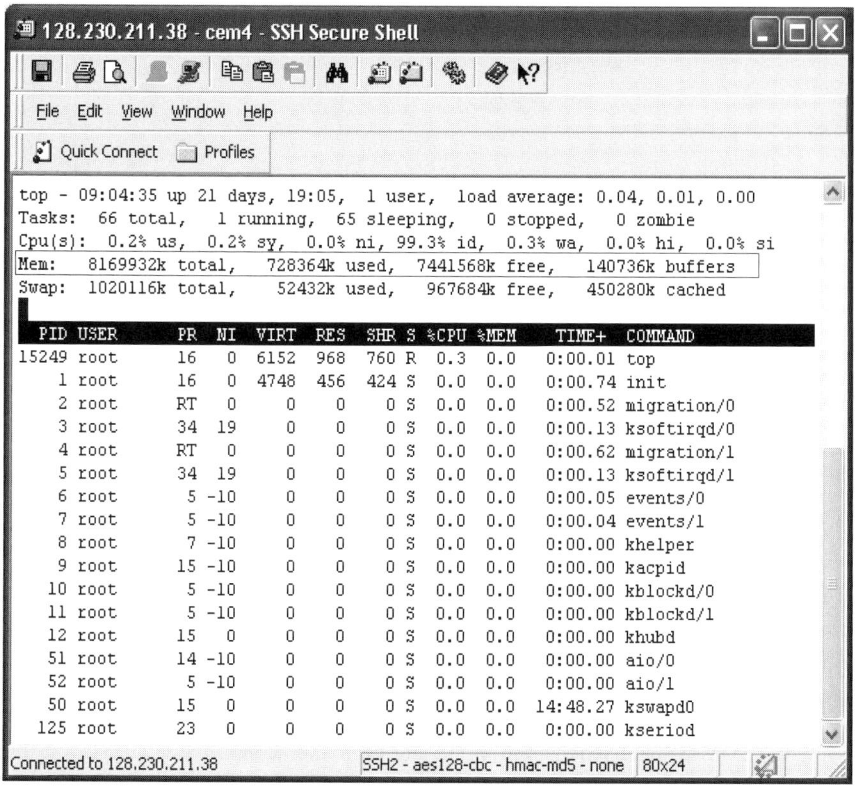

Figure 5.23. Display for the amount of free RAM when using the "top" command.

TABLE 5.12. Wall Time for the Complete Simulations

IASIZE (GB)	First Runtime (minutes)			Second Runtime (minutes)		
	Filling	Solving	Wall	Filling	Solving	Wall
3.0	33	379	412	31	380	411
3.1	33	380	413	31	380	411
3.2	32	394	426	32	394	426
3.3	32	366	398	31	365	396
3.4	31	369	400	31	370	401
3.5	32	368	400	31	368	399
3.6	40	387	427	38	384	422
3.7	46	419	465	45	417	462

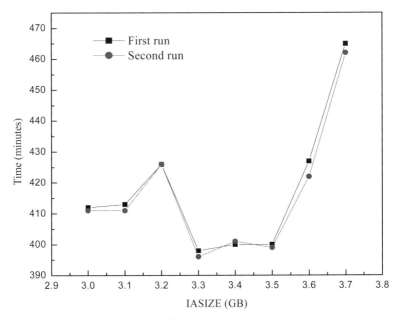

Figure 5.24. Wall time versus IASIZE on CEM-4.

For the second example, an EM64T workstation using a Windows operating system (CEM-8) is used as the platform to make a detailed analysis of the performance of the out-of-core solver. This computer has 32 GB RAM available and has two quad-core CPUs. The RCS of the airplane is calculated at 2.05 GHz using 60,653 unknowns, which requires 59 GB of hard disk storage. As shown in Figure 5.25, the RAM used by the operating system for this platform is 650 MB (this value is obtained from the Windows task manager), which is occupied before the execution of the parallel code starts. The measured time is plotted in Figure 5.26. and it shows that the optimum IASIZE for this application is 3.6 GB, which is 90% of the amount of free RAM for each process.

5.4.2   Optimizing IASIZE for a Parallel MoM Code Using RWG Basis Functions

All the previous analyses in Section 5.4.1 have been made for a parallel integral equation solver utilizing the higher-order basis functions. In this subsection, two different projects are simulated on CEM-7, using one node and two nodes, to check how the performance of a parallel code with the piecewise triangular patch (RWG) basis functions will be influenced by the choice of the IASIZE.

The first example deals with the scattering from an aircraft (introduced in Chapter 3) at 200 MHz. This project involves 79,986 unknowns and is run on one node of CEM-7. The observed wall times are listed in Table 5.13.

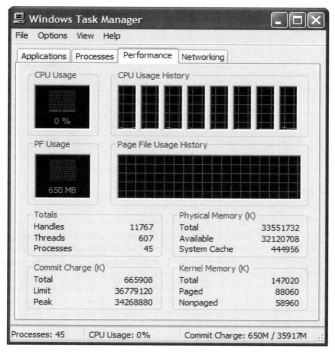

Figure 5.25. Amount of available RAM as seen on the Windows XP task manager.

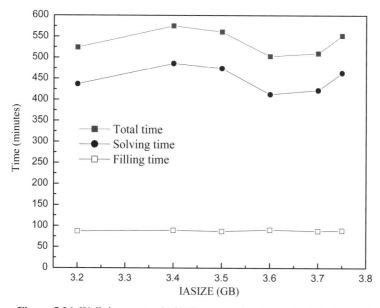

Figure 5.26. Wall time versus IASIZE on a workstation using Windows XP.

TABLE 5.13. Wall Times on One Node for Solving Problems with 79,986 Unknowns and Using Different IASIZE

IASIZE (GB)	Time (minutes)		
	Filling	Solving	Wall
1.5	27	702	729
1.6	28	692	720
1.7	36	680	716
1.8	70	853	923
1.9	74	842	916

The second example involves scattering from the same aircraft at a different frequency of 225 MHz. It involves 104,592 unknowns for this project and 2 nodes are used for the simulation. The observed wall times are listed in Table 5.14.

TABLE 5.14. Wall Times on Two Nodes for Solving Problems with 104,592 Unknowns and Using Different IASIZE

IASIZE (GB)	Time (minutes)		
	Filling	Solving	Wall
1.6	26	808	834
1.7	30	790	820
1.8	29	789	818
1.9	67	1107	1174

The wall times required to execute these two examples are plotted in Figures 5.27–5.29. The matrix filling time increases rapidly when the IASIZE is larger than 1.7 GB when using one node and 1.8 GB when using two nodes, as depicted in Figure 5.27. As shown in Figure 5.28, the matrix solving time increases in a similar fashion for each problem. The total time taken to solve the problem is plotted in Figure 5.29 and indicates that the optimum value of IASIZE for one node is 1.7 GB, while for the two node configuration it is 1.8 GB. These two examples together with the earlier discussions in this chapter illustrate that the IASIZE affects the performance of the parallel out-of-core integral equation solver, regardless of what kind of basis function is used.

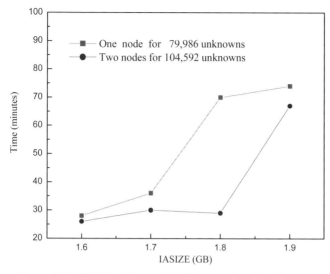

Figure 5.27. Wall time for matrix filling as a function of IASIZE.

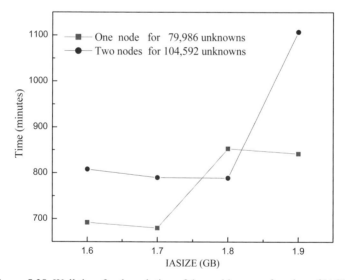

Figure 5.28. Wall time for the solution of the problem as a function of IASIZE.

Note that the optimum value of the IASIZE will vary slightly for the same platform when using different basis functions, or when running different projects with the same basis function. Also, the optimum value of the IASIZE may not be the same for different computational platforms. It is always advisable for the user to choose the proper IASIZE to avoid using virtual memory of the computer.

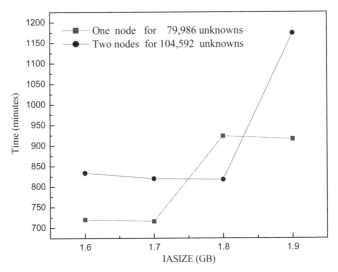

Figure 5.29. Total wall time as a function of IASIZE.

5.4.3   Influence of Physical RAM Size on Performance

The previous discussion shows that by properly choosing the size of the in-core buffer the performance can be optimized. In this context, it is important to address the following auxiliary issue: Does more available RAM necessarily benefit the application of the parallel out-of-core solver?

To address this question, scattering from the aircraft model using 93,091 unknowns, described in Section 4.7.2.3 of Chapter 4, is executed on four nodes of CEM-4 with 2 GB RAM/core and 4 GB RAM/core. This computer cluster has 8 GB RAM per node (with two single-core CPUs). 4 GB of RAM can be removed from each of the four nodes to yield 4 GB RAM/node. This configuration enables CEM-4 to simulate the cases using 2 GB or 4 GB of physical RAM per CPU. The observed results are listed in Table 5.15(*a*) and 5.15(*b*) for 2 GB and 4 GB of RAM size, respectively.

TABLE 5.15(*a*). Results on CEM-4 Using 2 GB RAM per Core and Different IASIZE

RAM per Core (GB)	IASIZE (GB)	Total Cores	Total Wall Time (minutes)	Wall Time for LU (seconds)
2	1.5	8	1071	53,736
2	1.6	8	1060	53,077
2	1.7	8	1076	53,790
2	1.8	8	1117	55,735

TABLE 5.15(*b*). Results on CEM-4 Using 4 GB RAM per Core and Different IASIZE

RAM per Core (GB)	IASIZE (GB)	Total Cores	Total Wall Time (minutes)	Wall Time for LU (seconds)
4	1.5	8	1063	53,428
4	1.6	8	1053	52,765
4	1.7	8	1042	52,286
4	1.8	8	1042	52,454
4	1.9	8	1029	51,662
4	2.0	8	1016	50,999
4	2.1	8	1017	51,071
4	2.2	8	1017	51,237
4	2.3	8	1007	50,531
4	2.4	8	1006	50,510
4	2.5	8	1006	50,659
4	2.6	8	993	49,879
4	2.7	8	993	49,922
4	2.8	8	995	50,024
4	2.9	8	996	50,169
4	3.0	8	979	49,188
4	3.1	8	983	49,381
4	3.2	8	986	49,519
4	3.3	8	989	49,636
4	3.4	8	989	49,858
4	3.5	8	990	49,546
4	3.6	8	1026	50,882
4	3.7	8	1088	53,709
4	3.8	8	1140	55,072

Figure 5.30 plots the total wall time as a function of the size of the in-core buffer, i.e. the IASIZE, which specifies how much of the RAM is used for each process. It illustrates that more time is consumed at the very low end, then there is a "sweet spot" for some intermediate values, and then there is a degradation in performance as the value approaches the amount of the available RAM. The curve typically has a signature shape with a plateau of acceptable values before the sharp degradation resulting in longer runtimes. It is seen that for several values of IASIZE toward the end region, using more RAM as the in-core buffer does not always improve the performance of the out-of-core solver.

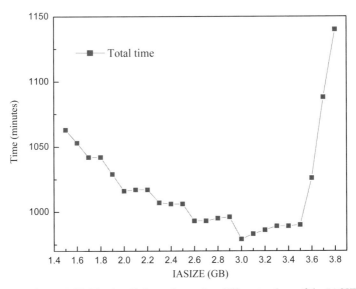

Figure 5.30. Total wall time when using different values of the IASIZE.

Next, the total time used for computation by the different computer nodes are compared, when using 2 GB RAM/core and 4 GB RAM/core. When a reasonable value of the IASIZE is chosen for the 2 GB RAM/core case, the total time obtained by using 2 GB of physical RAM/core is similar to the total time obtained using 4 GB of physical RAM/core with the same IASIZE, as demonstrated in Figure 5.31. There is a difference of less than 10 minutes between the two cases. The total time taken to solve the complete problem is more than 1000 minutes. Note that if sufficient caches are available for the 4 GB RAM/core case during the I/O operation, then IASIZE can be of the same value as in the 2 GB RAM/core case. This indicates that the I/O operation involved in the parallel out-of-core solver is a very small portion of the total time used for the complete solution for this problem.

The shortest time for the simulation using 2 GB of physical RAM per core is 1060 minutes. The shortest time for the simulation using 4 GB of physical RAM per core is 979 minutes. The difference between them is 81 minutes. This indicates that the performance can be improved by approximately 8%, if the computer's physical RAM is doubled and the code uses the optimum amount of that RAM as the in-core buffer. Note that these tests have been performed on a machine using a single-core CPU, a single hard disk (without RAID controller) and a Gigabit Ethernet. Therefore, the time difference between 2 GB/core and 4 GB/core can be even more insignificant for a general case.

These benchmarks serve to support the conclusion that more RAM can benefit the parallel out-of-core applications. However, the advantage realized will cost more with the purchase of more RAM, which may not be very cost-effective when compared to the similar storage requirements on a high-speed SAS hard drive.

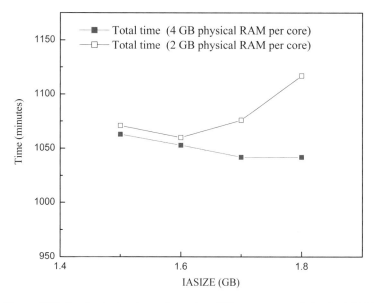

Figure 5.31. Total solution time when using different amount of physical RAM.

5.5   RELATIONSHIP BETWEEN SHAPE OF THE PROCESS GRID AND IN-CORE BUFFER SIZE

The discussions in Sections 5.3 and 5.4 indicate that the shape of the process grid and the size of the in-core buffer influence the performance of the parallel out-of-core integral equation solver significantly. Here we explore the relationship between these two important parameters.

If one of the parameters cannot be fixed during the optimization of the other, the best performance can never be obtained by just running several benchmarks. Fortunately, the way that the IASIZE parameter influences the performance of the code remains almost unchanged when using different shapes of the process grids. The following examples are presented to investigate the relationship between the shape of the process-grid and the size of the in-core buffer.

Three possible process grids, 1×16, 2×8, and 4×4, are first investigated using 16 processes on CEM-4. The wall time for each process grid is observed as a function of the IASIZE parameter changing from 3.0 GB to 3.9 GB. The results are plotted in Figure 5.32. Since a project using 69,921 unknowns is a relatively small problem for a parallel out-of-core solver, a simulation is also made for a larger project with 128,029 unknowns. The results are plotted in Figures 5.33 and 5.34. From these plots, it is seen that the shape of the process grid influences the performance when the problem size is large.

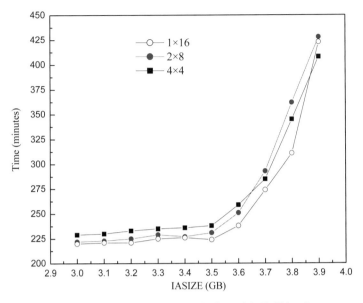

Figure 5.32. Total solution time when dealing with 69,921 unknowns.

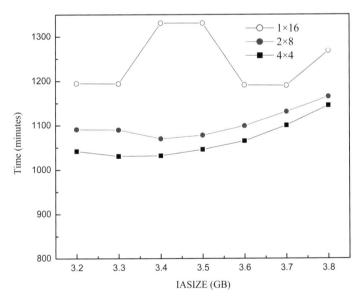

Figure 5.33. Wall time required for the LU decomposition on CEM-4 when using 128,029 unknowns.

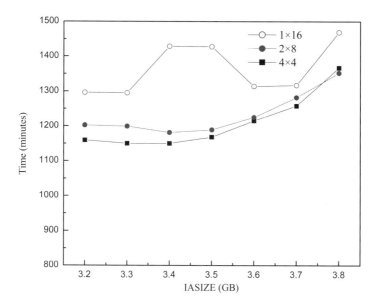

Figure 5.34. Total time taken to solve the problem on CEM-4 when using 128,029 unknowns.

There appears to be no change in the performance of the code when different process grids are used for the solution of problems with a smaller number of unknowns. For a larger problem, the influence of the process grid on the performance is very noticeable. For a given process grid, the performance deteriorates rapidly when the IASIZE exceeds the upper bound as discussed previously. When using different process grids, the IASIZE will vary slightly. The smallest value of the proper IASIZE obtained for different process grids can be considered as the upper bound. The plots also indicate that for larger sized problems, a square process grid is generally the best shape even though for smaller projects it may not bring any benefit.

The choice of three possible process grids is also investigated using another 16 processes on CEM-9. The wall time for obtaining the solution for each process grid is measured as a function of different values of the IASIZE. Simulations with 82,652 unknowns and 97,539 unknowns show that the best process grid is square-shaped, and the best IASIZE is around 2.4 GB (90% of the system free RAM for each process), as shown in Figures 5.35 and 5.36. In Figure 5.36, the size of the IASIZE was varied thrice for the 1×16 process grid. Poor performance indicated that no further testing should continue. Note that for smaller problems, using a square process grid on CEM-9 does not bring much benefit. These conclusions are similar to that obtained on CEM-4.

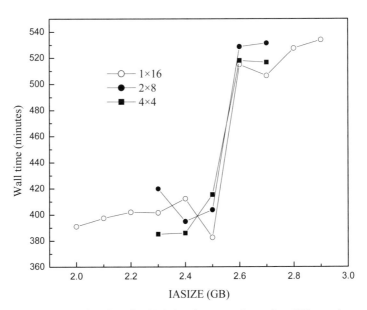

Figure 5.35. Total solution time for 82,652 unknowns when using different shapes of the process grid and the IASIZE on CEM-9.

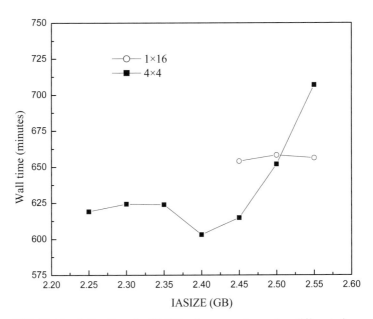

Figure 5.36. Total solution time for 97,539 unknowns when using different shapes of the process grid and the IASIZE on CEM-9.

5.6   OVERALL PERFORMANCE OF A PARALLEL OUT-OF-CORE SOLVER ON HPC CLUSTERS

It has now been established how to properly allocate memory and choose a suitable shape of the process grid for a parallel out-of-core solver. In this section, the performance of the parallel out-of-core solver on large high performance computing (HPC) clusters is investigated.

The performance of the out-of-core integral equation solver is observed on different HPC platforms. The same model airplane introduced in Chapter 4 is used for the analysis. Table 5.16 summarizes the 13 cases simulated and the wall times achieved for these problems. The corresponding process grid is listed in Table 5.16 for each case. The value of block size is taken to be 104 for all the cases, except for the last row where the value of block size is chosen to be 128. The value of IASIZE is selected as 1.7 GB.

The computation of the RCS for the aircraft at 2.8 GHz is completed in 164 minutes when using 80 cores of Cluster-1. The computational time is reduced to 72 minutes when using 160 cores of Cluster-1. The electrical dimensions of the airplane at this frequency are 108.3λ in length, 65.3λ in width, and 27.3λ in height.

TABLE 5.16. Summary of the Test Results Obtained on Six Different Clusters

Unknowns	Matrix Storage	Platform	Process Grid	Used Cores	Wall Time
110,994	197 GB	Cluster-1	8×10	80	164 minutes
		Cluster-1	10×16	160	72 minutes
		Cluster-2	8×10	80	171 minutes
		Cluster-2	10×16	160	85 minutes
		Cluster-3	8×10	80	213 minutes
		Cluster-3	10×16	160	103 minutes
232,705	866 GB	Cluster-1	10×16	160	628 minutes
		Cluster-2	16×16	256	418 minutes
		Cluster-4	10×16	160	707 minutes
		Cluster-5	10×16	160	706 minutes
		Cluster-6	12×20	240	399 minutes
500,469	4 TB	Cluster-5	10×16	160	125.65 hours
500,479	4 TB	Cluster-2	16×32	512	27.03 hours

The speedup in the solution time obtained by double the cores used on Cluster-1, Cluster-2 and Cluster-3 is better than the theoretically expected value of 2 (twice faster) as illustrated in Table 5.16. This is because this simulation needs only approximately 197 GB storage for 110,994 unknowns. When 160 cores are used, the system can provide 320 GB of RAM, which is more than required to solve the problem in-core. The out-of-core code is executed in an in-core mode in this case. When the simulation results from Cluster-2 and Cluster-3 are compared, it is seen that Cluster-2 is faster than Cluster-3. This is due to the different configurations of the CPU and the different type of network connections.

The solution for the RCS of the airplane at 4.1 GHz is completed in less than 12 hours using 20 nodes (160 cores) of Cluster-4. The time is reduced to less than 7 hours by using 60 nodes (240 cores) of Cluster-6 or 32 nodes (256 cores) of Cluster-2. The electrical dimensions of the airplane at this frequency are 158.5λ in length, 95.7λ in width, and 39.9λ in height.

Using the wall time for 110,994 unknowns as the benchmark, the estimated time for the solution of a 232,705 unknowns problem on Cluster-1 is calculated to be $[(232,705/110,994)^3 \times 72$ minutes$]$ = 664 minutes. The measured time is 628 minutes, which is 36 minutes shorter than the estimated one. This indicates that very good scalability can be provided by this cluster. Using the wall time for 110,994 unknowns as the benchmark, the estimated time for solving 232,705 unknowns on Cluster-2 is calculated to be $[(232,705/110,994)^3 \times (160/256) \times 85$ minutes$]$ = 490 minutes. The measured time is 418 minutes, which is 72 minutes shorter than estimated. This means very good scalability can also be obtained on Cluster-2.

Figure 5.37 plots (from top to bottom) the wall time, user time, the MPI usage time, and the system (Sys) time for the simulation executed on Cluster-4. As shown in Figure 5.37, the wall time taken to solve the problem for all the processes is essentially the same, as shown by the straight line at the top, since the slowest process determines the nature of the performance. The user, MPI, and Sys (system) times vary slightly with different processes because of the very small load unbalance. This figure shows that excellent load balancing of the TIDES software [6] for solving these classes of problems is achievable. It also indicates that the MPI usage time accounts for only 13% of the total time.

The analysis of scattering from the airplane at 6.15 GHz was completed in approximately 5 days using 40 nodes (160 cores total) of Cluster-5, which generates a matrix equation with over half a million unknowns and requires over 4 TB (terabytes) of storage. The electrical dimensions of the airplane at this frequency are: 237.8λ in length, 143.5λ in width, and 59.9λ in height. Using the wall time for 232,705 unknowns as the benchmark, the estimated time for the solution of 500,469 unknowns is calculated to be $[(500,469/232,705)^3 \times 706$ minutes$]$ = 117 hours. The measured time is 126 hours and is greater than the estimated one. The difference between the estimated and the measured time is 9 hours, which indicates that when the problem size increases from 866 GB to 4000 GB, the performance of the out-of-core code degrades by roughly 8%.

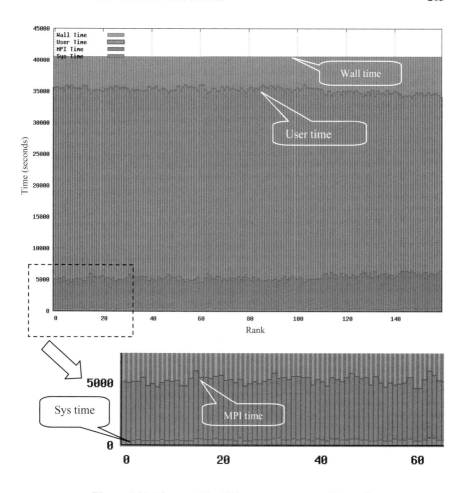

Figure 5.37. Time used by different processes on Cluster-4.

The analysis of scattering from the airplane at 6.15 GHz was completed in approximately 27.03 hours when using 64 nodes (512 cores total) of Cluster-2, which generates a matrix equation with over half a million unknowns and requires over 4 TB of storage. Note that the number of unknowns is different from the previous example executed on Cluster-5, since several modifications have been made at the nose of the aircraft model.

The storage space required on Cluster-2 is provided by the local hard disks installed on all the compute nodes rather than on an external file storage system as for Cluster-5. The load balance is shown in Figure 5.38, which plots (from top to bottom) the wall time, user time, the MPI time, and the system (Sys) time for the simulation executed on Cluster-2.

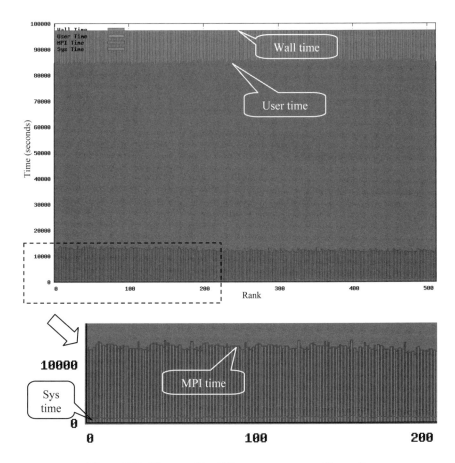

Figure 5.38. Time used by different processes on Cluster-2.

As shown in Figure 5.38, the wall time for all the processes is essentially the same, as shown by the straight line at the top, since the slowest process determines the performance. The user, MPI and Sys times vary slightly with the different processes because of the very small load imbalance. This figure, again, shows that excellent load balancing can be achieved when analyzing challenging problems using the TIDES software [6]. It also indicates that the MPI time accounts for only 13% of the total time in this case.

The results presented in this section indicate that the parallel codes have very good scalability on different types of clusters. The total solution time had been computed when the computer system was running with the optimum choice of the relevant parameters, such as the shape of the process grid and the in-core buffer size allocated for each process.

5.7   CONCLUSION

In this chapter, the factors that influence the performance of parallel in-core and parallel out-of-core solvers are investigated with emphasis on the latter. Block size, shape of the process grid, and size of the in-core buffer are the most important parameters that influence performance. It is recommended that the block size should be the optimum value obtained from various simulations or the value as suggested by the software vendors. When analyzing electrically large problems, setting the shape of the process grid as square as possible enables the parallel out-of-core solver to be most efficient. Using approximately 90% of the free RAM (rather than the complete RAM) as the in-core buffer size will ensure optimum performance. These are some of the guidelines that one can use to optimize the performance of a parallel solver.

REFERENCES

[1]   Ganglia Monitoring System, "What is Ganglia?", Planet Lab. Available at: http://ganglia.sourceforge.net/. Accessed Sept. 2008.

[2]   Intel®Software Network, *Intel®Math Kernel Library 10.0 — ScaLAPACK*. Available at:http://www.intel.com/cd/software/products/asmo-na/eng/266861.htm. Accessed Sept. 2008.

[3]   F. Desprez, S. Domas, and B. Tourancheau, "Optimization of an LU Factorization Routing Using Communication/Computation Overlap," *INRIA*, No. 3094, Feb. 1997. Available at: ftp://ftp.inria.fr/INRIA/publication/publi-pdf/RR/RR-3094.pdf. Accessed Sept. 2008.

[4]   L. S. Blackford, J. Choi, A. Cleary, J. Demmel, and I. Dhillon, "ScaLAPACK: A Portable Linear Algebra Library for Distributed Memory Computers — Design Issues and Performance," *Proceedings of the 1996 ACM/IEEE Conference on Supercomputing (SC'96)*, 1996.

[5]   Netlib Repository at UTK and ORNL, *Choice of Grid Size*. Available at: http://www.netlib.org/utk/papers/scalapack/node20.html. Accessed Sept. 2008.

[6]   Y. Zhang, J. Porter, M. Taylor, and T. K. Sarkar, "Solving Challenging Electromagnetic Problems Using MoM and a Parallel Out-of-Core Solver on High Performance Clusters," *2008 IEEE AP-S & USNC/URSI Symposium*, San Diego, CA, July 2008.

6

REFINEMENT OF THE SOLUTION USING THE ITERATIVE CONJUGATE GRADIENT METHOD

6.0 SUMMARY

In this chapter two different variants of the conjugate gradient (CG) method are presented for the solution of a matrix equation. One bottleneck of the CG method is the large computational burden of performing the matrix–vector products involved in the computations. To accelerate the CG method, a parallel computation framework is employed.

The other goal of this chapter is to provide a method of checking and refining the results obtained from the parallel solution of the LU decomposition. By following the conventional solver with the CG method, accuracy of the solution of the large systems to a prespecified degree can be guaranteed. For this reason, a parallel LU-CG scheme is also presented.

6.1 DEVELOPMENT OF THE CONJUGATE GRADIENT METHOD

A general approach is developed to iteratively solve the matrix equation [1–4]

$$[A][X]=[B],\tag{6.1}$$

where $[A]$ is a square matrix of $N \times N$ known elements, $[X]$ is a column matrix of $N \times 1$ unknowns to be solved for, and $[B]$ is a column matrix of $N \times 1$ known excitations. Using an iterative method, instead of solving the equation $[A][X] = [B]$ directly, a functional $F([X])$ through

$$F([X]) = \langle [S][R], [R] \rangle = [R]^{H}[S][R]\tag{6.2}$$

is minimized, where ^{H} denotes the conjugate transpose of a matrix and $[S]$ is a Hermitian positive definite operator, which is assumed to be known. The residual $[R]$ is given by

$$[R] = [B] - [A][X].$$ (6.3)

The inner product and the norms are defined as

$$\langle [R], [R] \rangle = [R]^{H}[R] = \| [R] \|^{2}.$$ (6.4)

In an iterative method the objective is to start from an initial guess $[X]_i$ and proceed along the direction $[P]_i$ a distance t_i to arrive at the updated point $[X]_{i+1}$. The distance t_i is selected in such a way that $F([X]_{i+1})$ is minimized. Hence

$$[X]_{i+1} = [X]_i + t_i [P]_i.$$ (6.5)

Since in this case $F([X])$ is a quadratic function, it has only one stationary point and that point is the unique global minimum that is sought. There are no other local minima at all. So t_i is selected such that

$$
\begin{aligned}
F([X]_{i+1}) &= \langle [S][R]_{i+1}, [R]_{i+1} \rangle \\
&= \langle [S]([R]_i - t_i [A][P]_i), [R]_i - t_i [A][P]_i \rangle
\end{aligned}
$$ (6.6)

is minimized. It is required that

$$\frac{\partial F([X]_{i+1})}{\partial a_i} = 0, \qquad \frac{\partial F([X]_{i+1})}{\partial b_i} = 0,$$ (6.7)

where a_i and b_i denote the real and imaginary parts of the length of the search directions t_i, respectively. Note that

$$
\begin{aligned}
F([X]_{i+1}) &= \langle [R]_{i+1}, [S][R]_{i+1} \rangle \\
&= \langle [R]_i, [S][R]_i \rangle - t_i \langle [A][P]_i, [S][R]_i \rangle \\
&\quad - t_i^* \langle [R]_i, [S][A][P]_i \rangle + |t_i|^2 \langle [A][P]_i, [S][A][P]_i \rangle,
\end{aligned}
$$ (6.8)

where * denotes the conjugate of a complex number and $|\ |$ denotes the magnitude of a complex number. Application of (6.7) yields

$$t_i = a_i + jb_i = \frac{\langle [R]_i , [S][A][P]_i \rangle}{\langle [A][P]_i , [S][A][P]_i \rangle}. \tag{6.9}$$

It is clear from Eqs. (6.3) and (6.5) that the residuals can be generated recursively as follows

$$[R]_{i+1} = [B] - [A][X]_{i+1} = [R]_i - t_i [A][P]_i, \tag{6.10}$$

and it is seen from Eqs. (6.9) and (6.10) that

$$\langle [R]_{i+1} , [S][A][P]_i \rangle = 0. \tag{6.11}$$

The search directions $[P]_i$ are also generated recursively from the gradient of $F([X])$ as

$$[P]_{i+1} = s_{i+1} \left([K][A]^H [S][R]_{i+1} + q_i [P]_i \right), \tag{6.12}$$

where $[A]^H$ is the adjoint operator (conjugate transpose for a matrix) and s_{i+1} is a scale factor chosen to make the vector $[P]_{i+1}$ a unit vector. Scaling can be quite effective for a certain class of problems. The matrix $[K]$ denotes another Hermitian positive definite operator to be specified. The constant q_i is chosen for the conjugate gradient method such that $[A][P]_i$ are orthogonal with respect to the following inner product

$$\langle [A][P]_i , [S][A][P]_j \rangle = 0. \text{ (for } i \neq j.) \tag{6.13}$$

The condition of Eq. (6.13) guarantees that the method will converge in a finite number of steps, barring roundoff error accumulated through the computation of $[A][P]$ or $[A]^H[R]$ and the numerical roundoff error propagated through the recursive computations of $[R]_i$ and $[P]_i$. Past experience has shown that numerical errors accumulated through recursive computations of $[R]_{i+1}$ and $[P]_{i+1}$ are negligible. There are also small numerical errors associated with the evaluation of $[A][P]$ and $[A]^H[R]$.

By solving for q_i using Eqs. (6.12) and (6.13), q_i is obtained as

$$q_i = -\frac{\langle [A][K][A]^H [S][R]_{i+1} , [S][A][P]_i \rangle}{\langle [A][P]_i , [S][A][P]_i \rangle}. \tag{6.14}$$

By solving for t_i using Eqs. (6.10)–(6.12), one obtains

$$t_i = \frac{\left\langle [R]_i, s_i [S][A][K][A]^H [S][R]_i \right\rangle}{\left\langle [A][P]_i, [S][A][P]_i \right\rangle}. \tag{6.15}$$

It can be shown that

$$\left\langle [R]_{i+1}, [S][A][K][A]^H [S][R]_i \right\rangle = 0. \tag{6.16}$$

Equations (6.15) and (6.16) describe a class of conjugate gradient methods. This class is generated by how one selects the possible Hermitian definite operators $[S]$ and $[K]$ and the scale factor s_{i+1}. It is easy to see that the functional $F([X])$ (the error function) decreases after each iteration. This is because

$$\begin{aligned} F\left([X]_i\right) - F\left([X]_{i+1}\right) &= \frac{\left| \left\langle [R]_i, [S][A][P]_i \right\rangle \right|^2}{\left\langle [A][P]_i, [S][A][P]_i \right\rangle} \\ &= |t_i|^2 \left\langle [A][P]_i, [S][A][P]_i \right\rangle \end{aligned} \tag{6.17}$$

is always positive since $[S]$ is a positive definite operator.

Two special cases of the conjugate gradient method are now considered.

Case 1: RCG (residual of CG). Choose

$$[S] = [K] = [I] \tag{6.18}$$

and

$$s_{i+1} = 1 \quad \text{for all } i. \tag{6.19}$$

This results in

$$F\left([X]_i\right) = \left\| [R]_i \right\|^2 \tag{6.20}$$

$$[X]_{i+1} = [X]_i + t_i [P]_i \tag{6.21}$$

$$[R]_{i+1} = [R]_i - t_i [A][P]_i \tag{6.22}$$

$$t_i = \frac{\left\| [A]^H [R]_i \right\|^2}{\left\| [A][P]_i \right\|^2} \tag{6.23}$$

$$[P]_{i+1} = [A]^H [R]_{i+1} + q_i [P]_i \tag{6.24}$$

with

$$q_i = \frac{\left\| [A]^H [R]_{i+1} \right\|^2}{\left\| [A]^H [R]_i \right\|^2} \tag{6.25}$$

$$[P]_0 = [A]^H [R]_0 = [A]^H \left([B] - [A][X]_0 \right). \tag{6.26}$$

For case 1, the residuals are minimized at each iteration. The RCG method yields slow convergence when $\left\| [A]^H [R] \right\|$ is small even though $\left\| [R] \right\|$ is large. A different way to approach the problem is to use a preconditioning matrix that derives the search directions from vectors $[K][A]^H [R]_i$ rather than from $[A]^H [R]_i$, as given by the algorithm RPCG to be described next.

Case 2: RPCG (residual of preconditioned CG). Choose

$$[S] = [I] \tag{6.27}$$

$$[K] = \text{ a Hermitian matrix} \tag{6.28}$$

(a good choice for $[K]$ is the square of the inverse diagonal elements of $[A]^H$) and

$$s_{i+1} = 1 \quad \text{for all } i. \tag{6.29}$$

This results in

$$F\left([X]_i \right) = \left\| [R]_i \right\|^2 \tag{6.30}$$

$$[X]_{i+1} = [X]_i + t_i [P]_i \tag{6.31}$$

$$[R]_{i+1} = [R]_i - t_i [A][P]_i \tag{6.32}$$

$$t_i = \frac{\left\langle [A]^H [R]_i, [K][A]^H [R]_i \right\rangle}{\left\| [A][P]_i \right\|^2} \tag{6.33}$$

$$[P]_{i+1} = [K][A]^H [R]_{i+1} + q_i [P]_i \quad \text{with } q_0 = 0. \tag{6.34}$$

Here, $[K]$ is a preconditioning operator, with

$$q_i = \frac{\left\langle [A]^H [R]_{i+1}, [K][A]^H [R]_{i+1} \right\rangle}{\left\langle [A]^H [R]_i, [K][A]^H [R]_i \right\rangle} \tag{6.35}$$

For case 2, the residuals are minimized at each iteration, and a preconditioning operator $[K]$ is introduced. If $[K]$ can be chosen judiciously, the pitfalls of RCG can be rectified. If a good choice for $[K]$ is not available, an alternative is to use the biconjugate or the augmented conjugate gradient algorithms [1,4].

6.2   THE ITERATIVE SOLUTION OF A MATRIX EQUATION

Here, a numerical example which uses the RCG algorithm to solve the matrix equation generated from the application of a MoM methodology is presented. The general procedure to solve the linear equation $[A][X] = [B]$ using the RCG method is displayed in Figure 6.1.

Initial step:

$$\text{Guess } [X]_0$$

$$[R]_0 = [B] - [A][X]_0$$

$$[P]_1 = [A]^H [R]_0$$

$$error_0 = \frac{\left\| [R]_0 \right\|}{\left\| B \right\|}$$

While ($error_k > error$), iterate $k=1, 2\ldots$

$$\alpha_k = \frac{\left\| [A]^H [R]_{k-1} \right\|^2}{\left\| [A][P]_k \right\|^2}$$

$$[X]_k = [X]_{k-1} + \alpha_k [P]_k$$

$$[R]_k = [R]_{k-1} - \alpha_k [A][P]_k$$

$$error_k = \frac{\left\| [R]_k \right\|}{\left\| B \right\|}$$

$$\beta_n = \frac{\left\| [A]^H [R]_k \right\|^2}{\left\| [A]^H [R]_{k-1} \right\|^2}$$

$$[P]_{k+1} = [A]^H [R]_k + \beta_k [P]_k$$

Figure 6.1. The flowchart for the iterative solution using the conjugate gradient method.

As shown in Figure 6.1, an initial guess $[X] = [X]_0$ is chosen and the residual error $[R]_0 = [B] - [A][X]_0$ is computed. The initial search direction vector is given by $[P]_1 = [A]^H [R]_0$. The estimated updates for the relevant parameters are obtained using iterations. The iteration is halted when the residual error $error_k = \| [R]_k \| / \| B \|$ falls below a predetermined value, but in no case will it require more iterations than the dimension of [A] in the total absence of roundoff errors.

The combined cube-and-sphere PEC model introduced in Section 3.9.1.2 (see Chapter 3) is used as an example to illustrate the application of this methodology. The magnitude of the residual as a function of the number of iterations for the RCG method is plotted in Figure 6.2. A total of 769 iterations are needed to guarantee a residual value of less than 10^{-3}.

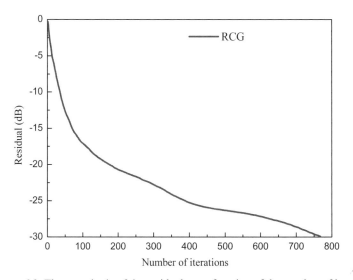

Figure 6.2. The magnitude of the residuals as a function of the number of iterations.

6.3 PARALLEL IMPLEMENTATION OF THE CG ALGORITHM

It can be seen that in the CG algorithms, the computational bottleneck lies in carrying out a matrix–vector product (MVP), such as $[A][P]$ and $[A]^H[R]$.

This computational bottleneck can be eliminated by using preconditioning techniques to accelerate the rate of convergence, and/or fast Fourier transform (FFT) to carry out the matrix–vector multiplication [5]. Previous research [6–8] has provided that a highly efficient preconditioner can be obtained, which takes the near field interactions into account when constructing the preconditioning matrix. Using this preconditioner, it is possible to speed up the convergence of the CG method by one order of magnitude. Typical examples of MoM methods using global sinusoidal basis functions, rooftop basis functions, or the RWG basis functions can be found in the literature [7].

The computational bottleneck can also be eliminated by using parallel computation to carry out the matrix–vector multiplication. This significantly enhances the performance of the CG algorithms. Designing a parallel CG code is quite straightforward. By rewriting the code available in the literature [9] for the matrix–vector multiplication using parallel computational methodology, an efficient parallel computer code can easily be generated.

The performance of the parallel RCG method is tested on CEM-4 with the calculation of the RCS of a sphere at 1.98 GHz, using 21,936 unknowns as discussed in Chapter 4. The measured times using different numbers of nodes (two processes per node) are listed in Table 6.1. In this simulation, the initial starting value of the solution in the RCG method is taken to be zero. A very good speedup is obtained on CEM-4 for this simulation, as shown in Figure 6.3. Note that each process has a small load when 18 processes are used; therefore, the speedup can be even better when a larger load is distributed over all the processors.

TABLE 6.1. Computational Times Required by the Parallel RCG Method on CEM-4

Number of Nodes	1	2	3	4	5	6	7	8	9
Number of Processes	2	4	6	8	10	12	14	16	18
Wall time (minutes)	267	136	94	72	58	49	43	40	35

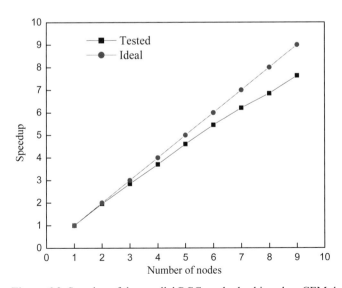

Figure 6.3. Speedup of the parallel RCG method achieved on CEM-4.

In this example, the total solution time taken by the CG method is longer than the time used by the LU decomposition method as described in Chapter 4. The residual error for different numbers of iterations is plotted in Figure 6.4. Also observe that the rate of convergence is slow. Readers who are interested in accelerating this method can try to implement the preconditioning method presented in the literature [6–8].

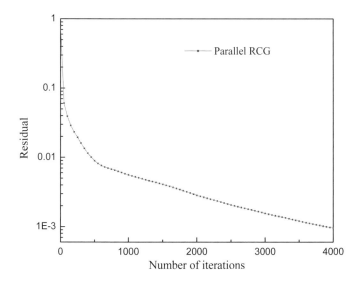

Figure 6.4. Residual error versus number of iterations for the solution related to a sphere.

6.4   A PARALLEL COMBINED LU-CG SCHEME TO REFINE THE LU SOLUTION

After the LU decomposition is employed to solve the matrix equation, it is beneficial to assess the accuracy of the final solution to ensure a high degree of confidence in the solution of very large problems. A combined LU-CG scheme is applied for this purpose.

After the LU decomposition, the original impedance matrix is overwritten with the LU factors. Therefore, it is very important to save the original impedance matrix to the hard disk before performing LU factorization (here we assume that the matrix equation obtained from MoM may use up the entire RAM). The method used to manage the I/O operations in the parallel out-of-core solver can also be used to read/write the impedance matrix from/to the hard disk.

With these considerations in mind, a flowchart of the combined LU-CG scheme is presented in Figure 6.5 for a parallel in-core solver. The steps with notation "may skip" are not required when the right-hand side (RHS) of the matrix equation can be handled all at once (when RHS is a vector or a matrix that can be fit into RAM along with impedance matrix).

1. Write impedance matrix to hard disk

2. Perform LU decomposition (only once)

3. Write results of LU decomposition to hard disk (may skip)

4. Read results ([L] and [U] submatrices) of the LU decomposition from hard disk (may skip)

5. Solve [X] using [L] and [U] factors of [A]

6. Read the impedance matrix from hard disk

7. Execute the parallel CG method (using the solution from the LU decomposition as an initial guess)

8. Postprocessing (using [X])

9. Go to step 4, until the solutions for all the excitations are calculated (may skip)

10. Delete the unnecessary files from hard disk

Figure 6.5. The parallel combined LU-CG scheme.

It is required that the data be written to the hard disk at least twice, and be read from it several times (depending on the number of independent excitations at which the problem needs to be solved). The scheme in Figure 6.5 can easily be extended to the parallel out-of-core solver, with the exception that the amount of I/O operation will be increased. Even though it is meaningful to check the residuals of the solution of the CG method to guarantee the accuracy, fortunately, the result obtained using the LU method with partial pivoting is quite accurate and so the CG method needs only a few iterations (in most cases one only).

As an example, the 60-element Vivaldi antenna array described in Chapter 4 (see Section 4.7.1.2.1) is simulated using the parallel combined LU-CG solver. When the solution from the LU decomposition is used as the initial guess for the CG method, a magnitude of the residual of 5.13×10^{-13} is obtained. If the parallel CG code uses an accuracy criteria of $\|\text{residual}\| \leq 10^{-8}$, then the code terminates after a single iteration!

6.5   CONCLUSION

In this chapter, two different conjugate gradient algorithms are reviewed. These algorithms have been found useful in solving matrix equations obtained from the application of the MoM methodology to a typical frequency-domain integral equation. A parallel CG scheme is implemented to accelerate the efficiency of the CG method. The parallel combined LU-CG scheme provides a way to further refine the solution of the LU decomposition method and thus guarantees the accuracy of the final solution to a prespecified degree in the solution of very large problems.

REFERENCES

[1]   M. Hestenes, *Conjugate Direction Methods in Optimization*, Springer-Verlag, New York, 1980.

[2]   J. M. Daniel, *The Conjugate Gradient Method for Linear and Nonlinear Operator Equations*, PhD dissertation, Stanford University, Stanford, CA, 1965.

[3]   T. K. Sarkar, "Application of Conjugate Gradient Method to Electromagnetics and Signal Analysis," *Progress in Electromagnetics Research*, Vol. 5, Elsevier, New York, 1991.

[4]   T. K. Sarkar, X. Yang, and E. Arvas, "A Limited Survey of Various Conjugate Gradient Methods for Solving Complex Matrix Equations Arising in Electromagnetic Wave Interactions," *Wave Motion*, Vol. 10, pp. 527–547, North-Holland, New York, 2000.

[5]   T. K. Sarkar, E. Arvas, and S. M. Rao, "Application of FFT and the Conjugate Gradient Method for the Solution of Electromagnetic Radiation from Electrically Large and Small Conducting Bodies," *IEEE Transactions on Antennas and Propagation,* Vol. 34, pp. 635–640, 1986.

[6]   Y. Zhang, Y. Xie, and C. H. Liang, "A Highly Effective Preconditioner for MoM Analysis of Large Slot Arrays," *IEEE Transactions on Antennas and Propagation,*, Vol. 52, No. 5, pp. 1379–1382, 2004.

[7]   Y. Zhang. *Study of Key Techniques of FDTD and MoM and Applications of Parallel Computational Electromagnetics*, PhD dissertation, Xidian University, Xi'an, China, June 2004.

[8]   Y. Zhang, J. Y. Li, C. H. Liang, and Y. Xie, "A Fast Analysis of Large Finite-Slot Phased Arrays Fed by Rectangular Waveguides," *Journal of Electromagnetic Waves and Applications*, Vol. 18, No. 6, pp. 715–727, 2004.

[9]   T. K. Sarkar, M. C. Wicks, M. Salazar-Palma, and R. J. Bonneau, *Smart Antennas*, Wiley-IEEE Press, Piscataway, NJ, 2003.

7

A PARALLEL MOM CODE USING HIGHER-ORDER BASIS FUNCTIONS AND PLAPACK-BASED IN-CORE AND OUT-OF-CORE SOLVERS

7.0 SUMMARY

To reduce the number of unknowns in a MoM code, the higher-order polynomial functions were introduced in Chapter 4 as the basis functions. To extend the capability of the MoM code, this chapter describes a newly developed out-of-core solver based on the library package PLAPACK (*parallel linear algebra package*) rather than ScaLAPACK (*scalable linear algebra package*). The method presented here can also extend beyond the physical memory size while maintaining the efficiency of an in-core implementation. Results on a typical computer platform for solutions of challenging electromagnetic problems indicate that the out-of-core solver can be as efficient as an in-core solver in solving large matrix equations.

7.1 INTRODUCTION

In Chapters 3 and 4, the parallel in-core and out-of-core MoM solvers using ScaLAPACK were discussed. As a prototype of a more flexible alternative to ScaLAPACK, PLAPACK [1,2], which contains a number of parallel linear algebra solvers, is discussed next.

PLAPACK is an MPI-based parallel linear algebra package designed to provide a user-friendly infrastructure for building parallel dense linear algebra libraries. PLAPACK has the following three unique features [1,2]:

1. A matrix distribution based on the natural distribution of an application (physically based matrix distribution)
2. An application interface for filling and querying matrices and vectors, which uses one-sided communication requiring only a standard compliant MPI implementation

3. A programming interface allowing the code to be written in a way that closely resembles the way algorithms are naturally explained, using object-based (MPI-like) programming

By taking this approach to parallelization, more sophisticated tasks (e.g., restarting the job after a computer system crashes) can be implemented. PLAPACK is not as well commercialized as ScaLAPACK. Users must have a thorough knowledge of PLAPACK in order to use it, since commercial support is not readily available. With the necessary concerns for the implementation of parallel computation using PLAPACK taken into account, this chapter delves into the development of a parallel MoM code using high-order basis functions.

7.2   FACTORS THAT AFFECT A PARALLEL IN-CORE AND OUT-OF-CORE MATRIX FILLING ALGORITHM

Application program interface (API) functions of PLAPACK can be used to compute the elements of the MoM impedance matrix in parallel. Earlier, a parallel implementation of a MoM code, using the RWG basis functions, was carried out by Dr. Robert A. van de Geijn, the author of PLAPACK. It was implemented as an in-core solution process, where the matrix was distributed to different processes by one column at a time using the PLAPACK API functions [3].

Rather than using API functions, this chapter presents another method to fill the impedance matrix. This alternative method is similar to the methods using ScaLAPACK introduced in earlier chapters. By simply introducing a flag to each geometric element involved with the source, as previously done in Chapter 4, the computation of all the redundant integrals can be avoided on any given process.

When developing the matrix filling codes, subroutines are needed to return the coordinates of a process that holds a given matrix element, and/or the local index for a given global index of a matrix element. These subroutines are provided by the ScaLAPACK library package but are not explicitly included in the current version of the PLAPACK library. Therefore, the FORTRAN subroutine "PLAG2LnP" is provided here and described in Figure 7.1.

The input parameters of the code are: block size nb_distr, number of process rows $nprows$, number of process columns $npcols$, global matrix row index i (start indexing at 1), and global matrix column index j (start indexing at 1). The output parameters of the code are: index of process row $irow$ (start indexing at 0) that owns the matrix element (i, j), index of process column $icol$ (start indexing at 0) that owns the matrix element (i, j), local matrix row index $iloc$ (start indexing at 1), and local matrix column index $jloc$ (start indexing at 1).

The matrix is split into blocks of size $nb\_distr \times nb\_distr$. Once the process to which the global matrix elements belong and the corresponding local matrix indices of the global matrix elements are known, it is not difficult to implement a parallel in-core matrix filling scheme in this MoM formulation. The framework is exactly the same as that introduced in Section 4.6.1 of Chapter 4.

```
Subroutine PLAG2LnP (nprows, npcols, nb_distr, i, j, iloc, jloc, irow, icol)
!       this code returns the local matrix index (iloc, jloc)
!       and the process coordinates (irow, icol)
!       for a given global matrix index (i, j)
        integer nprows, npcols, nb_distr, i, j, iloc, jloc, irow, icol
        integer ic, jc, ir, jr, ib, jb, iitemp, itemp, jtemp
!       convert to indexing starting at 0, like C language
        ic = i-1;
        jc = j-1;
!       ic, jc resides in block ib, jb (start indexing at 0)
        ib = int( ic/nb_distr );
        jb = int( jc/(nprows * nb_distr ) );
!       compute irow and icol by looking for which process row
!       and column own block ib, jb
        irow = mod( ib, nprows );
        icol = mod( jb, npcols );
!!!     compute iloc
        ir = mod( ic, nb_distr );        ! remainder
!       number of blocks on irow, not counting last block
        itemp = int( ib / nprows );
!       number of elements on irow, not counting last block
        iloc = itemp * nb_distr;
        iloc = iloc + ir;
!       convert to FORTRAN indexing
        iloc = iloc + 1;
!!!     compute jloc
        iitemp=nprows * nb_distr
        jr = mod( jc, iitemp ) ;         ! remainder
!       number of blocks on icol, not counting last block
        jtemp = int( jb/ npcols );
!       number of elements on icol, not counting last block
        jloc = jtemp * (nb_distr*nprows);
        jloc = jloc + jr;
!       convert to FORTRAN indexing
        jloc = jloc + 1;
        End Subroutine
```

Figure 7.1. Description of the FORTRAN subroutine dealing with the local information.

Once the parallel in-core matrix solution algorithm is obtained, it is necessary to reflect on how it will map to a parallel out-of-core matrix filling scheme as the data are being decomposed into different slabs. Subroutine "PLAG2LnP" provides the local index for an in-core matrix element. However, when the matrix is divided into different slabs, it cannot provide the correct local index for the matrix element of different slabs.

For the purpose of load balancing, it is necessary to keep the width of each slab the same, being a multiple of $n_b \times P_r \times P_c$, except for the fringe slab. Recall that n_b is the block size, and P_r and P_c denote the number of rows and columns of the process grid respectively. As an example, the 2×3 process grid in Figure 2.5 (see Chapter 2) is used to distribute the elements of the matrix as indicated in Figure 7.2. The slab width is taken as $(n_b \times P_r \times P_c)$, and the whole matrix consists of two complete slabs and a fringe slab.

To resolve the index alignment problem of each process for different slabs, the width of each slab is evaluated using the PLAPACK subroutine PLA_Obj_local_width [2]. For the first slab, the local index of a matrix element can be directly obtained by calling "PLAG2LnP". Starting with the second slab, the local index of a matrix element on a given process returned by "PLAG2LnP" should subtract the summation of the local widths of all the slabs to its left on this particular process. For example, to calculate the local index of a matrix element in Slab 3 for processes whose column coordinate in the process grid is 0, as shown in Figure 7.2, first add up the widths of the portion of the matrix that these processes hold for Slab 1 and Slab 2 (left of Slab 3), as shown by the shaded regions in Slab 1 and Slab 2 in Figure 7.2. Then, take this summed value off the local index returned by "PLAG2LnP" on the processes with 0 column coordinates and get the correct local index.

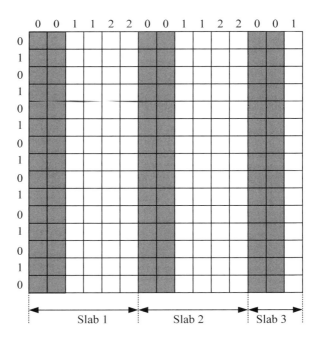

Figure 7.2. The distribution of the matrix and the definition of the slabs.

Once the local index is properly handled, the method used to avoid redundant calculation during the parallel out-of-core matrix filling is the same as introduced in Chapter 4. Figure 7.3 is an example flowchart for a PLAPACK matrix filling algorithm.

PLAPACK Parallel Out-of-Core Matrix Filling Demonstration Code

```
! Partition matrix into left part and right part
        ifrom = 0;
        iend = 0;
        icountx = 0;
        call PLA_Obj_view_all_f(A_ooc,A_ooc_cur , ierror);
100     call PLA_Obj_global_width_f(A_ooc_cur, size, ierror);
        size =min(size, nb_slab);
! While the width of the right portion of the whole matrix, size, is greater than 0
        if ( size > 0 ) then
        icountx=icountx+1;
        iend=ifrom+size;
! Separate one slab A_ooc_1 from the remaining right portion of matrix A_ooc_cur
        call PLA_Obj_vert_split_2_f(A_ooc_cur, size,
    &           A_ooc_1, A_ooc_cur, ierror);
        A_in_1(1:2)=0;
        call PLA_Matrix_create_conf_to_without_buffer_f(A_ooc_1,
    &           A_in_1, ierror);
        call PLA_Obj_local_width_f(A_in_1, n_A, ierror);
        nacounter(icountx)=n_A+nacounter(icountx-1);
        call PLA_Obj_local_length_f(A_in_1, m_A, ierror);

! Fill this slab
        if ( m_A > 0 .and. n_A > 0 ) then
        allocate(buffer_temp_to(1:m_A,1:n_A));
        buffer_temp_to(1:m_A,1:n_A) = 0;
        if ( n > ifrom .and. n <=ifrom+nb_slab ) then
        call PLAG2LnP (nprows, npcols, nb_distr, m, n, iloc, jloc, irow, icol);
        if ( myrow == irow .and. mycol == icol ) then
        icount = (n-1)/nb_slab + 1;

        ! Align data of different slabs for each process
        jloc = jloc-nacounter(icount);
        buffer_temp_to(iloc,jloc) = buffer_temp_to(iloc,jloc) + delta(m,n);
        ! delta(m,n) is from the calculation of the (m,n) element of the impedance matrix
        endif
        endif

! Submit this slab to out-of-core matrix
        call PLA_Obj_attach_buffer(A_in_1,buffer_temp_to,m_A, ierror);
        call PLA_Copy_f(A_in_1, A_ooc_1, ierror);
        if ( ierror == 0 ) deallocate(buffer_temp_to)
        endif
        ifrom = ifrom + size;
        go to 100
        else
        endif
```

Figure 7.3. A flowchart for out-of-core filling of the impedance matrix using PLAPACK.

Together with the out-of-core matrix equation solver, and using PLAPACK as discussed in Chapter 2, a parallel out-of-core MoM code can now be implemented. Next we analyze the performance of such a code.

7.3   NUMERICAL RESULTS

Two different models are chosen to illustrate the solution of electrically large EM problems on a parallel computer platform. For a composite problem, a large array of Vivaldi antennas is used. For an electromagnetic scattering problem, the RCS from a full-scale electrically large airplane is computed at three different frequencies.

7.3.1   Radiation from an Array of Vivaldi Antennas

The analysis of radiation from an array of Vivaldi antennas is presented in this section. A 60-element and a 112-element dual-polarized Vivaldi antenna array are shown in Figure 4.15 (*a*) (see Chapter 4) and Figure 7.4, respectively. The dimensions of a single element are 76 mm × 25.6 mm × 6 mm. The dielectric constant of the substrate is 2.2. The operating frequency of this Vivaldi array varies from 1 to 5 GHz, where it behaves as a traveling-wave antenna. The structure is analyzed at the highest frequency of interest as the model will generate the largest number of unknowns.

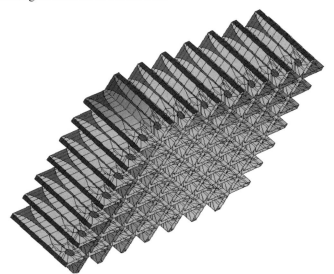

Figure 7.4. A 112-element Vivaldi array.

Modeling a 60-element array requires 34,390 geometric elements, which are composed of wires, plates, and junctions. The PMCHW formulation yields 45,500 unknowns for the electric currents and 4320 unknowns for the magnetic currents to calculate the current distribution on the array. The total number of unknowns for this array is 49,820, which requires approximately 40 GB of RAM when performing computation using double-precision (complex) arithmetic.

Modeling a 112-element array requires 63,952 geometric elements. The total number of unknowns for this array is 92,729, which requires approximately 137.6 GB of RAM when using double-precision (complex) arithmetic. The two Vivaldi antenna arrays are simulated on CEM-4 and the results are presented in Figures 7.5 and 7.6. Figure 7.5 (*a*) and (*b*) show the radiation pattern of the 60-element Vivaldi array in the azimuth and the elevation planes, respectively. Figure 7.5 indicates that the computed results from a serial MoM code using a ScaLAPACK based out-of-core solver (OCS) agree well with that obtained from a parallel MoM code using the PLAPACK in-core solver. Figure 7.6 (*a*) and (*b*) show the radiation pattern of the 112-element Vivaldi array in the azimuth and elevation planes, respectively.

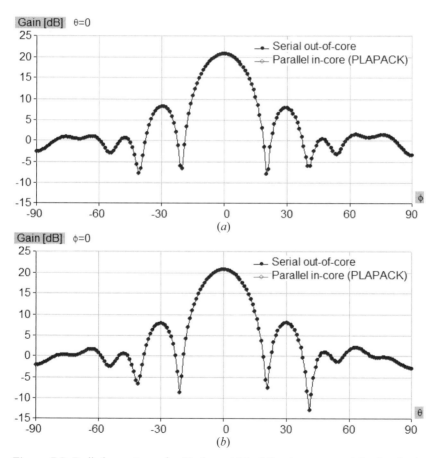

Figure 7.5. Radiation pattern of a 60-element Vivaldi antenna array: (*a*) azimuth plane pattern (0° starts from *x* axis in the *xoy* plane); (*b*) elevation plane pattern (0° starts from *x* axis in the *xoz* plane).

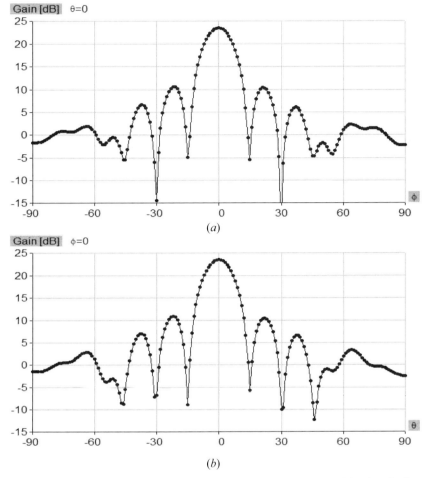

Figure 7.6. Radiation pattern of a 112-element Vivaldi antenna array: (*a*) azimuth plane pattern (0° starts from *x* axis in the *xoy* plane); (*b*) elevation plane pattern (0° starts from *x* axis in the *xoz* plane).

The total computation time taken to solve the problem using nine compute nodes (18 processes) is listed in Table 7.1. Although simulation of these complicated structures is easily achieved with the higher-order basis functions as presented in Chapter 4, the number of geometric elements in these examples is comparable with the number of unknowns. Therefore, the benefit of using higher-order basis functions cannot be fully demonstrated here. However, the benefits of using the higher-order basis will be apparent in the second example. The analysis of the performance of the parallel code for these examples is limited primarily to the solution of the matrix equations and does not deal with the matrix filling operation. This is because the computation taken to perform the LU decomposition in the matrix solution process typically takes significantly more time than does evaluation of the elements of the matrix.

TABLE 7.1. Simulations for Two Large Vivaldi Antenna Arrays

Vivaldi Antenna Array	Case 1-1 (60-element)	Case 1-2 (60-element)	Case 1-3 (60-element)	Case 1-4 (112-element)
PLAPACK Code	In-core	Out-of-core	Out-of-core	Out-of-core
Frequency (GHz)	5.0	5.0	5.0	5.0
Unknowns	$N_1 = 49{,}820$	$N_2 = 49{,}820$	$N_3 = 49{,}820$	$N_4 = 92{,}729$
In-core buffer	N/A	2.4 GB/process	1.7 GB/process	1.7 GB/process
Hard disk (GB)	N/A	39.7	39.7	137.6
Filling (minutes)	51	51	73	491
Solving (minutes)	76	77	91	578
Total (minutes)	$T_1 = 127$	$T_2 = 128$	$T_3 = 164$	$T_4 = 1069$

In Table 7.1, case 1-1 uses the PLAPACK in-core solver, and the other three cases use the PLAPACK out-of-core solver. For case 1-2, the in-core buffer used per process is 2400 MB, which is slightly larger than the 2205 MB per process required for an in-core solution. In this case the out-of-core problem is solved essentially as an in-core problem with one additional operation of writing to the hard disk. For case 1-1, the time required for solution of the matrix equations is 76 minutes, and for case 1-2 it is 77 minutes. These results illustrate that the CPU times spent on solving the matrix equations in cases 1-1 and 1-2 are similar. For both cases, the matrix filling time is 51 minutes. This is due to the fact that there is sufficient additional cache memory available on the system and thus the I/O overhead is reduced for this simulation.

For case 1-3, the RAM per process used for one slab is set to be 1700 MB. The matrix is split into two slabs because the in-core buffer size for each process is less than 2205 MB, which is required per process for an in-core solution. The matrix solving time is 91 minutes, which is longer than required for case 1-2 (out-of-core) where the slab is big enough to fit all the data using one slab. When compared to case 1-1 (in-core), the matrix solving time required using the out-of-core solver for case 1-3 is approximately 20% more.

Using the clock time of case 1-1 as the benchmark, case 1-4 will require $(92{,}729/49{,}820)^3 \times 76$ minutes = 490 minutes to solve the problem. The actual computation time is 578 minutes, which is 88 minutes longer than the predicted value. This indicates that for a larger job, the PLAPACK out-of-core solver incurs less than 18% penalty for solving a large matrix equation when compared with the PLAPACK in-core solver.

Considering the high cost of RAM, this degradation in performance by using the cheaper hard disks rather than expensive RAM provides a good price–performance scaling and thus will be very cost-effective. Performance of the solver will be further analyzed by simulating the scattering from an electrically large aircraft.

7.3.2   Scattering from an Electrically Large Aircraft

Scattering from a full-size aircraft is presented in the following section to further demonstrate the high efficiency achieved using the parallel implementation of the integral equation solver using PLAPACK. The aircraft structure is 11.6 m long, 7.0 m wide, and 2.92 m high. It is modeled as a PEC surface with symmetry along the *yoz* plane. The plane wave is incident from the negative *y*-axis direction and is polarized along the *z*-axis direction, as shown in Figure 4.17 (*a*) (see Chapter 4). The structure is analyzed at three different frequencies on CEM-4 so that one can observe how the solution time for these problems scale with the increase in problem size.

Figure 7.7 plots the bistatic RCS in dB (decibels) along the *yoz* plane at 2.20 GHz. This simulation requires 70,995 unknowns to accurately calculate the currents. At 2.20 GHz, the airplane is 85λ long, 51.3λ wide, and 21.4λ high. Next, the structure is analyzed at 3.0 GHz. The corresponding wavelength at 3.0 GHz is 0.01 m, and the structure is 116λ long. In this case, 271.1 GB RAM is needed for the in-core solver using double-precision (complex) arithmetic for computation when using 130,165 unknowns. Here, 18 processes are used with just 36 GB RAM and 271.1 GB hard-disk storage.

The ratio of the wall time for the solution of the different numbers of unknowns, N_2 and N_1, should be within the range of $(N_2/N_1)^2$ and $(N_2/N_1)^3$. This is because the computational complexity of computing all the elements of the matrix scales as $O(N^2)$, which sets the lower limit for the speed of the solution. The matrix solution process, which uses the LU decomposition, has a computational complexity of $O(N^3)$, which sets the upper limit for the total time. The estimated time to complete the matrix solving of the problems with different number of unknowns is listed in Table 7.2. Indeed, the measured wall times are not more than the estimated for all the cases that have been analyzed.

As illustrated in the previous analysis of the Vivaldi antenna array, the focus here will also be on the performance of the matrix solution methodology. The ratio of the time between cases 2-1 and 2-3 in Table 7.2 demonstrates the applicability of the $O(N^3)$ relationship. Using the relationship that T_3/T_1 is proportional to $(N_3/N_1)^3$, one can predict the maximum solution time for case 2-3 using case 2-1 as a benchmark. The matrix solution time for case 2-3 is predicted to be $[(229 \text{ minutes}) \times (130,165/70,995)^3] = 1411$ minutes. Since the actual computational time taken is 1317 minutes and is less than the value that is predicted, it demonstrates the expected behavior in the solution methodology. Similar analysis and conclusions can also be applied to case 2-2 using case 2-1 as the benchmark. If case 2-2 is used as a benchmark, the predicted time for the solution of case 2-3 will be 1353 minutes. The actual computational time taken is 1317 minutes, which is shorter and it compares with the estimated time quite well.

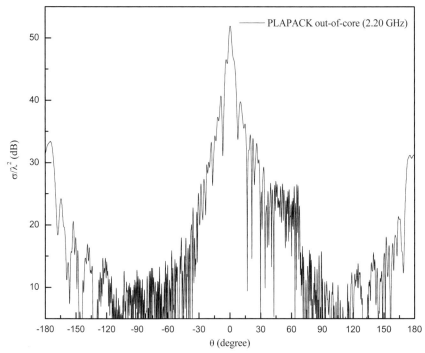

Figure 7.7. RCS of the airplane computed using the PLAPACK out-of-core solver (0° starts from *y* axis in the *yoz* plane).

TABLE 7.2. Simulations for the Airplane at Three Different Frequencies

	Case 2-1	Case 2-2	Case 2-3
Frequency (GHz)	2.2	2.5	3.0
Unknowns	$N_1 = 70,995$	$N_2 = 93,091$	$N_3 = 130,165$
Memory (GB)	36.0	36.0	36.0
Hard disk (GB)	80.6	138.7	271.1
Matrix filling (minutes)	43	85	251
Matrix solving (minutes)	229	495	1317
Estimate based on case 1	—	516	1411
Estimate based on case 2	—	—	1353
Total time (minutes)	$T_1 = 272$	$T_2 = 580$	$T_3 = 1568$

It is interesting to estimate the simulation time for the analysis of the aircraft by using the in-core solution time of case 1-1 related to the solution of the Vivaldi array as the benchmark. The estimated solution time is $[(130,165/49,820)^3 \times (76 \text{ minutes})] = 1355$ minutes. Compared to the actual time of 1317 minutes, the total solution time is again shown to be faster than estimated. This indicates that the out-of-core solver offers very good performance when the cheaper storage on the hard disk is used as opposed to using the expensive RAM memory.

7.3.3   Discussion of the Computational FLOPS Achieved

The GFLOPS, which is 10^9 floating-point operations per second (FLOPS) is now estimated for the parallel in-core and out-of-core LU decomposition methodology when solving the problems listed in Table 7.3. Except for case 1-4, the numbers of GFLOPS of the parallel in-core and the parallel out-of-core solvers are similar. The worst penalty for the parallel out-of-core solver one has to pay is a longer CPU time of approximately 20% than the in-core solver. Considering the prices of hard disk and RAM of comparable size, this degradation of performance is reasonable and is quite acceptable in the solution of electrically large problems.

TABLE 7.3. GFLOPS Achieved for the Parallel Solvers When Solving the Problems Mentioned Earlier

Case	Parallel Solver	Unknowns	CPU Time (seconds)	GFLOPS
Case 1-1	In-core	49,820	3554.9	92.8
Case 1-2	Out-of-core	49,820	3559.6	92.6
Case 1-3	Out-of-core	49,820	3710.6	88.9
Case 1-4	Out-of-core	92,729	28483.5	74.6
Case 2-1	Out-of-core	70,995	10755.1	88.7
Case 2-2	Out-of-core	93,091	23626.9	91.0
Case 2-3	Out-of-core	130,165	66152.4	88.9

Note that for all these examples, the various parameters of interest have not been completely optimized as yet [4,5]. To improve the performance of the out-of-core solver, the size of the in-core buffer for the process and the shape of the process grid should be optimized, as has been done for the ScaLAPACK-based parallel in-core and out-of-core solvers described in

Chapter 5. This additional tuning of the PLAPACK code will result in further improvement of overall performance.

7.4 CONCLUSION

This chapter presents the key concepts for designing efficient parallel in-core and parallel out-of-core MoM solvers using the PLAPACK library. Several challenging examples have been solved using a representative parallel computing platform to illustrate the scalability and the efficiency of the techniques. With the methodology described in this chapter, electrically large problems that are difficult to solve on a single processor can be solved in reasonable time when using the parallel mode, and the more important advantage is that the size of the RAM no longer remains a limitation to the application of MoM. The advantage of PLAPACK over ScaLAPACK is that checkpointing can be implemented much more easily. This implies that the computing process can be restarted from the point where the system failed rather than restarting the solution process from the beginning, thereby saving significant time.

REFERENCES

[1]   G. Morrow and R. A. van de Geijn, *Half-day Tutorial on: Using PLAPACK: Parallel Linear Algebra PACKage*, Texas Institute for Computational and Applied Mathematics, University of Texas at Austin, Austin, Texas. Available at: http://www.cs.utexas.edu/users/plapack/tutorial/SC98/. Accessed Aug. 2008.

[2]   R. A. van de Geijn, *Using PLAPACK: Parallel Linear Algebra Package*, MIT Press, Cambridge, MA, 1997.

[3]   T. Cwik, R. A. van de Geijn, and J. E. Patterson, "The Application of Parallel Computation to Integral Equation Models of Electromagnetic Scattering," *Journal of the Optical Society of America A* , Vol. 11, No. 4, pp. 1538–1545, April 1994.

[4]   Y. Zhang, T. K. Sarkar, R. A. van de Geijn, and M. C. Taylor, "Parallel MoM Using Higher Order Basis Function and PLAPACK In-Core and Out-of-Core Solvers for Challenging EM Simulations," *IEEE Antennas and Propagation Society Symposium*, San Diego, 2008.

[5]   Y. Zhang, R. A. van de Geijn, M. C. Taylor, and T. K. Sarkar, "Parallel MoM Using Higher-Order Basis Functions and PLAPACK In-Core and Out-of-Core Solvers for Challenging EM Simulations," *IEEE Antennas and Propagation Magazine* (to be published).

8

APPLICATIONS OF THE PARALLEL FREQUENCY-DOMAIN INTEGRAL EQUATION SOLVER — TIDES

8.0   SUMMARY

In this chapter, the parallel electromagnetic (EM) computer code *t*arget *ide*ntification *s*oftware (TIDES) presented in the earlier chapters will be used to analyze challenging electromagnetic problems in frequency domain. Performance comparisons between TIDES and a commercially available software using the piecewise RWG basis functions are also presented. The results of these comparisons indicate that less computer resources are required by TIDES, due to the employment of higher-order basis functions as opposed to the use of the conventional subdomain basis functions.

A unique feature of TIDES is the parallel out-of-core MoM-based solver; it can handle problems that cannot fit into the RAM of the computer systems. The parallel out-of-core solution feature of this code is a unique capability and is not readily available for most commercial codes. The application examples presented in this chapter illustrate the efficiency and the accuracy of TIDES, which can be used as a versatile tool for analyzing challenging real-world electromagnetic problems.

At the end of this chapter, the computational results of some complex composite structures using TIDES are compared with the experimental measurements. The complex composite structure to be studied consists of an L-band antenna array and a dielectric radome with seven large structural ribs for supporting the radome shell. It is demonstrated that using this electromagnetic software — TIDES, one can compute results for the radiation pattern which are within several tenths of a decibel (dB) when compared with measurements for the grating lobe amplitudes. The difference between theory and experiment falls within the resolution of the measurements. The results demonstrate that TIDES can be used to accurately predict the interaction of an electrically large array with its surrounding dielectric structures.

8.1   PERFORMANCE COMPARISON BETWEEN TIDES AND A COMMERCIAL EM ANALYSIS SOFTWARE

In this section, numerical results are presented that are obtained from two different electromagnetic analysis codes based on an integral equation method. One of the two codes (TIDES), which uses GiD software tool for generating meshes [1], is the out-of-core parallel integral equation solver, based on the MoM using higher-order basis functions. The other is a commercial software that solves the integral equations using MoM and/or MLFMA using the piecewise RWG basis functions. To investigate the performance of these two EM software, simulations were made at the National Key Laboratory of Antennas and Microwave Technology, Xidian University, Xi'an, China in 2007, and the results have been presented in [2], and are included here for completeness.

The two codes are executed on a PC with a Pentium P4 chip of 3.0 GHz CPU, 1.0 GB RAM, and using the Microsoft Windows XP operating system. The numerical results indicate that the TIDES software using the higher-order basis functions requires less than 40% RAM of the other software as it solves the same problem using fewer unknowns. Hence, the execution time is significantly reduced when compared with the use of piecewise triangular patch basis functions. Even when the speedup of the computations is achieved in the commercial software using the MLFMA algorithm, it still needs much more execution time and RAM than TIDES. These results seem to indicate that the use of higher-order basis functions is much more efficient than the piecewise RWG basis functions in the traditional integral equation solvers.

8.1.1   Analysis of a Scattering Problem

For the first problem, the bistatic RCS is calculated for a PEC sphere of 1.0λ radius. The structure is illuminated by an incident plane wave propagating along the $-x$ axis. The plots of the structure used in the two codes, are shown in Figure 8.1 for the case using TIDES and in Figure 8.2 using the commercial software.

In TIDES, which uses the GiD software tool for preprocessing and postprocessing [1], the surface of the sphere is meshed into bilinear quadrilateral patches and a quarter of the circle is subdivided into 12 subsections as illustrated in Figure 8.1. In the commercial EM software, the surface is meshed into triangular patches with the edge length of the triangular facets being approximately one-tenth the wavelength in size as seen in Figure 8.2.

The VV (vertical–vertical) and HH (horizontal–horizontal) polarization components of the bistatic RCS along the xoy plane are calculated and plotted in Figures 8.3 and 8.4, respectively. As indicated in these two figures, the VV component of the RCS obtained from TIDES and that from the commercial software are in excellent agreement with each other. The HH polarization component of the RCS computed by the two codes is also in excellent agreement with each other.

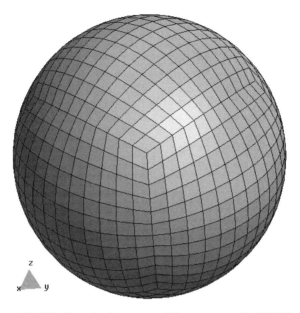

Figure 8.1. The discretized structure of the sphere used in TIDES.

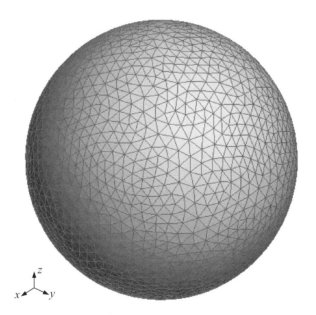

Figure 8.2. The discretized model of the sphere generated by the commercial software.

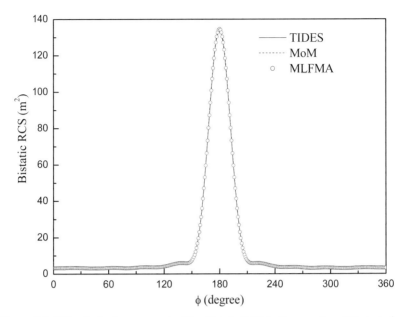

Figure 8.3. VV polarized component of the bistatic RCS in the *xoy* plane (0° starts from *x* axis in the *xoy* plane).

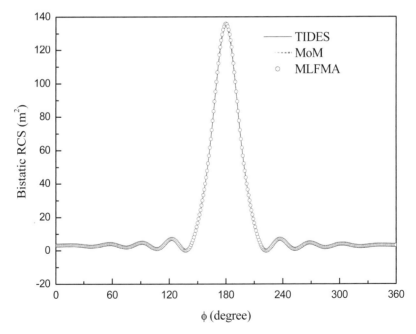

Figure 8.4. HH polarized component of the bistatic RCS in the *xoy* plane (0° starts from *x* axis in the *xoy* plane).

Note that both the EFIE and the CFIE (combined field integral equation) yield results that are in very good agreement with each other, when either MoM or MoM incorporating MLFMA is used as in the commercial software for this simulation. As such, only the results from the MLFMA using the EFIE are presented in Figures 8.3 and 8.4.

The execution time listed in Table 8.1 is the total elapsed time for the computation of the vertically and the horizontally polarized cases. In this example, TIDES needs only about 30% of the total number of unknowns (NUN) as compared to that used in the piecewise triangular patch model. Also as Table 8.1 illustrates, the amount of RAM and the computation time required by TIDES are much less than those by the commercial software for analyzing the same problem [2].

Note that in the commercial software, the MLFMA/EFIE formulation requires much larger RAM than the MLFMA/CFIE formulation, because of the use of default ILU (incomplete LU) preconditioner in the computations. The preconditioner needs about 65.4 MB RAM in the MLFMA/EFIE formulation, while it requires only 1.15 MB in the MLFMA/CFIE formulation [2].

TABLE 8.1. Performance Comparison for the RCS Calculation

Code	Algorithm /Equation	Basis Function	NUN	Solver (Preconditioner)	RAM (MB)	Time (seconds)
TIDES	MoM /EFIE	Higher-order	1728	LU (no preconditioner)	22.78	62.0
Commercial software	MoM /EFIE	RWG	5607	LU (no preconditioner)	242.37	348.11
	MLFMA /EFIE	RWG	5607	BiCGStab (ILU)	108.10	167.39
	MLFMA /CFIE ($\alpha = 0.5$)	RWG	5607	BiCGStab (ILU)	42.16	152.69

8.1.2 Analysis of a Radiation Problem

In the second example, two monopole antennas are placed on a flat PEC square plate. The plate lies in the xoy plane and the length of the edge is 10 m. The origin of the coordinate system is set to the center of the plate. The length of the monopole antennas are taken as 0.25 m with a radius of 0.002 m. With a 300 MHz working frequency, the antennas are fed with 1.0 V injected at the wire junction connecting to the plates. The coordinates of the antenna–plate junctions are (2,0,0) and (–2,0,0), respectively. The radiation patterns along the *xoz* and *yoz* planes are calculated. The simulation model used in TIDES is shown in Figure 8.5. Figure 8.6 displays the corresponding geometry used in the commercial software. The mesh used in TIDES is shown in Figure 8.5. It can be seen from this figure that the mesh size in this example is approximately 1.0λ. In the commercial software, the plate is meshed into the triangular patches with an edge length of 0.1λ. The mesh is not shown in Figure 8.6 because the amount of triangles is too large to be plotted clearly.

This example deals with an electrically large open body, therefore, only the MLFMA/EFIE formulation is employed in the commercial software. Normalized radiation patterns along the *xoz* and *yoz* planes are plotted in Figures 8.7 and 8.8, respectively. The results from TIDES and the commercial software agree very well with each other.

Note that the default preconditioner in the commercial software is the ILU preconditioner, which occupies relatively large RAM and in totality needs more than 1 GB of RAM. Therefore, the *sp*arse *a*pproximate *i*nverse (SPAI) preconditioner is used in this simulation.

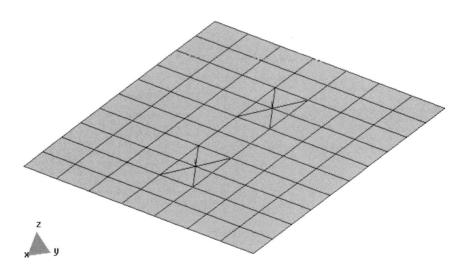

Figure 8.5. Simulation model used in TIDES.

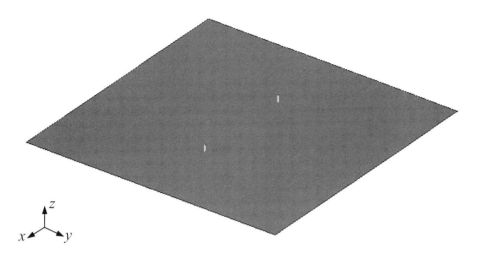

Figure 8.6. Simulation model used in the commercial software.

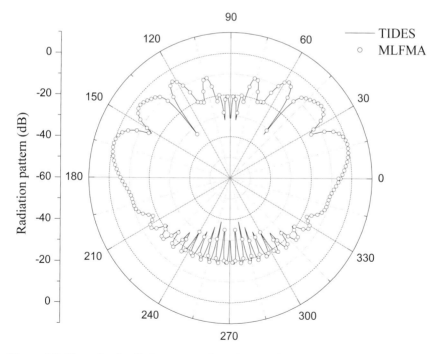

Figure 8.7. Normalized radiation pattern along the *xoz* plane (0° starts from *x* axis in the *xoz* plane).

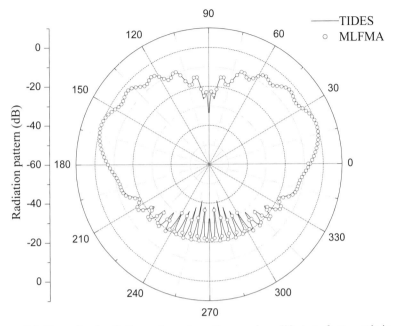

Figure 8.8. Normalized radiation pattern along the *yoz* plane (0° starts from *y* axis in the *yoz* plane).

In this example, TIDES needs only about 10.5% of the total number of unknowns (NUN) than that used in the commercial software. The RAM required by TIDES is only 56.7% of that required by the MLFMA/EFIE formulation used in the commercial software, and the computation time for TIDES is only 33.1% of that for the commercial software using the MLFMA/EFIE formulation. All the details are listed in Table 8.2 [2]. In this example, the SPAI preconditioner in the commercial software uses about 69 MB of RAM out of the total amount of 287.11 MB used.

TABLE 8.2. Performance Comparison for a Radiation Problem

Code	Algorithm /Equation	Basis Function	NUN	Solver (Preconditioner)	RAM (MB)	Time (seconds)
TIDES	MoM /EFIE	Higher-order	4,618	LU (no preconditioner)	162.70	231.0
Commercial software	MLFMA /EFIE	RWG	44,170	BiCGStab (SPAI)	287.11	697.4

8.1.3   Analysis of a Coupling Problem

For the third example, two monopoles are mounted on a PEC cylinder oriented along the z axis. The length and radius of the cylinder are 5.0 m and 1.0 m, respectively. One of the two antennas is directed along the x axis, and the other along the y axis. The length and radius of each monopole antenna are 0.25 m and 0.002 m, respectively. The antennas are fed at the junction between the wires and the plates with 1.0 V and the coordinates of the antenna–cylinder junctions are (1,0,0) and (0,1,0), respectively.

The surface of the cylinder is meshed into bilinear surfaces in TIDES and into the triangular patches with an edge length of 0.125λ in the commercial software at 400 MHz. Again, only the MLFMA/EFIE formulation is used in the commercial software for this simulation.

The S parameters of the antenna network are calculated in the frequency band of 200–400 MHz using 21 sampling frequencies. The electromagnetic models used in TIDES and in the commercial software are given in Figure 8.9 (a) and (b) [2].

The magnitude and phase of the S parameters are calculated and plotted in Figures 8.10 and 8.11, using the two methods. As indicated by the figures, the results obtained using the two different codes agree very well with each other.

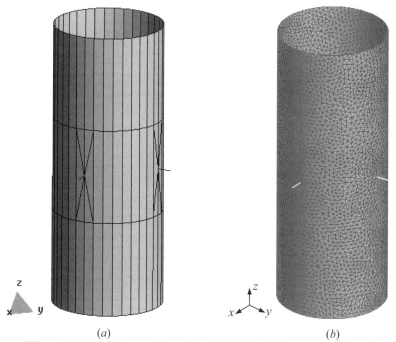

(a)   (b)

Figure 8.9. Simulation model used in: (a) TIDES; (b) the commercial software.

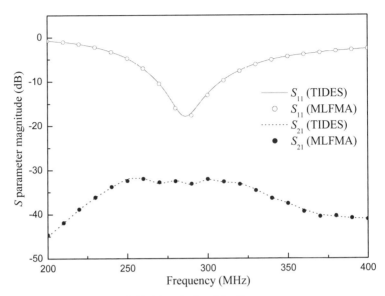

Figure 8.10. Magnitude of the *S* parameters.

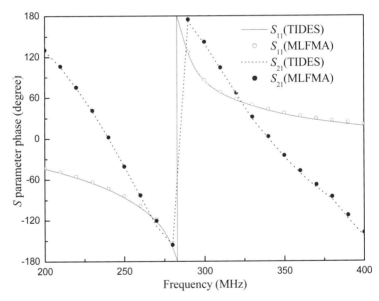

Figure 8.11. Phase of the *S* parameters.

In this example, the number of unknowns (NUN) and the RAM required by TIDES are 30.7% and 17.8% of those required by the commercial software, respectively. The execution time used by TIDES is only 39.6% of that used by the commercial software using the MLFMA/EFIE formulation. Detailed information about this simulation is listed in Table 8.3. For simulations executed at different frequency samples, the RAM required by TIDES and the commercial software are different. Only the maximum amount of RAM used is listed in Table 8.3.

TABLE 8.3. Performance Comparison for the Coupling Calculations

Code	Algorithm /Equation	Basis Function	NUN	Solver (Preconditioner)	RAM (MB)	Time (seconds)
TIDES	MoM /EFIE	Higher-order	3,660	LU (no preconditioner)	102.20	2334
Commercial software	MLFMA /EFIE	RWG	15,508	BiCGStab (ILU)	574.24	5889.06

The TIDES and the commercial software have been used in the solution of three typical types of electromagnetic problems. It can be seen that the obtained results using the two different software agree very well with each other, but TIDES requires less than 40% of the RAM used in the commercial software. The execution time in TIDES is also significantly less than that of the commercial software [2].

The use of higher-order basis functions substantially reduces the number of unknowns, the problem size and also the total solution time. Benchmark tests have proven TIDES to outperform the commercial software (MoM and MLFMA solvers) for these three problems without even using the parallel processing methodology [2].

Next, we illustrate the applications of the parallel codes.

8.2   EMC PREDICTION FOR MULTIPLE ANTENNAS MOUNTED ON AN ELECTRICALLY LARGE PLATFORM

Antennas are generally designed assuming that they are radiating in a free space environment without interference from the other antennas or between them and the platform on which the antennas are mounted. In practice, antennas are often placed in close proximity to one another and are mounted on large physical structures, such as the mast of a ship, the fuselage of an aircraft, and the like, which significantly influence their radiation and coupling characteristics.

Antenna-to-antenna couplings have been difficult to solve for because of the complexity of the related analysis. Measurements of the radiation and coupling characteristics of the antennas mounted on a large platform are difficult or even impossible to perform. Therefore, the challenge exists to accurately simulate the interaction between antennas mounted on electrically large platforms. In this section, we use TIDES to calculate the coupling between antennas in order to estimate the EMI/EMC coupling. The isolations between the four antennas mounted on a full-size aircraft model, shown in Figure 8.12, are calculated. The four circles in Figure 8.12 are used to show the junction between the four wire antennas and the aircraft structure.

The full-size aircraft model is meshed at 1.5 GHz and is 58 wavelengths long at this frequency. This simulation employed 70,853 unknowns to discretize the structure into quadrilaterals. It took 3.5 hours to calculate the isolation at each frequency by running the parallel in-core code on the 20 CPUs of CEM-4. The results for the isolation between the various antennas on the aircraft structure are plotted in Figures 8.13−8.16.

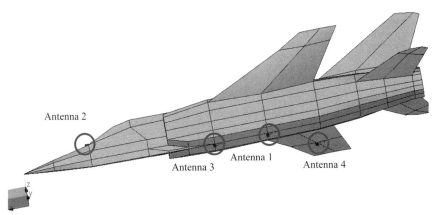

Figure 8.12. Four antennas mounted on a full-size airplane model.

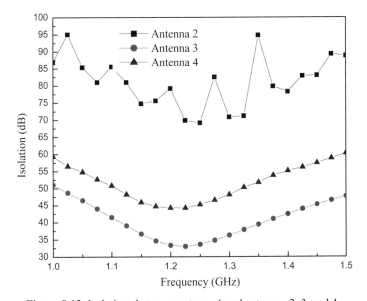

Figure 8.13. Isolations between antenna 1 and antennas 2, 3, and 4.

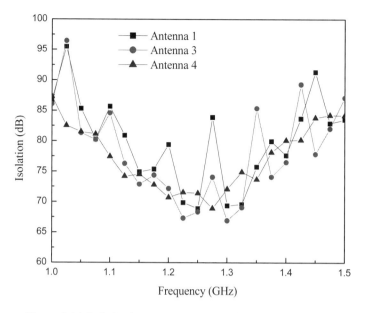

Figure 8.14. Isolation between antenna 2 and antennas 1, 3, and 4.

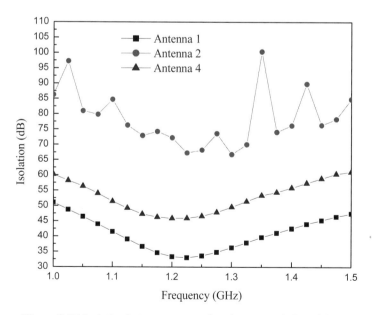

Figure 8.15. Isolation between antenna 3 and antennas 1, 2, and 4.

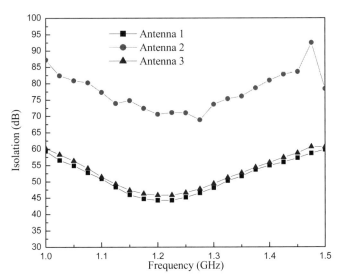

Figure 8.16. Isolation between antenna 4 and antennas 1, 2, and 3.

To assess the accuracy of the numerical results, isolations between three pair of antennas are plotted together in Figure 8.17. In the legend of Figure 8.17, the first number in each pair represents the antenna acting as a transmitter, and the second, represents the antenna used as a receiver.

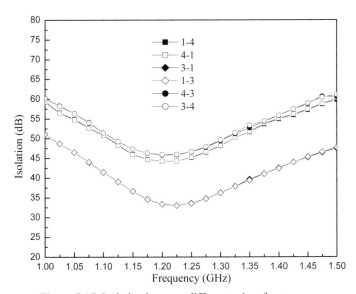

Figure 8.17. Isolation between different pairs of antennas.

Coupling between antenna 1 and antenna 3 is the strongest and the numerical isolation between them exactly meets the reciprocity relationship of any passive network, as shown by the two lowest curves in Figure 8.17.

As shown in Figure 8.17, the isolation between antennas 1 and 4 and antennas 3 and 4 is larger than 42.5 dB over the frequency band. The isolation between antennas 1 and 4 (or antennas 3 and 4) also satisfies the reciprocity theorem, no matter which one is used as the transmitter (and the other as the receiver). A similar situation exists for the isolation results between antennas 2 and 4, where the isolation is greater than 70 dB and is much higher than the isolation between antennas 1 and 4 (or antennas 3 and 4).

The radiation pattern for these monopole antennas mounted on the airplane can also be calculated. Figure 8.18 shows the gain (in dB) in the xoz plane of antenna 2 at 1.25 GHz. The airplane icon in this figure is used to indicate the orientation of the aircraft in the xoz plane.

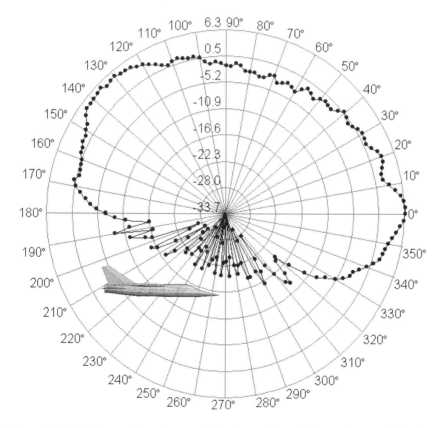

Figure 8.18. Radiation pattern of antenna 2 in the xoz plane (0° starts from x axis in the xoz plane).

8.3   ANALYSIS OF COMPLEX COMPOSITE ANTENNA ARRAY

The model considered is a 112-element Vivaldi antenna array as shown in Figure 7.4 (see Chapter 7). This large composite PEC and dielectric radiating structure is a challenging antenna analysis problem, which is presented here to demonstrate the capability of the TIDES code.

Modeling a 112-element Vivaldi array at 4.7 GHz requires 63,938 geometric elements. The total number of unknowns for this array is 89,124, which requires approximately 127.1 GB of storage to analyze the problem using double precision (complex) arithmetic.

The simulation time on CEM-3, using four single-core CPUs, and on CEM-9, using four quad-core CPUs, is listed in Table 8.4. The wall time required by CEM-3 is 9421 minutes when executing the parallel in-core MoM code of TIDES. The wall time can be reduced to 614 minutes by running the parallel out-of-core MoM code of TIDES on CEM-9.

The azimuth pattern and elevation pattern of the Vivaldi antenna array are plotted in Figures 8.19 and 8.20, respectively, when all the elements are excited with an amplitude of 1.0 V. As shown in these figures, the results of using the parallel in-core MoM code overlay with those obtained by the parallel out-of-core MoM code.

TABLE 8.4. Simulation Times on CEM-3 and CEM-9

Platform	CEM-3	CEM-9
Code	Parallel in-core	Parallel out-of-core
Wall time (minutes)	9421	614

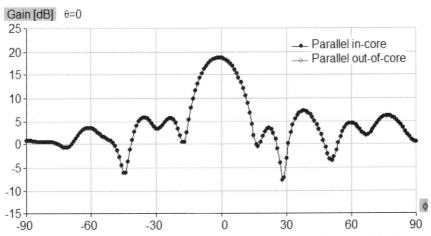

Figure 8.19. Azimuth pattern of the Vivaldi antenna array (0° starts from x axis in the *xoy* plane).

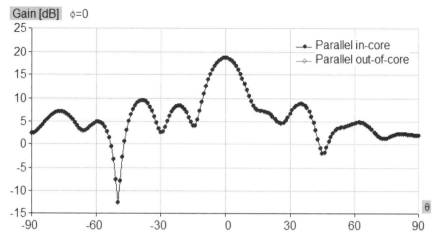

Figure 8.20. Elevation pattern of the Vivaldi antenna array (0° starts from x axis in the *xoz* plane).

8.4   ARRAY CALIBRATION FOR DIRECTION-OF-ARRIVAL ESTIMATION

It is well known that mutual coupling between antenna elements in an array and nearby scatterers or the platform it is mounted on degrades the performance of direction-of-arrival (DOA) estimation. The effect of mutual coupling can be reduced by a proper array calibration.

In this example, waves incident from known directions are simulated and the currents (or voltages) received on each antenna are computed and used to determine a calibration matrix. This matrix transforms a real steering vector, corrupted by mutual coupling, into an ideal steering vector. The DOA estimation is then processed using these transformed signals. The calibration angles in this example include those ranging from 60° to 120° in azimuth (ϕ) and $-30°$ to 30° in elevation (θ).

In this application, 176 dipole antennas are put on each side of the Global Hawk unmanned aerial vehicle (UAV) (http://images.google.com/images?um=1&hl=en&q=global+hawk) for DOA estimation, as shown in Figure 8.21. The length and the radius of each dipole antenna are 0.47λ and 0.9493 mm, respectively. The structure is analyzed at 1.0 GHz.

Figure 8.22 shows the layout of the antennas at one side of the aircraft. There are 11 rows and 16 columns of antennas on each side of the aircraft. The distances between the rows and the columns are both 0.5λ. The distance between the antenna array and the aircraft model is 0.25λ.

Figure 8.21. Model of the UAV and the antenna array along with the directions of the incident waves.

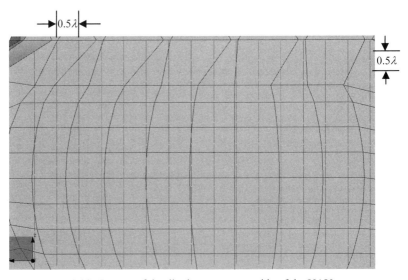

Figure 8.22. Layout of the dipole array at one side of the UAV.

The array calibration is performed at 1.0 GHz, which creates 68,133 unknowns using the MoM procedure. Calculation times on CEM-3 using 1 CPU and on CEM-4 using 20 CPUs are listed in Table 8.5. It indicates that a problem run in the serial mode, that previously took 3 weeks to complete, can now be finished in 9 hours by using the parallel code on a high-performance computer.

TABLE 8.5. Simulation Time on CEM-3 and CEM-4

Platform	CEM-3	CEM-4
Code	Serial	Parallel in-core
Wall time	21 days 6 hours 36 minutes	9 hours 7 minutes

Once the calibration matrix is obtained, the 2D Matrix Pencil method [3] is used to estimate the DOAs and amplitudes of the signals. Figure 8.23 shows two incident signals received by the array. The arrows in the figure represent the polarization of the electric field and the direction of the wave propagation. In this case, one of the signals is incident from $\phi = 70°$ and $\theta = 20°$ and the other arrives from $\phi = 110°$ and $\theta = -10°$ (θ coordinate is measured from xoy plane to z axis). Both signals have their complex amplitude (α) equal to 1. Table 8.6 shows the exact and estimated directions of arrival along with the amplitudes of the signal.

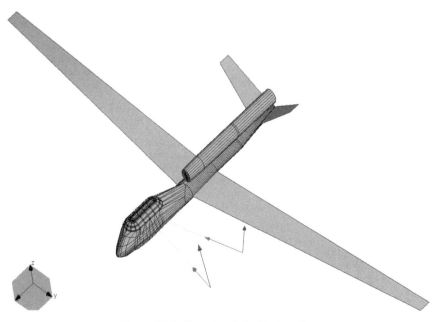

Figure 8.23. Two signals incident on the array.

TABLE 8.6. Estimated Direction of Arrival of the Signal and Their Amplitudes

	ϕ_{exact}	$\phi_{est.}$	θ_{exact}	$\theta_{est.}$	α_{exact}	$\alpha_{est.}$
Signal 1	70°	69.97°	20°	20.01°	1+j0	1.005+j0.004
Signal 2	110°	110.03°	−10°	−9.92°	1+j0	1.000+j0.033

8.5   RADAR CROSS SECTION (RCS) CALCULATION OF COMPLEX TARGETS

8.5.1   RCS Calculation of a Squadron of Tanks

In this example, simulation of the RCS of several tanks (http://en.wikipedia.org/wiki/Tank) is performed. The dimensions of the tanks are 8.0 m × 3.7 m × 2.75 m, as shown in Figure 8.24. The arrows show the plane wave propagating direction and its polarization. The RCS from a five-tank formation, which is shown in Figure 8.25, is also simulated and the results are plotted together in Figures 8.26 and 8.27. Both of the problems are considered to be in free space (without any ground plane) for demonstration purposes. The ground plane will be taken into account in the next example. The simulation for the five tanks at 100 MHz leads to 15,673 unknowns. It took 1.7 hours on a Dell Precision workstation 470 using the parallel out-of-core solver to complete the simulation. This PC has 1 dual-core Xeon 2.8 GHz CPU. The distances between the tanks are 5.0 m each along both the length and width of the tank, respectively.

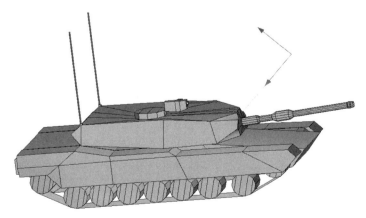

Figure 8.24. A tank illuminated by a plane wave.

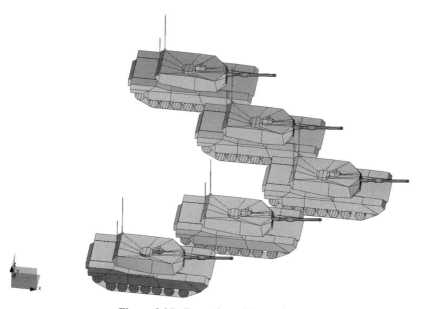

Figure 8.25. Formation of five tanks.

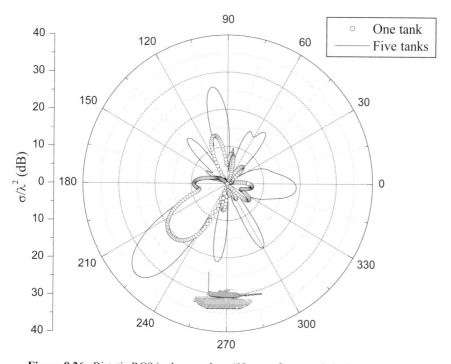

Figure 8.26. Bistatic RCS in the *xoz* plane (0° starts from *x* axis in the *xoz* plane).

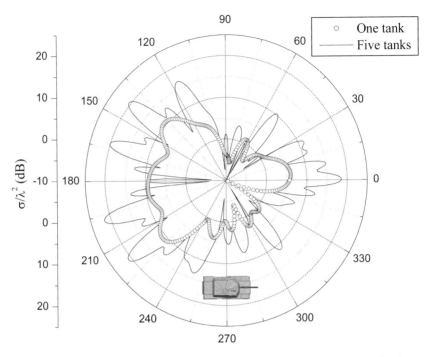

Figure 8.27. Bistatic RCS in the *xoy* plane (0° starts from *x* axis in the *xoy* plane).

8.5.2   RCS of the Tanks inside a Forest Environment

Scattering of EM waves from trees and foliage as well as the propagation of EM waves in the presence of forests plays an important role in many civilian and military applications; for example, in the usage of foliage penetrating radar for detecting potential targets in the forest. If all the trees are modeled as dielectric rods and plates, the time needed for a rigorous MoM simulation is unacceptable in the solution of real-life problems. Computationally efficient modeling of trees and foliage can be done with metallic wires (for branches) and metallic plates (for leaves) with distributed loadings over them. The error made in this approximation for the trees is negligible if all the branches have a radius $a \leq \lambda/8$, where a is the radius and λ is the wavelength at the operating frequency. The branches that do not satisfy this condition have to be modeled rigorously as dielectric rods. The model of a tree and four tanks inside a segment of the forest are shown in Figures 8.28 and 8.29, respectively.

The geometric model of the forest consists of 63,750 wires and 10,325 plates, which results in 97,175 unknowns. The model for the four tanks under the tree canopy generates 109,713 unknowns. The monostatic RCS of the forest and tanks inside the forest is calculated at 100 MHz using 16 nodes of Cluster-2, and it took approximately 6 seconds for each excitation after the matrix had been LU factored. The monostatic RCS in ϕ and θ cut planes obtained using the incident wave $1.0\,\hat{\theta}+1.0\,\hat{\phi}$ are plotted in Figures 8.30 and 8.31. It is interesting to observe that the radar returns from the tree canopy can sometimes be larger than the radar returns when the tanks are inside the tree canopy. Note that there is an infinite PEC ground plane for each model in this example.

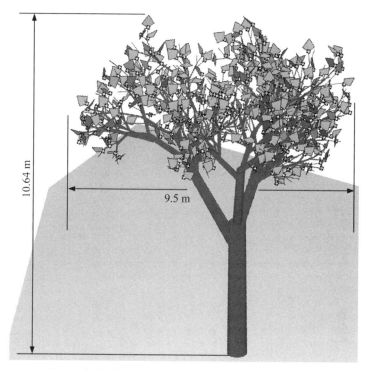

Figure 8.28. Tree modeled using wires and loaded plates.

Figure 8.29. Perspective view of four tanks inside a small forest of size 50 m × 50 m (the center of each tank is located at (-15,-15,0), (-5,-5,0), (5,5,0), and (-15,15,0), and the distance between the neighboring trees is 10.0 m in both x and y directions).

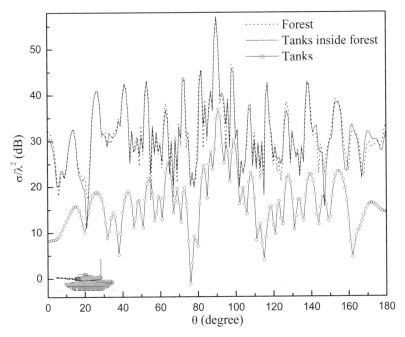

Figure 8.30. Monostatic RCS at ϕ cut plane (0° starts from x axis in the xoz plane).

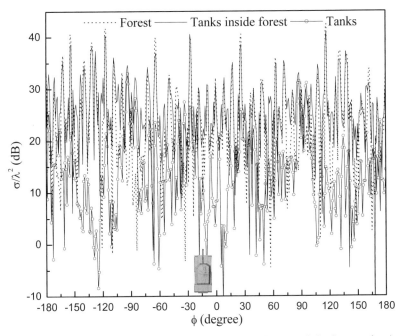

Figure 8.31. Monostatic RCS at θ cut plane (0° starts from x axis in the xoy plane).

8.5.3 RCS from an Aircraft and a Formation of Aircraft

In this example, the RCS of a "V" shape aircraft formation is calculated. The arrows in the Figure 8.32 represent the polarization of the electric field and the direction of the wave propagation with respect to the lead aircraft. The distances between any two neighboring aircraft are $\Delta x = 10.0$ m along the head direction, $\Delta y = 10.0$ m along the wing direction, and $\Delta z = 0.0$ m along the height direction.

The aircraft formation is shown in Figure 8.33. The RCS is calculated at 1.25 GHz using 32 nodes of Cluster-2 and plotted in Figures 8.34 and 8.35. The number of unknowns for the aircraft formation is 351,071 and it took 18.5 hours using 32 nodes of Cluster-2 for this simulation.

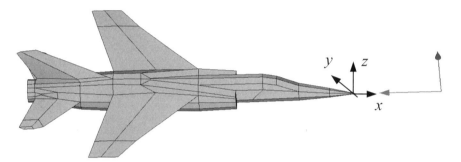

Figure 8.32. An aircraft with a plane wave excitation.

Figure 8.33. Aircraft flying in a V formation.

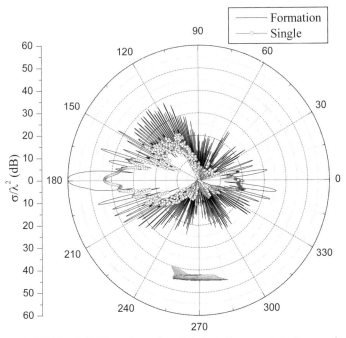

Figure 8.34. Bistatic RCS at ϕ cut plane (0° starts from x axis in the xoz plane).

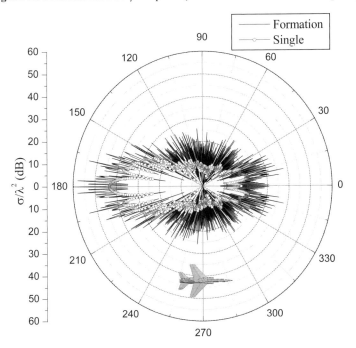

Figure 8.35. Bistatic RCS at θ cut plane (0° starts from x axis in the xoy plane).

8.5.4 RCS Simulation with Million Level Unknowns

To validate the robustness and the stability of the code, the bistatic RCS of a single aircraft is calculated at 6.15 GHz, the full size aircraft is shown in Figure 8.32. The numbers of unknowns in this case is 954,618 (approximately one million unknowns) when the structure is meshed at a frequency of 5.95 GHz. The meshed airplane is shown in Figure 8.36. The bistatic RCS is given in Figure 8.37.

This project was simulated in 177 hours using 64 nodes (512 cores) of Cluster-2, and it required approximately all the available hard disk space on every compute node of the cluster. This simulation demonstrates that the code is stable and can be used for even more unknowns as long as sufficient hard disk space is available.

Figure 8.36. The meshed airplane model (mesh frequency is 5.95 GHz).

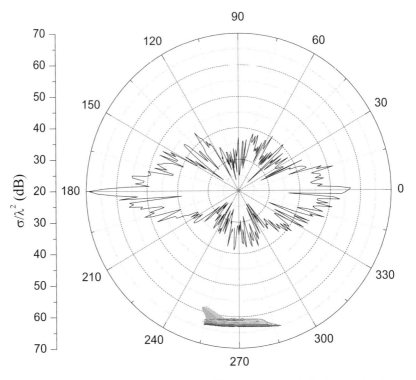

Figure 8.37. Bistatic RCS at ϕ cut plane (0° starts from x axis in the xoz plane).

8.5.5 RCS of an Aircraft Carrier

An aircraft carrier is a warship designed with a primary mission of deploying and recovering aircraft, thus acting as a seagoing airbase. Aircraft carriers allow a naval force to project air power from great distances without having to depend on local bases for staging aircraft operations. The carrier model used here is generated according to http://en.wikipedia.org/wiki/Aircraft_carrier. The model is about 265 m long, 66 m wide, and 47 m high. Sixty-one aircraft and six helicopters have been laid out along the deck of the carrier as shown in Figure 8.38 for demonstrating the analysis capability of the code. The aircraft model is the same as that described in Section 8.5.3. The helicopter is shown in Figure 8.39 (http://en.wikipedia.org/wiki/Ka-50). The body of the helicopter is modeled as the material with parameters $\varepsilon_r = 2$ and $\mu_r = 2$, whereas the rotating blades are modeled as metals. The top, front, and the side views of the model are given in Figure 8.40. The structure is considered in free space (not over an ocean surface) for demonstration purposes. The simulation for this model at 150 MHz requires a total of 559,059 unknowns. It took 38 hours and 45 minutes using 64 nodes (512 cores) of Cluster-2 to complete the simulation. The RCS of the structure is plotted in Figures 8.41 and 8.42.

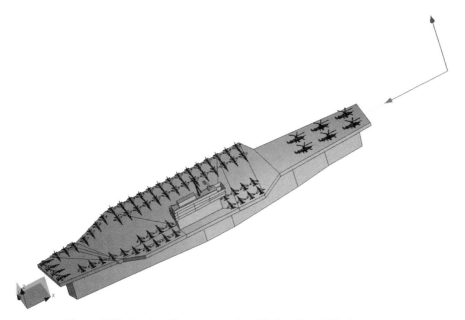

Figure 8.38. An aircraft carrier carrying 61 aircraft and 6 helicopters.

Figure 8.39. A model of a helicopter.

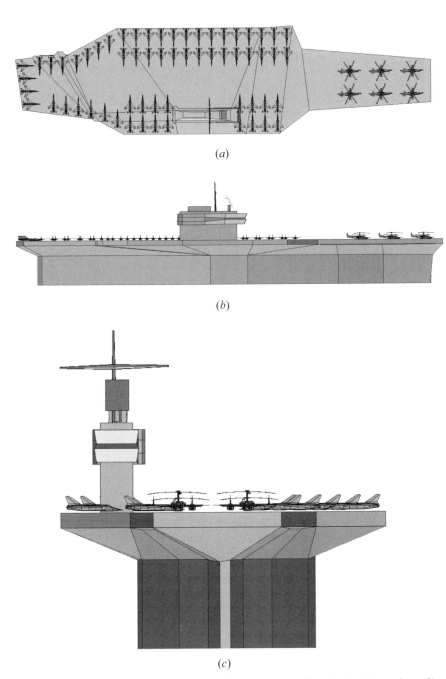

Figure 8.40. Layout of the aircraft carrier with relevant aircraft on deck: (*a*) top view; (*b*) side view; (*c*) front view.

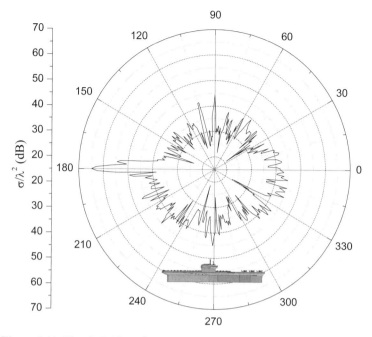

Figure 8.41. Bistatic RCS at ϕ cut plane (0° starts from x axis in the xoz plane).

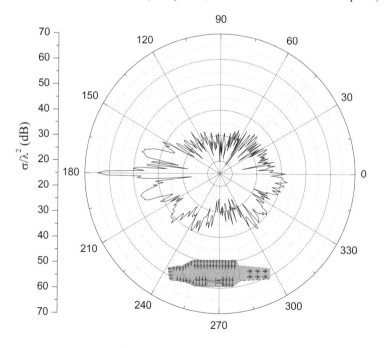

Figure 8.42. Bistatic RCS at θ cut plane (0° starts from x axis in the xoy plane).

8.6   ANALYSIS OF RADIATION PATTERNS OF ANTENNAS OPERATING INSIDE A RADOME ALONG WITH THE PLATFORM ON WHICH IT IS MOUNTED

Modern aircraft utilize electromagnetically transparent radome structures to protect antennas from environmental stresses while preserving the aerodynamic integrity of the vehicle's superstructure. This section shows that TIDES can be used to design antennas encased by a radome and operating in the presence of the platform on which they are mounted.

The *Yagi–Uda antenna*, commonly known simply as the *Yagi antenna* or *Yagi*, is a directional antenna system consisting of an array of dipoles and additional closely coupled parasitic elements (usually a reflector and one or more directors). In this example, a Yagi array and its radome are mounted on an aircraft model. The antenna array consists of

- 41 Yagi antennas
- Space between each Yagi pair: 0.6λ

The parameters of the Yagi elements of the array are shown in Figure 8.43. Figure 8.43 (*a*) shows a single Yagi antenna, in which the radius of each element antenna is 5.1 mm, and the radius of the feed wire of the folded element is 2.21739 mm. Figure 8.43 (*b*) shows the whole Yagi array backed by a metal plate.

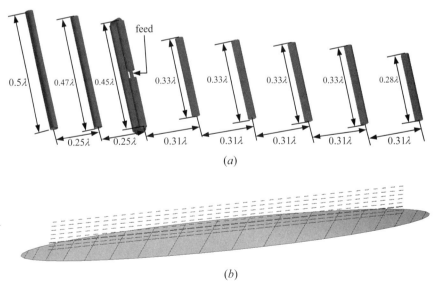

Figure 8.43. Layout of the Yagi array: (*a*) dimensions of a single Yagi antenna; (*b*) Yagi array and the reflection plate.

A model is developed for the radome inside which the array is placed. The parameters of the radome are

- Relative permittivity $\varepsilon_r = 2$
- Height of radome : 1 m
- Diameter of radome : 8.56 m
- Thickness of radome : 0.05λ

The placement of the antenna array inside the radome is shown in Figure 8.44.

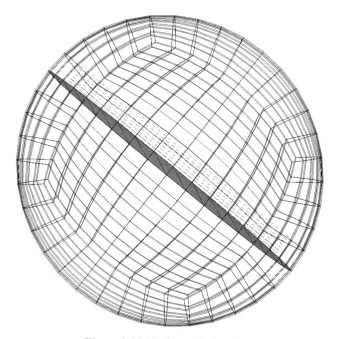

Figure 8.44. Yagi array inside the radome.

Figure 8.45 shows the perspective drawing of the whole structure consisting of the antenna, the radome, and the aircraft platform. It is important to understand how the radome affects the antenna array properties such as the gain and the beamwidth. These parameters can be investigated through simulations, if sufficient computer resources are available. A total of 611,318 unknowns are required to model the antenna array along with the radome, and the aircraft frame at 1 GHz. The aircraft model is 36 m long, 40 m wide, and 10.5 m high. It corresponds to 120λ, 133.3λ, and 35λ. The array shown in this example is for demonstration purpose only. The model of the aircraft is made according to http://en.wikipedia.org/wiki/Image:Tu-126_Moss.jpg.

The radiation patterns of the Yagi antenna array directed towards the tail are calculated using the parallel out-of-core solver. The simulation of the array inside the radome took 12.5 hours using 32 nodes of Cluster-2. When this array was placed inside the radome, and mounted on the airplane, the simulation took 49.4 hours using 64 nodes of Cluster-2. The overview of the model is shown in Figure 8.46. In the radiation patterns plotted in Figures 8.47 and 8.48, the small icons of the aircraft are used to show the orientation of the aircraft in the radiation plane. It is seen from Figure 8.47 that the level of the side and back lobes in the azimuth plane are increased because of the disturbance produced by the fuselage, tail, and wings. In Figure 8.48, the downward radiation in the elevation plane is decreased because of the effects of shielding by the fuselage and wing of the airplane platform. There is a slight asymmetry in the pattern of the array (without the platform) due to the presence of the Yagi folded element.

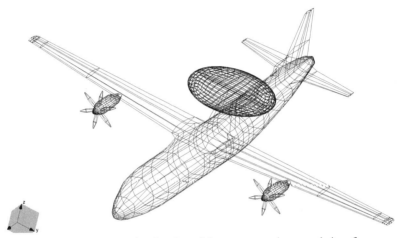

Figure 8.45. Perspective drawing of the antenna, radome, and aircraft.

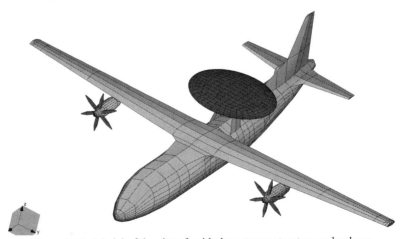

Figure 8.46. Model of the aircraft with the antenna structure and radome.

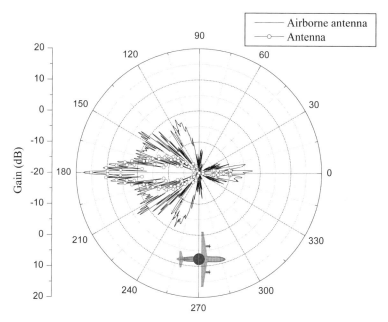

Figure 8.47. Azimuth radiation pattern (0° starts from *x* axis in the *xoy* plane).

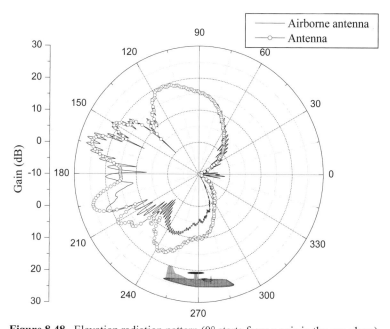

Figure 8.48. Elevation radiation pattern (0° starts from *x* axis in the *xoz* plane).

8.7   ELECTROMAGNETIC INTERFERENCE (EMI) ANALYSIS OF A COMMUNICATION SYSTEM

Electronic components and subsystems (e.g., microprocessor boards) are essential parts of modern civilian and military systems. Due to the impact of electromagnetic fields on the functionality of electronic components, microwave applications are of great interest in the analysis of undesired field penetration onto a structure which may interfere with a communication system. The capabilities of high-power microwaves (HPM) and Ultrawideband (UWB) components are the reason for intensive research activities in the area of microwave sources, and antennas. High-power RF/microwave energy can affect anything that responds to electromagnetically induced voltages and currents, which include electronics, materials and personnel.

Most aircraft and precision weapons have sensors, flight control surfaces, and other apertures that allow entry of radio frequency (RF) radiation. RF radiation uses two mechanisms to penetrate into a target, called *frontdoor coupling* and *backdoor coupling*.

Frontdoor coupling occurs when RF energy enters the system through a sensor or antenna designed to receive this type of radiation. Backdoor coupling encompasses every other way RF energy can penetrate a system. RF energy may enter through cracks, seams, seals, conduits, cable runs, apertures, solar cells, optical sensors, etc. UWB radiation is particularly effective at backdoor coupling because of the wide range of wavelengths involved, from 1 mm (gigahertz frequencies) to 3 m (megahertz frequencies), which allows them to penetrate from multiple points.

In this example, a scenario is given where an impulse radiating antenna (IRA) is assumed to be a pulsed power system. The IRA is a highly directive antenna and it can radiate an approximate impulse in the far field, when a rising step input is given. The IRA model used in this example is based on a Farr research design, using a $D = 1.83$ m diameter reflector and 45° feed arms with a 200 Ω load connecting each feed arm to the reflector, as shown in Figure 8.49 [4].

An electromagnetic communication system using the IRA is shown in Figure 8.50. The IRA is fed with 20,000 V at 2 GHz. Here, the IRA is mounted on one of the three trucks, which are used for supplying power. The trucks have identical dimensions of 7.0 m long, 2.6 m wide and 2.47 m high. The distance between the aircraft fuselage and the focus of the IRA is taken as 61.094 m, which is approximately $2\sqrt{2}\, D^2/\lambda$, where λ is the free space wavelength at 2 GHz, and D is the diameter of the reflector of the IRA antenna. It is notable that this distance is just taken for demonstration purposes, in fact, it can be any distance in the real case, and the varying distance will not result in an increase of the number of unknowns. The number of unknowns in this project is 541,512 and the simulation took 33 hours using all the 64 compute nodes (512 cores) of Cluster-2. The field distributions near the IRA antenna and aircraft are plotted in Figures 8.51 and 8.52, where the unit is kV/m.

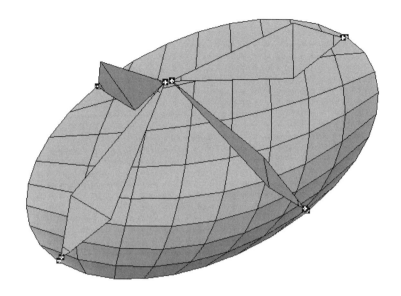

Figure 8.49. IRA model.

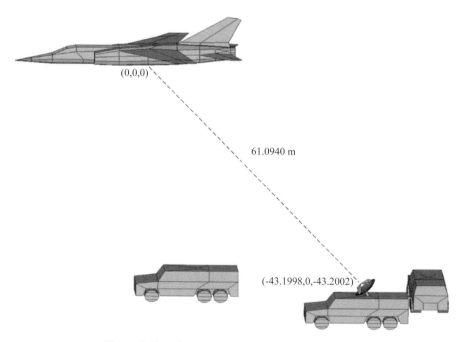

(0,0,0)

61.0940 m

(-43.1998,0,-43.2002)

Figure 8.50. Diagram of a communication system.

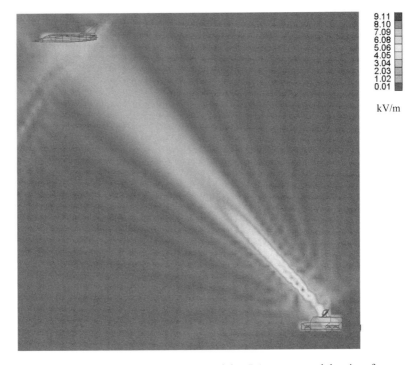

Figure 8.51. Field distribution around the IRA antenna and the aircraft.

Figure 8.52. Field distribution around the aircraft.

8.8 COMPARISON BETWEEN COMPUTATIONS USING TIDES AND MEASUREMENT DATA FOR COMPLEX COMPOSITE STRUCTURES

In this example, the numerical results from TIDES are compared with the measured data for complex composite structures. The complex structure consists of a dielectric radome containing seven large structural ribs that support an aircraft radome shell along with an L-band antenna array. The goal of this study is to examine the suppression of radome induced grating lobes as illustrated in [5]. The grating lobes are generated by seven large structural ribs that support an aircraft radome shell in front of an L-band antenna array. The structural ribs form a diffraction grating that introduces grating lobes in the array radiation pattern [5]. Our goal here is to demonstrate that using the computational electromagnetic code — TIDES, one can compute results for the radiation pattern which are within several tenths of a dB when compared with measurements for the grating lobe amplitudes. The difference between theory and experiment falls within the resolution of the measurements [5].

The complex composite structure considered here consists of the seven structural ribs bounding the L-band array, as shown in Figure 8.53, which is taken from Figure 3 of [5].

Figure 8.53. L-band antenna array with seven ribs.

The L-band array feeds a diffraction grating comprising seven large ribs. The center rib is the largest while the rest gets progressively smaller but in a symmetric fashion about the largest. These ribs generate grating lobes in the radiation pattern. The tapering of the diffraction grating has an effect on the grating lobes. The grating lobes appear because the distance between consecutive ribs is larger than the wavelength. A rib of a certain size, however, scatters a wave differently from a rib of another size. Thus, each rib has its own element pattern that influences the location and shape of a grating lobe. These structural ribs are fiber-reinforced composite sandwich structures using a honeycomb core. An effective dielectric constant for these structures was computed so that they could be modeled as homogenous structures in TIDES.

The mainbeam to first grating lobe amplitude ratio is the metric of interest in this study. The mainbeam to grating lobe ratio computed by TIDES was within 0.2 dB for all cases evaluated [5]. The measured and calculated patterns were nearly indistinguishable down to levels nearly 30 dB below the peak of the mainbeam. Figure 8.54 shows an example of the excellent match that was achieved.

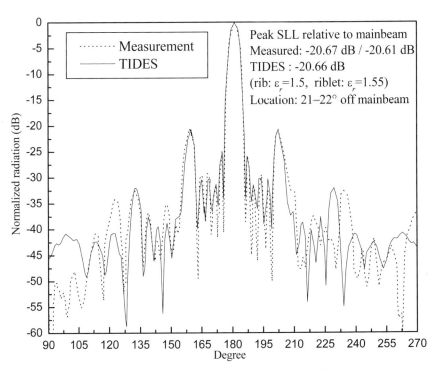

Figure 8.54. Comparison of measured array pattern and TIDES calculation.

8.9   CONCLUSION

In this chapter, several examples have been presented to provide an overview of the various applications of the code, TIDES, developed for this research. Numerical results from these examples, along with the comparison between TIDES and experimental measurements, show that the parallel in-core and out-of-core integral equation solver provides a very accurate and cost-effective solution for analyzing challenging real-life EM problems in the frequency domain.

REFERENCES

[1]   CIMNE, *GiD — The personal pre and post processor.* Available at: http://gid.cimne.upc.es/. Accessed Nov. 2008.

[2]   X. Zhao and C. Liang, "Performance Comparison between Two Commercial EM Software Using Higher Order and Piecewise RWG Basis Functions," *Microwave and Optical Technology Letters*, Vol. 51, No. 5, pp. 1219–1225, 2009.

[3]   T. K. Sarkar, M. Salazar Palma, and E. L. Mokole, *Physics of Multiantenna Systems and Broadband Processing*, Wiley Press, Piscataway, NJ, 2007.

[4]   M. C. Taylor and T. K. Sarkar, "Deconvolution of Target Signature for a Scene with Bistatic Impulse Radiating Antennas and a Sphere," *IEEE Conference on Radar*, Rome, NY, April 2006.

[5]   S. N. Tabet, J. S. Asvestas, and O. E. Allen. "Suppression of Radome Induced Grating Lobes," *24th Annual Review of Progress in Applied Computational Electromagnetics*, Niagara Falls, Canada, pp. 714–718, April 2008.

APPENDIX A

A SUMMARY OF THE COMPUTER PLATFORMS USED IN THIS BOOK

A.0 SUMMARY

Different computing platforms have been used in this book to observe how the parallel in-core and the out-of-core solvers operate in various hardware and software environments. In this appendix, a comprehensive summary is provided of the various computing platforms that have been used during the development of the various parallel codes for the solution of a frequency-domain integral equation.

A.1 DESCRIPTION OF THE PLATFORMS USED IN THIS BOOK

The parallel codes have been executed on different computer platforms. Table A.1 lists all the platforms that have been used along with a summary of the various hardware and software configurations for each platform. Note that when referring to the storage capacity of the hard disks, we use the abbreviations GB to represent 1,000,000,000 bytes and TB to represent 1,000,000,000,000 bytes. It is quite important to note that for the RAM a gigabyte represents 2^{30} bytes = 1,073,741,824 bytes which is actually bigger than 10^9 bytes, whereas for the hard disk a gigabyte of storage is actually 10^9 bytes. Therefore, the accessible capacity of the hard disk is less than the amount of RAM when the same terminology is used to characterize them. This distinction in this characterization of the same abbreviation is important when calculating the size of the problem that one can solve on a particular computer.

The various computer platforms that have been used for the parallel solution of the frequency-domain integral equation cover a range of computer architectures, including single-core, dual-core, and quad-core CPUs from both of the primary CPU manufacturers, Intel and AMD. The two common operating systems for the computational platforms, Linux and Windows, and three system manufacturers have been well represented. The various systems are presented in Table A.1 and described next for completeness. The various platforms were assembled at the Computational Electromagnetics Laboratory (CEM-LAB) of Syracuse University.

TABLE A.1. Summary of the 10 Computer Platforms Used in This Book

Platform (Architecture)	Operating System	Model	Processor	RAM(GB)
CEM-1 (IA-32)	MS-Windows 2000 Advanced Server	Custom-built	Intel Pentium III, 8 CPUs, single-core, 550 MHz	8
CEM-2 (EM64T)	MS-Windows XP Professional x64	Dell Precision workstation 670	Intel Xeon, 2 CPUs, dual-core, 2.8 GHz	8
CEM-3 (IA-64)	MS-Windows (64-bit) Advanced Server 2003	HP Integrity Server rx4640	Intel Itanium 2, 4 CPUs, single-core, 1.5 GHz	128
CEM-4 (EM64T)	Linux 2.6.9-11.ELsmp (x86_64)	Dell PowerEdge 1855 blade	Intel Xeon, 20 CPUs, single-core, 3.6 GHz	80
CEM-5 (IA-32)	Linux 2.6.9-42.0.2.ELsmp (x86)	CEM-LAB PC cluster	Intel Pentium 4, 6 CPUs, single-core, 3 GHz	12
CEM-6 (AMD64)	Linux 2.6.9-11.ELsmp (x86_64)	IBM BladeCenter LS41	AMD Opteron, 8 CPUs, dual-core, 2.80 GHz	16
CEM-7 (EM64T)	Linux 2.6.9-11.ELsmp (x86_64)	Dell PowerEdge 1955 blade	Intel Xeon, 4 CPUs, quad-core, 2.33 GHz	32
CEM-8 (EM64T)	MS-Windows XP Professional x64	Dell workstation	Intel Xeon, 2 CPUs, quad-core, 2.66 GHz	32
CEM-9 (EM64T)	Linux 2.6.18-53. ELsmp (x86_64)	IBM system x3500	Intel Xeon, 4 CPUs, quad-core, 3.0 GHz	48
CEM-10 (EM64T)	Windows Vista Ultimate SP1	Lenovo T61p laptop	Intel Core 2, 1 CPU, dual-core, 2.6 GHz	4

The first machine used is an IA-32 server called CEM-1 as shown in Figure A.1. In IA-32, each CPU cannot access more than 4 GB RAM. The machine has eight 550 MHz Intel Pentium III (single-core) processors (512 kB L2 cache and a 100-MHz frontside bus (FSB) [1]), 8 GB of RAM and a 70 GB hard disk. It uses a 32-bit Windows operating system.

Figure A.1. CEM-1 located at CEM-LAB.

The second machine, CEM-2, is a Dell Precision workstation 670 with two dual-core Intel Xeon 2.8 GHz EM64T processors (2 MB L2 cache with a 800-MHz FSB), 8 GB of RAM, and a 80-GB SATA (serial advanced technology attachment) [2] hard disk, shown in Figure A.2. The SATA hard drives are cheaper than the SAS [serial attached SCSI (small computer system interface)] drives but have a reduced performance level. The SATA computer bus has the primary function of transferring data between the motherboard and mass storage devices (such as hard-disk drives and optical drives) inside a computer. The SAS, on the other hand, is a data transfer technology designed to move data to and from computer storage devices such as hard drives and tape drives. It is a point-to-point serial protocol that replaces the parallel SCSI bus technology that first appeared in the mid-1980s in corporate data centers, and uses the standard SCSI command set. Construction of very high-capacity storage arrays with SAS will be possible. SAS hardware allows multipath I/O to devices, while SATA (prior to SATA II) does not.

The multicore technology allows a single processor to efficiently process multiple threads with each of the processors' cores executing one thread at full speed. Either four serial jobs or a single four-way parallel job can be executed on CEM-2. It uses a 64-bit Windows operating system.

Figure A.2. CEM-2 located at CEM-LAB.

The third machine, CEM-3, is a Hewlett-Packard Integrity rx4640 server, as shown in Figure A.3. It has four 1.5 GHz 64-bit Itanium 2 processors (256 kB L2 cache/4 MB L3 cache), 128 GB of RAM, and two 300 GB SCSI drives (RAID 1 [3]). This system is best suited for executing either serial or parallel jobs that require large shared memory. Since no commercial ScaLAPACK routines for Itanium 2 running on the 64-bit Windows operating system were available at the time of this research, a custom ScaLAPACK solver was developed. This is a true 64-bit machine.

Figure A.3. CEM-3 located at CEM-LAB.

The fourth machine, CEM-4, is a Dell blade cluster with 10 PowerEdge 1855 servers, as shown in Figure A.4. The blade cluster is composed of one head node and 9 compute nodes, all with the same configuration. Each blade has two Intel Xeon 3.6 GHz EM64T processors (2 MB L2 cache/800 MHz FSB), 8 GB of RAM, and one 146 GB, 15,000-rpm SCSI hard-disk drive, and another 300 GB, 10,000-rpm SCSI hard-disk drive. It uses a Linux operating system. All the nodes are interconnected with a Dell 5324 24-port gigabit Ethernet switch.

Figure A.4. CEM-4 located at CEM-LAB.

The fifth machine, CEM-5, is a Linux PC cluster. The PC cluster is composed of six PCs with the same configuration. Each PC has one 3.0 GHz Pentium 4 processor (512 kB cache/800 MHz FSB), 2 GB of RAM, and a 70-GB 7200-rpm SATA hard-disk drive. This machine is shown in Figure A.5. All the nodes are interconnected with a 10/100-Mbps adaptive switch.

The sixth machine, CEM-6, is an IBM blade cluster composed of two blades, as shown in Figure A.6. Each blade has four AMD dual-core Opteron 2.80 GHz AMD64 processors (2 MB L2 cache/1000 MHz Hyper transport), 16 GB of RAM and two 146-GB, 10,000-rpm SAS hard-disk drives (RAID 0). It also uses a Linux operating system. All the nodes are interconnected with a Dell 2708 8-port gigabit Ethernet switch.

Figure A.5 CEM-5 located at CEM-LAB.

Figure A.6. CEM-6 located at CEM-LAB.

The seventh machine, CEM-7, is also a Dell blade cluster with two PowerEdge 1955 servers, as shown in Figure A.7. The blade cluster is composed of one master node and one compute node, both with the same configuration. Each blade has two Intel quad-core Xeon 2.33 GHz EM64T processors (2×4 MB L2 cache/1333 MHz FSB), 16 GB RAM, and two 73-GB, 10,000-rpm SAS hard-disk drives (RAID 0). All the nodes are interconnected with a Dell 2708 8-port gigabit Ethernet switch. It uses a Linux operating system.

The eighth machine, CEM-8, is a Dell workstation. The workstation contains one node that has two Intel quad-core Xeon 2.66 GHz EM64T processors (2×4 MB L2 cache/1333 MHz FSB), 32 GB RAM, and three 300-GB, 10,000-rpm SAS hard-disk drives (RAID 5). A picture of CEM-8 is given in Figure A.8. It uses a Windows operating system.

The ninth machine, shown in Figure A.9, CEM-9, is a custom-built IBM cluster. The cluster is composed of two nodes, each having two Intel quad-core Xeon 3.0 GHz EM64T processors (12 MB L2 cache/1333 FSB), 24 GB RAM, and three 300-GB, 15,000-rpm SAS hard-disk drives (RAID 0). It uses a Linux operating system. All the nodes are interconnected with a DELL 2708 8-port gigabit Ethernet switch.

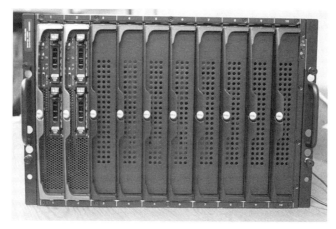

Figure A.7. CEM-7 located at CEM-LAB.

Figure A.8. CEM-8 located at CEM-LAB.

Figure A.9 CEM-9 located at CEM-LAB.

The tenth machine, CEM-10, is a Lenovo laptop as shown in Figure A.10. The laptop has one Intel Core 2 Duo T9500 2.66 GHz EM64T processor (3 MB L2 cache/800 MHz FSB), 4 GB RAM, and 200-GB, 7200-rpm hard-disk drives. Note that the user account control (UAC) must be turned off before installing MPICH2 under the Windows Vista operating system. Figure A.10 shows the computer labeled CEM-10.

Figure A.10. CEM-10 located at CEM-LAB.

Table A.2 provides a summary of several typical HP platforms that were also used but were located, except for the second one (Cluster-2), at the vendor's site. Cluster-2 is located at the CEM-LAB. Clusters labeled 1−5 are Hewlett-Packard c-Class blade systems with BL460c blades and Infiniband interconnect [4]. Cluster-6 is a HP ProLiant DL140 G3 packaged cluster. Cluster-4–Cluster-6 uses an HP Scalable File Share (SFS [5]) high-bandwidth storage appliance rather than the local hard disks.

Table A.2. A Summary of Six Typical HP Cluster Platforms

Cluster	Intel Processor	Model	Clock (GHz)	CPUs/ Cores per Node	RAM per Core (GB)	Total Cores	Disk Storage
Cluster-1	Xeon 5160	Dual-core	3.0	2/4	2	160	Local
Cluster-2	Xeon E5450	Quad-core	3.0	2/8	2	520	Local
Cluster-3	Xeon X5365	Quad-core	3.0	2/8	2	160	Local
Cluster-4	Xeon X5365	Quad-core	3.0	2/8	2	160	SFS
Cluster-5	Xeon 5160	Dual-core	3.0	2/4	2	160	SFS
Cluster-6	Xeon 5160	Dual-core	3.0	2/4	2	240	SFS

Cluster-1 has 40 nodes, each with two dual-core Intel Xeon 5160 3.0 GHz (2×2 MB L2 cache/1333 MHz FSB) EM64T processors, 8 GB of RAM, and 25 GB local hard disk drive capacity available for use. The nodes are interconnected via an Infiniband [4×DDR (double data rate)] switch.

Cluster-2 has one head node and 64 compute nodes, as shown in Figure A.11. The head node has two quad-core Intel Xeon E5450 3.0 GHz (2×6 MB L2 cache/1333 MHz FSB) EM64T processors, 16 GB of RAM, and four 146-GB 10,000-rpm SAS hard disks (two hard-disk drives are configured as RAID 1 and the other two as RAID 0). Each compute node has two quad-core Intel Xeon E5450 3.0 GHz (2×6 MB L2 cache/1333 MHz FSB) EM64T processors, 16 GB of RAM, and two 146-GB 10,000-rpm SAS hard disks (RAID 0). The platform is a BL460c blade system with three Infiniband (ConnectX IB DDR) switches.

Figure A.11. Cluster-2 located at CEM-LAB during the research.

Cluster-3 has 20 nodes, each with two quad-core Intel Xeon X5365 3.0 GHz (2×4 MB L2 cache/1333 MHz FSB) EM64T processors, 16 GB of RAM, and a 25-GB local hard-disk space. The nodes are interconnected via an Infiniband (4×DDR) switch.

The configuration of Cluster-4 is the same as that of Cluster-3 except that Cluster-4 uses an SFS, which is a high-bandwidth storage appliance.

The configuration of Cluster-5 is the same as that of Cluster-1 except that Cluster-5 uses an SFS storage appliance rather than the local hard disks.

Cluster-6 has 60 blades, each with two dual-core Intel Xeon 5160 3.0 GHz (2×2 MB L2 cache/1333 MHz FSB) EM64T processors, with 8 GB of RAM. The nodes are interconnected via an Infiniband (4×DDR) switch, and the cluster has an SFS high-bandwidth storage appliance.

Table A.3 summarizes the specifications of two additional IBM clusters (Cluster-7 and Cluster-8, which are IBM BladeCenter HS21 blade severs [6]) used in this research, but were located at the vendor's site.

Cluster-7 has 14 nodes, each with two quad-core Intel Xeon E5345 2.33 GHz (2×4 MB L2 cache/1333 MHz FSB) EM64T processors, 16 GB of RAM, and two 146-GB, 10,000-rpm SAS hard-disk drives. All the nodes are interconnected via an Infiniband (4×DDR) switch.

Cluster-8 has 8 nodes, each with two quad-core Intel Xeon E5450 3.0 GHz (12 MB L2 cache/1333 MHz FSB) EM64T processors, 16 GB of RAM, and two 73 GB, 10,000-rpm SAS hard disk drives. All the nodes are interconnected via an Infiniband (4×DDR) switch.

Table A.3. Summary of the Two IBM Cluster Platforms

Cluster	Intel Processor	Model	Clock (GHz)	CPUs/ Cores per Node	RAM per Core (GB)	Total Cores	Disk
Cluster-7	Xeon E5345	Quad-core	2.33	2/8	2	112	Local
Cluster-8	Xeon E5450	Quad-core	3.0	2/8	2	64	Local

A.2   CONCLUSION

In this appendix, several computer platforms have been introduced, including desktop PCs, PC clusters and high-performance clusters, which involve single-core, dual-core, and quad-core CPUs from both the primary CPU manufacturers, Intel and AMD. The two common operating systems for computational platforms, Linux and Windows (including Vista), have also been used in these representative computers. The portability of the parallel code presented in this book has been tested on a wide range of platforms, and the performance of the codes on each specific platform of interest has been tuned.

REFERENCES

[1] Wikipedia contributors, "Front Side Bus," *Wikipedia, The Free Encyclopedia.* Available at http://en.wikipedia.org/w/index.php?title=Front_side_bus&oldid= 220015743. Accessed June 2008.

[2] Wikipedia contributors, "Hard Disk Drive," *Wikipedia, The Free Encyclopedia.* Available at: http://en.wikipedia.org/w/index.php?title=Hard_disk_drive&oldid= 221422514. Accessed June 2008.

[3] Wikipedia contributors, "RAID," *Wikipedia, The Free Encyclopedia.* Available at: http://en.wikipedia.org/w/index.php?title=RAID&oldid=220543027.   Accessed June 2008.

[4] Hewlett-Packard, HP Cluster Platforms, HP Development Company, 2008. Available at: http://h20219.www2.hp.com/HPC/cache/276446-0-0-0-121.html. Accessed Oct. 2008.

[5] Hewlett-Packard, *HP Storage Works Scalable File Share*, HP Development Company, 2008. Available at: http://h20338.www2.hp.com/HPC/cache /276636-0-0-0-121.html. Accessed Oct. 2008.

[6] IBM, IBM BladeCenter HS21, IBM Website, 2008. Available at: http:// www-03.ibm.com/systems/bladecenter/hardware/servers/hs21/.   Accessed   Oct. 2008.

APPENDIX B

AN EFFICIENT CROSS-PLATFORM COMPILATION OF THE SCALAPACK AND PLAPACK ROUTINES

B.0   SUMMARY

There are many libraries available to support high-performance computation using the Linux and Windows operating systems. CMKL, as an example, supports both Windows and Linux. However, in some cases there is still the need to generate libraries for some particular computing platforms. Hence, there is a need to discuss how to tailor a particular library package to a specific platform. There are some commercial parallel libraries available for Itanium 2 with the Linux operating system. Even though CMKL now supports the library for Windows operating system, during this work, such a library was not available, and it was quite necessary to develop such a parallel library supporting the true 64-bit Itanium 2 systems. This appendix will give a detailed guide on how to develop such libraries. The Windows operating system will be used as an example. In the following, libraries of ScaLAPACK and PLAPACK will be generated for both 32-bit and 64-bit Windows operating system.

B.1   TOOLS FOR COMPILING BOTH SCALAPACK AND PLAPACK

Some software tools are needed for compiling ScaLAPACK and PLAPACK in Windows operating system. These software tools need to match the operating systems. For example, 64-bit (rather than 32-bit) Python should be used for a 64-bit Windows operating system. A list of these tools is as follows:

1. Intel FORTRAN compiler [1]
2. Intel C/C++ compiler [1]
3. Python (http://opsrc.org/download/releases/2.5/) [2]
4. SCons (http://www.scons.org/download.php) [3]
5. MPICH2 [4]

The Intel FORTRAN and Intel C/C++ compiler deliver outstanding performance by optimizing applications for Intel processors.

Python is a dynamic object-oriented programming language that can be used for many kinds of software development.

SCons is an open source software construction tool and is a next-generation build tool.

MPICH2 is an implementation of the message passing interface (MPI).

B.2   GENERATING THE SCALAPACK LIBRARY

B.2.1   Source Codes for Compiling ScaLAPACK

Before generating ScaLAPACK for Windows operating system, a list of source codes should be first downloaded. They are: blas.gz, mpiblacs.gz, mpiblacs-patch03.gz, and scalapack-1.7.4.gz. Note that it is not necessary for these source codes to be the latest version, and they are used here only for demonstration purpose.

B.2.2 Steps for Compiling ScaLAPACK

Suppose Intel C/C++ and FORTRAN compiler have already been installed, then we need to:

1. Install Python (e.g., install Python Version 2.5 into C:\Python2.5).
2. Install SCons (e.g., install SCons Version 0.96.1 into C:\Python2.5) by using the following commands:

    cd scons-0.96.1
    C:\Python2.5\python.exe setup.py install

3. Install MPICH2 for Microsoft Windows (e.g., install MPICH2 Version 1.0.3 into C:\Project\MPICH2).

Then, for BLAS, BLACS, and ScaLAPACK, follow these procedures:

1. Create directory for BLAS, BLACS, and ScaLAPACK. (e.g., C:\Project\BLAS; C:\Project\BLACS; C:\Project\ScaLAPACK-1.7.4)
2. Unzip ".gz" files into their corresponding directories.
3. Create the "SConstruct" and "SConscript" script files and put them in the corresponding directories.
4. Go into each directory and type "C:\Python2.5\scons-0.96.1.bat". Take BLAS as an example, type "C:\Python2.5\scons-0.96.1.bat >scons.log".
5. If you want to rebuild it, type "scons –c" to clean out the build and then execute step 4.

A user can directly compile the source codes by typing "scons-0.96.1.bat", and using the proper script files. How to create these script files will be described next. Note that the script files for the 32-bit Windows operating system may not work well for the FORTRAN files. In this case, the reader can use the method introduced for 64-bit Windows operating system to generate the libraries for 32-bit Windows system.

B.2.3   Script Files for 32-bit Windows Operating System

B.2.3.1. Script Files for BLAS. For BLAS, the SCons file list is

```
|+SConstruct
|+SConscript
```

The content of the file "SConstruct" is

SConscript('SConscript', build_dir = 'bld', duplicate = 0)

The content of the file "SConscript" is

```
import os
env = Environment(FORTRAN = 'ifort -nologo', ENV = os.environ)
files = Split('''
src/caxpy.f
src/ccopy.f
src/cdotc.f
src/cdotu.f
src/cgbmv.f
src/cgemm.f
src/cgemv.f
src/cgerc.f
src/cgeru.f
src/chbmv.f
src/chemm.f
src/chemv.f
src/cher.f
src/cher2.f
src/cher2k.f
src/cherk.f
src/chpmv.f
src/chpr.f
src/chpr2.f
src/crotg.f
src/cscal.f
src/csrot.f
```

src/csscal.f
src/cswap.f
src/csymm.f
src/csyr2k.f
src/csyrk.f
src/ctbmv.f
src/ctbsv.f
src/ctpmv.f
src/ctpsv.f
src/ctrmm.f
src/ctrmv.f
src/ctrsm.f
src/ctrsv.f
src/dasum.f
src/daxpy.f
src/dcabs1.f
src/dcopy.f
src/ddot.f
src/dgbmv.f
src/dgemm.f
src/dgemv.f
src/dger.f
src/dnrm2.f
src/drot.f
src/drotg.f
src/drotm.f
src/drotmg.f
src/dsbmv.f
src/dscal.f
src/dsdot.f
src/dspmv.f
src/dspr.f
src/dspr2.f
src/dswap.f
src/dsymm.f
src/dsymv.f
src/dsyr.f
src/dsyr2.f
src/dsyr2k.f
src/dsyrk.f
src/dtbmv.f
src/dtbsv.f
src/dtpmv.f
src/dtpsv.f
src/dtrmm.f
src/dtrmv.f

src/dtrsm.f
src/dtrsv.f
src/dzasum.f
src/dznrm2.f
src/icamax.f
src/idamax.f
src/isamax.f
src/izamax.f
src/lsame.f
src/sasum.f
src/saxpy.f
src/scasum.f
src/scnrm2.f
src/scopy.f
src/sdot.f
src/sdsdot.f
src/sgbmv.f
src/sgemm.f
src/sgemv.f
src/sger.f
src/snrm2.f
src/srot.f
src/srotg.f
src/srotm.f
src/srotmg.f
src/ssbmv.f
src/sscal.f
src/sspmv.f
src/sspr.f
src/sspr2.f
src/sswap.f
src/ssymm.f
src/ssymv.f
src/ssyr.f
src/ssyr2.f
src/ssyr2k.f
src/ssyrk.f
src/stbmv.f
src/stbsv.f
src/stpmv.f
src/stpsv.f
src/strmm.f
src/strmv.f
src/strsm.f
src/strsv.f
src/xerbla.f

src/zaxpy.f
src/zcopy.f
src/zdotc.f
src/zdotu.f
src/zdrot.f
src/zdscal.f
src/zgbmv.f
src/zgemm.f
src/zgemv.f
src/zgerc.f
src/zgeru.f
src/zhbmv.f
src/zhemm.f
src/zhemv.f
src/zher.f
src/zher2.f
src/zher2k.f
src/zherk.f
src/zhpmv.f
src/zhpr.f
src/zhpr2.f
src/zrotg.f
src/zscal.f
src/zswap.f
src/zsymm.f
src/zsyr2k.f
src/zsyrk.f
src/ztbmv.f
src/ztbsv.f
src/ztpmv.f
src/ztpsv.f
src/ztrmm.f
src/ztrmv.f
src/ztrsm.f
src/ztrsv.f
''')
blaslib = env.Library('blas.lib', files)

End of the script file for BLAS

Note that all the FORTRAN source codes within one directory are denoted by "*.f" in the following for convenience. The reader should list the names of all the codes instead (e.g., one can use "dir *.f > namelist.txt" to obtain the list), as we did above for BLAS. Similarly, all the C source codes within one directory are denoted by "*.c" in the following SConscript files for different components of ScaLAPACK.

B.2.3.2. Script Files for BLACS. For BLACS, SCons file list is

```
|+SConstruct
|+SConscriptBlacsBase
|+SConscriptBlacsCBase
|+SConscriptBlacsFortranBase
|+SConscriptCBlacs
|+SConscriptFortranBlacs
```

The content of the file "SConstruct" is

```
SConscript('SConscriptBlacsBase', build_dir = 'bldBase')
SConscript('SConscriptCBlacs', build_dir = 'bldC')
SConscript('SConscriptFortranBlacs', build_dir = 'bldFort')
```

The content of the file "SConscriptBlacsBase" is

```
#this is SConscriptBlacsBase
#builds the BLACS base lib from the INTERNAL codes
cinclude = Split('''
C:/Project/MPICH2/include
SRC/MPI
''')
fortinclude = Split('''
C:/Project/MPICH2/include
''')
import os
env = Environment(
ENV = os.environ,
#FORTRAN = 'ifort –nologo –I fortinclude',
#C = 'icl',
CPPPATH = cinclude,
FORTRANPATH = fortinclude,
CPPDEFINES = ['UpCase', 'UseMpi2']
)
internalfiles = Split('''
SRC/MPI/INTERNAL/*.c
SRC/MPI/INTERNAL/*.f
''')
extralibs = []
fortlib = SConscript('SConscriptBlacsFortranBase', build_dir = 'bldFortBase')
extralibs.append(fortlib)
clib = SConscript('SConscriptBlacsCBase', build_dir = 'bldCBase')
extralibs.append(clib)
env.Library('blacsbase', internalfiles+extralibs)
```

The content of the file "SConscriptBlacsCBase" is

```
#this is SConscriptBlacsCBase
#builds the C-compatible part of the BLACS base lib
cinclude = Split('"
C:/Project/MPICH2/include
"')
fortinclude = Split('"
C:/Project/MPICH2/include
"')
import os
env = Environment(
ENV = os.environ,
#FORTRAN = 'ifort -nologo',
#C = 'icl',
CPPPATH = cinclude,
FORTRANPATH = fortinclude,
CPPDEFINES = ['UpCase', 'UseMpi2']
)
libfiles= Split('"
SRC/MPI/*.c
"')

clib = env.Library('blacsCbase', libfiles)
Return('clib')
```

The content of the file "SConscriptBlacsFortranBase" is

```
#this is SConscriptBlacsFortranBase
#builds the Fortran-compatible part of the BLACS base lib

cinclude = Split('"
C:/Project/MPICH2/include
"')

fortinclude = Split('"
C:/Project/MPICH2/include
"')

import os
env = Environment(
ENV = os.environ,
#FORTRAN = 'ifort -nologo',
#C = 'icl',
CPPPATH = cinclude,
FORTRANPATH = fortinclude,
```

```
CPPDEFINES = ['UpCase', 'UseMpi2', 'CallFromC']
)
libfiles= Split('''
```
SRC/MPI/*.c
```
''')

fortlib = env.Library('blacsFortbase', libfiles)
Return('fortlib')
```

The content of the file " SConscriptCBlacs" is

```
#this is SConscriptCBlacs
#builds C interface to the BLACS

cinclude = Split('''
C:/Project/MPICH2/include
''')

fortinclude = Split('''
C:/Project/MPICH2/include
''')

import os
env = Environment(
ENV = os.environ,
#FORTRAN = 'ifort -nologo',
#C = 'icl',
CPPPATH = cinclude,
FORTRANPATH = fortinclude,
CPPDEFINES = ['UpCase', 'UseMpi2']
)
obj1defines = ['UpCase', 'UseMpi2', 'CallFromC', 'MainInC']
obj1 = env.Object('Cblacs_pinfo', 'SRC/MPI/blacs_pinfo_.c', CPPDEFINES =
obj1defines)

obj2defines = ['UpCase', 'UseMpi2', 'MainInC']
obj2 = env.Object('blacs_pinfo', 'SRC/MPI/blacs_pinfo_.c', CPPDEFINES =
obj2defines)

env.Library('c_blacs.lib', obj2+obj1)
```

The content of the file "SConscriptFortranBlacs" is

```
#this is SConscriptFortranBlacs
#builds fortran interface to the BLACS
cinclude = Split('''
```

```
SRC/MPI
C:/Project/MPICH2/include
''')

fortinclude = Split('''
C:/Project/MPICH2/include
''')

import os
env = Environment(
ENV = os.environ,
#FORTRAN = 'ifort -nologo',
#C = 'icl',
CPPPATH = cinclude,
FORTRANPATH = fortinclude,
CPPDEFINES = ['UpCase', 'UseMpi2']
)
obj1defines = ['UpCase', 'UseMpi2', 'CallFromC', 'MainInF77']
obj1 = env.Object('Cblacs_pinfo', 'SRC/MPI/blacs_pinfo_.c', CPPDEFINES = obj1defines)
obj2defines = ['UpCase', 'UseMpi2', 'MainInF77']
obj2 = env.Object('blacs_pinfo', 'SRC/MPI/blacs_pinfo_.c', CPPDEFINES = obj2defines)

env.Library('fortran_blacs.lib', obj2+obj1)
```

B.2.3.3. Script Files for ScaLAPACK. For ScaLAPACK, SCons File list is

```
|+SConstruct
|+SConscript
```

The content of the file "SConstruct" is

```
import os
pwd = os.getcwd()
SConscript('SConscript', build_dir = 'bld', duplicate = 0)
```

The content of the file "SConscript" is

```
cinclude = Split('''
C:/Project/MPICH2/include
''')
fortinclude = Split('''
C:/Project/MPICH2/include
''')
```

```
import os
env = Environment(
ENV = os.environ,
#FORTRAN = 'ifort -nologo',
#C = 'icl',
CPPPATH = cinclude,
FORTRANPATH = fortinclude,
CPPDEFINES = ['UpCase'],
)
libfiles = Split('''
PBLAS/SRC/*.c
PBLAS/SRC/*.f
PBLAS/SRC/PTOOLS/*.c
PBLAS/SRC/PTZBLAS/*.f
SRC/*.c
SRC/*.f
TOOLS/*.f
TOOLS/*.c
C:/Project/blas/bld/blas.lib
C:/Project/blacs/bldFort/fortran_blacs.lib
#C:/Project/blacs/bldC/c_blacs.lib
C:/Project/MPICH2/lib/fmpich2.lib
''')
env.Library('scalapack', libfiles)
```

B.2.4 Script Files for 64-bit Windows Operating System

The method outlined above works for the 32-bit Windows operating systems. However, for the 64-bit Windows operating systems, a slight change has to be made. The "*.f " need to be changed into "*.obj " in the SConscript files. Keep in mind that the reader should list the names of all the ".obj" files. All the FORTRAN files listed in the different script files in Section B.2.3 can be compiled directly by entering into the corresponding folder and typing

$$ifort -c *.f$$

to generate the .obj files (note that the compiler option "–I " may be needed to include the head file of MPICH2).

Once the SConscript files are updated for the FORTRAN files, we can directly use these modified SConscript files to generate the libraries by executing "C:\Python2.5\scons-0.96.1.bat", with all the C files being taken into account automatically.

Note that the method introduced here works well for both 64-bit (including EM64T and IA-64) and 32-bit Windows operating systems, although the method introduced in Section B.2.3 offers a simpler way for 32-bit Windows system.

B.3   GENERATING THE PLAPACK LIBRARY

B.3.1   Source Codes for Compiling PLAPACK

The source codes for generating PLAPACK for the Windows operating system need to be downloaded first. The source codes are contained in PLAPACKR32.tgz at http://www.cs.utexas.edu/users/plapack/source.html. Note that it is not necessary for the source codes to be the latest version, and they are used here only for demonstration purposes.

Note that the "header" files of the source codes may need some modifications to meet the compatibility requirement of the operating system.

B.3.2   Script Files for PLAPACK

For PLAPACK, the SCons file list is

```
|+SConstruct
|+SConscript
```

The content of the file "SConstruct" is

```
import os
pwd = os.getcwd()
SConscript('SConscript', build_dir = 'bld', duplicate = 0)
```

The content of the file "SConscript" is

```
cinclude = Split('''
C:/Project/MPICH2/include
C:/Python25/PLAPACKR32/INCLUDE
''')

fortinclude = Split('''
C:/Project/MPICH2/include
''')

#import os user's environment variables into SCons,
#and set extra SCons environment variables

import os
env = Environment(
ENV = os.environ,
C = 'icl',
FORTRAN = 'ifort -nologo',
CPPPATH = cinclude,
```

```
FORTRANPATH = fortinclude,
CPPDEFINES = ['UpCase'],
)

libfiles = Split('''
TEMPLATE/*.c
API/*.c
BLAS1/*.c
BLAS2/*.c
BLAS3/*.c
COPY/*.c
Local_BLAS/*.c
OBJ/*.c
REDUCE/*.c
UTIL/*.c
IO/*.c
SOLVERS/LU/*.c
SOLVERS/Invert/*.c
FORTRAN_interface/pla_type_conversion_f.f
FORTRAN_interface/PLA_Conversion.c
lib/fmpich2.lib
''')
env.Library('plapack', libfiles)
```

End of the script file for PLAPACK

Note that "*.c" in the SConscript file denotes the name list for all the source code files with the extension ".c" in one folder. As an example

TEMPLATE/*.c

should be replaced with the following:

TEMPLATE/PLA_Environ.c
TEMPLATE/PLA_Init.c
TEMPLATE/PLA_Temp.c
TEMPLATE/PLA_Temp_distant_length.c

The "*.c" in the SConscript file for other folders need to be replaced in the same way.

B.4   TUNING THE PERFORMANCE BY TURNING ON PROPER FLAGS

Intel tunes its compilers to optimize for its hardware platforms to minimize stalls and to produce a code that executes in the smallest number of cycles. Different compiler flags will influence the performance [5]. A short list of different flags is given in Table B.1.

As a demonstration, Table B.2 lists the CPU time with different compiler flags. The PEC sphere model in Chapter 4 is simulated with 253 geometric elements and 12,012 unknowns. The results in Table B.2 indicate that using "-O2" compiler flags, the speed of computation is increased 2 times. Therefore, it is suggested that the user try different flags when compiling ScaLAPACK and PLAPACK library on their platforms.

TABLE B.1 Several Widely Used Flags for the Intel Compiler

Windows	Linux	Comment
/Od	-O0	No optimization; useful for a quick compiling
/O1	-O1	Optimize for size
/O2	-O2	Optimize for speed and enable some optimization
/O3	-O3	Enable all optimizations as O2, and intensive loop optimizations
/fast	-fast	Shorthand; on Windows equates to "/O3 /Qipo"; on Linux equates to "-O3 -ipo -static"; useful when compiling a program for release

TABLE B.2 CPU Time When Using Different Flags

Flags	Process			
	Process 0	Process 1	Process 2	Process 3
Matrix Filling CPU Time (seconds)				
-u -w -O	77.01	76.39	70.4	69.06
-u -w -O2	31.94	31.67	29.57	29.04
Far-Field Calculation CPU Time (seconds)				
-u -w -O	8.998E-003	8.998E-003	7.998E-003	6.999E-003
-u -w -O2	4.999E-003	4.999E-003	4.999E-003	2.9999E-003

B.5   CONCLUSION

In this appendix, a cross-platform compilation of the various components of the code is provided for generating ScaLAPACK and PLAPACK library packages using open-source codes. The method described here is quite effective in demonstrating the portability of the parallel codes.

REFERENCES

[1]   Intel® Software Network, "Intel® Compilers." Available at: http://software.intel.com/en-us/intel-compilers. Accessed April 2009.

[2]   Python™, "Download > Releases > 2.5 (Beta)." Available at: http://opsrc.org/download/releases/2.5/. Accessed Oct. 2008.

[3]   S. Knight, "1.3 Building and Installing SCons on Any System," *SCons User Guide*, 2007. Available at: http://www.scons.org/doc/HTML/scons-user/x166.html. Accessed Oct. 2008.

[4]   MPICH2, "High Performance and Widely Portable Message Passing Interface (MPI)." Available at: http://www.mcs.anl.gov/research/projects/mpich2. Accessed Oct. 2008.

[5]   Intel, "Intel® C++ Compiler Optimizing Applications," Intel Support, Performance Tools. Available at: ftp://download.intel.com/support/performancetools/c/linux/v9/optaps_cls.pdf.

APPENDIX C

AN EXAMPLE OF A PARALLEL MOM SOURCE CODE FOR ANALYSIS OF 2D EM SCATTERING

C.0   SUMMARY

This appendix presents an example for the implementation of a parallel 2D MoM code. First, the basic idea and procedures of MoM are introduced together with the solution of the functional equations. Then the analysis of a 2D scattering problem using the EFIE is illustrated. At the end of this appendix, the parallel code is provided to solve the 2D scattering problem. The parallel code presented here is suitable only for the solution of this problem. This demonstration code of this appendix can be considered as a starting point for interested readers to develop their own parallel codes. In that case, it is only necessary to modify the corresponding part to get their parallel MoM code for other applications. Codes for the solution of 3D problems are also available from the authors.

C.1   INTRODUCTION OF MOM

Consider the solution of the inhomogeneous equation

$$L(f) = g, \tag{C.1}$$

where L is a given linear operator, g is the known excitation, and f is the unknown response to be solved for. Here it is assumed that Eq. (C.1) is a *deterministic* problem, in other words, only one f is associated with a given g.

Let the unknown f be expanded in a series of known basis functions $f_1, f_2, f_3, \ldots$ in the domain of L so that

$$f = \sum_n \alpha_n f_n, \tag{C.2}$$

where α_n are the unknown constant coefficients to be solved for. f_n are the known expansion functions or basis functions. To obtain an exact solution, Eq. (C.2) usually requires an infinite summation and the functions $L(f_n)$ should form a complete set, so that it spans g. To obtain an approximate solution, Eq. (C.2) usually uses a finite summation. Substitute Eq. (C.2) into Eq. (C.1), and use the linearity of L, to obtain

$$\sum_n \alpha_n L(f_n) = g .$$
(C.3)

Assume that a suitable inner product $\langle f, g \rangle$ is defined for the problem. Now define a set of weighting functions, or testing functions, $w_1, w_2, w_3, \ldots$ in the range of L, and take the inner product of Eq. (C.3) with each w_m ($m = 1, 2, 3, \ldots$). The result is

$$\sum_n \alpha_n \langle w_m, L(f_n) \rangle = \langle w_m, g \rangle .$$
(C.4)

This set of equations can be written in matrix form as

$$[l_{mn}][\alpha_n] = [g_m],$$
(C.5)

where

$$[l_{mn}] = \begin{bmatrix} \langle w_1, L(f_1) \rangle & \langle w_1, L(f_2) \rangle & \langle w_1, L(f_3) \rangle & \cdots \\ \langle w_2, L(f_1) \rangle & \langle w_2, L(f_2) \rangle & \langle w_2, L(f_3) \rangle & \cdots \\ \langle w_3, L(f_1) \rangle & \langle w_3, L(f_2) \rangle & \langle w_3, L(f_3) \rangle & \cdots \\ \cdots & \cdots & \cdots & \cdots \end{bmatrix},$$
(C.6)

$$[\alpha_n] = \begin{bmatrix} \alpha_1 \\ \alpha_2 \\ \alpha_3 \\ \vdots \end{bmatrix},$$
(C.7)

$$[g_m] = \begin{bmatrix} \langle w_1, g \rangle \\ \langle w_2, g \rangle \\ \langle w_3, g \rangle \\ \vdots \end{bmatrix}.$$
(C.8)

The quality of the approximation provided by this solution depends on the choice of the functions f_n and w_m. The general procedure described above for solving these equations is called the *method of moments* (MoM) [1]. The particular choice $w_n = f_n$ leads to the *Galerkin's method*. The basic idea of MoM is to reduce a functional equation to a matrix equation, and then solve the matrix equation using available matrix techniques.

One of the main tasks in the solution of any particular problem is the choice of f_n and w_n. The function $L(f_n)$ must be linearly independent and should be chosen so that this superposition approximates g reasonably well. The testing function w_n must also be linearly independent. Certain rules that should be followed in the choice of the testing functions are provided in reference [2] from a mathematical standpoint.

C.2 SOLUTION OF A TWO-DIMENSIONAL SCATTERING PROBLEM

The MoM methodology is now used to calculate the scattering from a 2D perfectly conducting cylinder so that it is infinitely long along one direction and is illuminated by a transverse magnetic (TM)-polarized plane wave.

C.2.1 Development of the Integral Equation and the MoM Solution

Consider a perfectly conducting cylinder excited by an impressed electric field E_z^i, as shown in Figure C.1.

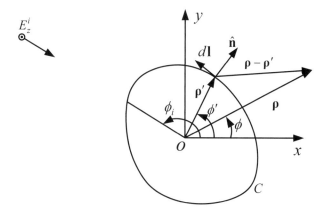

Figure C.1. A 2D PEC scattering problem.

The impressed field E_z^i induces a surface current J_z on the conducting cylinder, which produces a scattered field E_z^s, which is given by

$$E_z^s(\mathbf{\rho}) = \frac{-k\eta}{4} \int_C J_z(\mathbf{\rho}') H_0^{(2)}\left(k|\mathbf{\rho}-\mathbf{\rho}'|\right) dl', \qquad (C.9)$$

where $k = \omega\sqrt{\mu\varepsilon}$ is the wavenumber, $\eta = \sqrt{\mu/\varepsilon} \approx 120\pi$ is the intrinsic impedance of free space, and $H_0^{(2)}$ is the Hankel function of the second kind (zero order). Equation (C.9) is obtained by solving the two-dimensional Helmholtz equation:

$$\nabla^2 E_z + k^2 E_z = j\omega\mu J_z. \qquad (C.10)$$

The boundary condition is given by

$$E_z^{\text{total}} = E_z^i + E_z^s = 0. \qquad (C.11)$$

In other words, the total tangential electric field is zero on the conducting boundary C. Combining Eqs. (C.9) and (C.11) leads to the following integral equation

$$\frac{k\eta}{4} \int_C J_z(\mathbf{\rho}') H_0^{(2)}\left(k|\mathbf{\rho}-\mathbf{\rho}'|\right) dl' = E_z^i(\mathbf{\rho}), \quad \mathbf{\rho} \text{ on } C, \qquad (C.12)$$

where $E_z^i(\mathbf{\rho})$ is known and J_z is the unknown to be solved for. In this scattering problem the incident field is a TM plane-wave that is arriving from the ϕ_i direction. It is given by

$$E_z^i(x, y) = e^{jk(x\cos\phi_i + y\sin\phi_i)}. \qquad (C.13)$$

The simplest numerical solution of Eq. (C.12) consists of using the pulse functions as the basis function and using point matching for testing. To accomplish this, the contour of the scatterer C is divided into N segments of length ΔC_n, which are characterized by the N set of points (x_n, y_n) ($n = 1, \ldots, N$) on the PEC surface, as illustrated in Figure C.2. The midpoints of these segments are

$$x_n^c \approx \frac{x_{n+1} + x_n}{2}, \quad y_n^c \approx \frac{y_{n+1} + y_n}{2}. \quad (x_{N+1} = x_1, y_{N+1} = y_1.) \qquad (C.14)$$

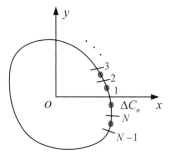

Figure C.2. Subdivision performed along the conducting boundary.

The distance between these points are given by

$$\rho_{mn} = \sqrt{\left(x_m^c - x_n^c\right)^2 + \left(y_m^c - y_n^c\right)^2}, \quad m,n = 1,2,\ldots,N. \tag{C.15}$$

First, the pulse basis functions, which are simple to deal with, are introduced as follows:

$$f_n(\boldsymbol{\rho}) = \begin{cases} 1 & \boldsymbol{\rho} \in \Delta C_n \\ 0 & \text{else} \end{cases}, \quad n = 1,2,\ldots,N. \tag{C.16}$$

Then, the testing functions are defined as

$$w_m(x,y) = \delta(x - x_m^c, y - y_m^c), \quad m = 1,2,\ldots,N, \tag{C.17}$$

which implies that Eq. (C.12) is satisfied at the midpoint (x_m^c, y_m^c) of each ΔC_m.
Now let

$$J_z = \sum_{n=1}^{N} \alpha_n f_n. \tag{C.18}$$

Substitute it into Eq. (C.12), and form the following inner product on both sides of the equation to yield

$$\left\langle w_m, \frac{k\eta}{4} \sum_{n=1}^{N} \alpha_n \int_C f_n(\boldsymbol{\rho}') H_0^{(2)}\left(k|\boldsymbol{\rho} - \boldsymbol{\rho}'|\right) dl' \right\rangle = \left\langle w_m, E_z^i(\boldsymbol{\rho}) \right\rangle. \tag{C.19}$$

This results in the following matrix equation:

$$[l_{mn}][\alpha_n] = [g_m], \tag{C.20}$$

where $[\alpha_n]$ is the coefficient vector and $[g_m]$ is the excitation vector associated with each element

$$g_m = E_z^i(x_m^c, y_m^c).$$ (C.21)

The elements of $[l_{mn}]$ are given by

$$l_{mn} = \frac{k\eta}{4} \int_{\Delta C_n} H_0^{(2)}\left(k\sqrt{(x_m^c - x)^2 + (y_m^c - y)^2}\right) dl.$$ (C.22)

Once the solution of Eq. (C.20) is obtained, the current can be found using Eq. (C.18).

There is no simple analytical expression for Eq. (C.22), but it can be evaluated by employing suitable approximations. Approximate the complete surface current over the nth segment as a line source J_{zn} located at $\left(x_n^c, y_n^c\right)$ and thus obtain the crudest approximation by treating an element $J_z \Delta C_n$ as a filament of current when the testing point is not on ΔC_n. This will result in

$$l_{mn} \approx \frac{k\eta}{4} \Delta C_n H_0^{(2)}\left(k\rho_{mn}\right), \ m \neq n,$$ (C.23)

where ρ_{mn} is given by Eq. (C.15) and the segment length ΔC_n is

$$\Delta C_n \approx \sqrt{\left(x_{n+1} - x_n\right)^2 + \left(y_{n+1} - y_n\right)^2}, \ n = 1, 2, ..., N.$$ (C.24)

For the self-interaction term represented by diagonal elements l_{mn}, the Hankel function has an integrable singularity, and the integral must be evaluated analytically. We now approximate ΔC_n by a straight line and use the small argument approximation for the Hankel function to yield

$$H_0^{(2)}(z) \approx 1 - j \frac{2}{\pi} \ln\left(\frac{\gamma z}{2}\right),$$ (C.25)

where $\gamma \approx 1.781$ is Euler's constant. An evaluation of Eq. (C.22) then yields

$$l_{mn} \approx \frac{k\eta}{4} \Delta C_n \left[1 - j\frac{2}{\pi}\left(\ln\left(\frac{\gamma k \Delta C_n}{4}\right) - 1\right)\right], \ m = n.$$ (C.26)

C.2.2   Evaluation of the Parameter of Interest

The scattering cross section is defined as [1]

$$\sigma(\phi) = \lim_{\rho \to \infty} 2\pi\rho \left|\frac{E^s(\phi)}{E^i}\right|^2 = \frac{k\eta^2}{4}\left|\left[V_n^s\right]^T \left[Z_{mn}\right]^{-1}\left[V_m^i\right]\right|^2,$$ (C.27)

where $\mathbf{E}^s(\phi)$ is the field at angle ϕ radiating from J_z, $\left[V_n^s\right]$ is the scattering matrix given by

$$\left[V_n^s\right] = \left[\Delta C_n \exp\left(jk(x_n \cos\phi + y_n \sin\phi)\right)\right], \tag{C.28}$$

$\left[Z_{mn}\right]$ is the impedance matrix given by

$$\left[Z_{mn}\right] = \left[\Delta C_m l_{mn}\right], \tag{C.29}$$

and $\left[V_m^i\right]$ is the voltage matrix given by

$$\left[V_m^i\right] = \left[\Delta C_m \exp\left(jk(x_m \cos\phi_i + y_m \sin\phi_i)\right)\right]. \tag{C.30}$$

C.3   IMPLEMENTATION OF A SERIAL MOM CODE

Since all the relevant expressions are now available for the calculation of MoM impedance matrix and excitation vector, we present the flowchart of the serial MoM algorithm and the associated source code written in Matlab as below.

C.3.1   Flowchart and Results of a Serial MoM Code

From the discussions in Section C.2, it is seen that the code will consist mainly of four parts, namely, geometry initialization, evaluation of the elements of the impedance and voltage matrices, solution of the matrix equation, and finally, performing some postprocessing to obtain the near/far fields or other parameters of interest. The flowchart of a typical serial MoM program is given in Figure C.3.

The code needs the geometry information for calculating the inner product between the basis and the testing functions. Once this is obtained, the code proceeds to calculate all the elements of the impedance and voltage matrices, followed by solving the matrix equation. After solutions for the current coefficients are obtained, postprocessing of the results is started to calculate the parameters of interest.

A Matlab code is given in the next section to implement the flowchart of the serial MoM code. The incident wave, which is propagating along the $+x$ axis (horizontal axis), the scattered field, and the total field near a PEC circular cylinder ($ka=1$, k is wavenumber, a is the radius of the PEC cylinder) are presented. An animated screenshot for each one is given in Figures C.4, C.5, and C.6, where the unit of the value shown in colormap is V/m. The numbers in the vertical and the horizontal axes in these figures represent the value of the sample. The width of the square sampling window is 8λ.

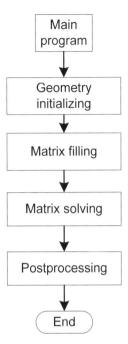

Figure C.3. The flowchart of a serial MoM code.

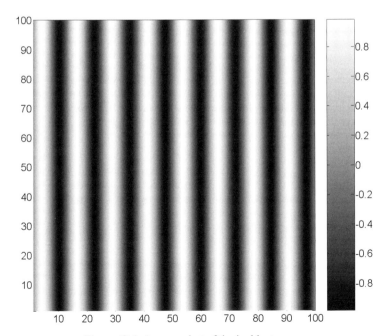

Figure C.4. A screenshot of the incident wave.

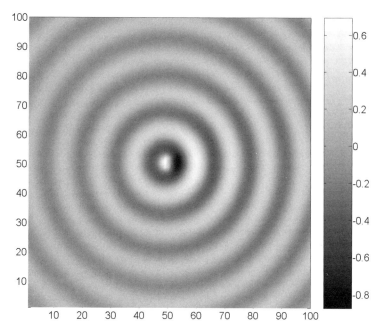

Figure C.5. A screenshot of the scattered field distributed around the circular cylinder.

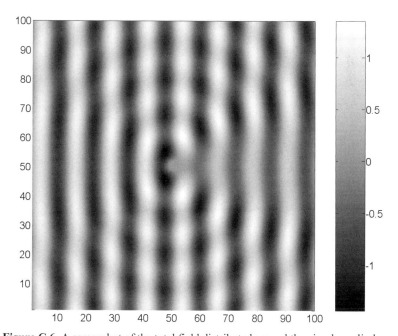

Figure C.6. A screenshot of the total field distributed around the circular cylinder.

C.3.2 A Serial MoM Source Code for the 2D Scattering Problem

```
% Matlab code for the MoM solution to 2D scattering problems
% TMz field, parametric PEC surface
function mom
wavek = 2*pi;
wavelength=2*pi/wavek;
a=1.0/wavek;
eta = 377;
c = (eta*wavek/4);  %global constant
N = 20; %number of segments on surface
Z_mn = zeros(N,N);  %"impedance" matrix
V_m = zeros(N,1);   %"voltage" matrix
w =4*wavelength;  %window goes from -w to w
Nx = 100;      %number of near field points across window in x direction
Ny = 100;      %number of near field points across window in y direction

%% geometry initializing
for n=1:N+1 % calculate points on PEC surface
   x(n) = a*cos(2.0*pi*(n-1.0)/N);
   y(n) = a*sin(2.0*pi*(n-1.0)/N);
end
for n=1:N
   xc(n) = (x(n+1)+x(n))/2; %mid point of segment
   yc(n) = (y(n+1)+y(n))/2;
   L(n) = sqrt((x(n+1)-x(n))^2+(y(n+1)-y(n))^2);   % segment length
end

%% filling the matrix
for m=1:N
   V_m(m) = Einc(xc(m),yc(m),wavek);
   for n=1:N
   if (m==n)
   Z_mn(m,n) = c*L(n)*(1-j*(2/pi)*(log(0.455*wavek*L(n))-1));
   else
   R = sqrt((xc(m)-xc(n))^2+(yc(m)-yc(n))^2);
   Z_mn(m,n) = c*L(n)*besselh(0,2,wavek*R);
   end
   end
end

%% solve the matrix equation for surface current...
I = inv(Z_mn)*V_m

%% post processing...
% calculate total field
```

```
A = zeros(Nx,Ny);
for ix=1:Nx
   xo = (2*((ix-1)/(Nx-1))-1)*w;
   for iy=1:Ny
   yo = (2*((iy-1)/(Ny-1))-1)*w;
   for n=1:N
   R = sqrt((xo-xc(n))^2+(yo-yc(n))^2);
   A(ix,iy) = A(ix,iy)-c*I(n)*L(n)*besselh(0,2,wavek*R);
   end
   A(ix,iy) = A(ix,iy)+Einc(xo,yo,wavek); % total field
   end
end
figure
Aoutput=abs(A);
save('nearf.dat','Aoutput','-ASCII')
pcolor(Aoutput');shading interp;axis equal;xlim([1 100]);

%% animating the field
figure
while (1)
   for k=1:20
   phi = 2*pi*k/20
   B = real(A*exp(j*phi));
   pcolor(B'); shading interp; axis equal; xlim([1 100]);
   pause(0.1);
   end
end

%% incident wave
function E = Einc(x,y,wavek)
%plane wave incident at angle phi
phi = 180*pi/180.0;
E = exp(j*wavek*(x*cos(phi)+y*sin(phi)));
```

C.4   IMPLEMENTATION OF A PARALLEL MOM CODE

In this section, the method for parallelizing the serial MoM code is described in detail.

C.4.1   Flowchart and Results of a Parallel MoM Code

The flowchart of a typical parallel MoM code is given in Figure. C.7. The code can be considered to be composed of seven parts, which include initialing and finalizing the MPI, setting up the ScaLAPACK environment, and the other four parts of the parallel version of the serial code illustrated in Figure C.3.

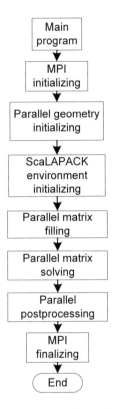

Figure C.7. The flowchart of a parallel MoM code.

MPI initializing. MPI must be initialized at the beginning of the program. MPI includes an initialization routine MPI_INIT. This routine must be called before any other MPI routine (apart from MPI_INITIALIZED).

Parallel geometry initializing. One possible strategy to initialize the geometric data for all the processes is to input it to the root process. After the root process computes the whole dataset, it then broadcasts the information to all the other processes, so that each process would have all the necessary geometric data ready for the matrix filling subroutine. As the electrical size of the problem grows, the time taken for processing the data on the root process also increases. Therefore, an alterative way for initializing is to distribute the input geometric dataset to all the processes. Once each process finishes its own computation, the information would then be communicated among all the processes to complete the parallel initialization.

ScaLAPACK environment initializing. Once the number of unknowns for each process is known, one can use the ScaLAPACK subroutine DESCINIT to initialize a descriptor vector. This vector stores the information necessary to establish the mapping between an object element and its corresponding process and memory location. The descriptor vector for the impedance matrix, the voltage matrix, and the coefficient vector of unknowns in MoM are also created at this stage.

Parallel matrix filling. The time to perform the calculations needed to fill the matrix increases as the problem size increases. For a given number of CPUs, the time for filling the elements of the matrix typically scales well as N^2, where N is the dimension of the matrix. Parallel matrix filling schemes may vary for different types of basis functions to obtain an optimum performance. The key is to avoid any redundant calculations and keep the load in every process balanced. For example, if the results of one integral over a geometric surface can be used for computing different impedance elements (such as for the RWG basis functions), then a proper flag needs to be set to avoid repeatedly calculating it. If Green's function can be reused for different orders of basis functions (such as for the higher-order polynomial basis functions), then the code needs to be designed accordingly so that the relevant calculation will be only performed once.

Parallel matrix solving. Solving a matrix equation involves the LU factorization and the substitutions. The time to factor the double precision, complex, nonsymmetric matrix into lower and upper triangular matrices, with partial pivoting included, increases with problem size and scales as N^3. The parallel efficiency typically depends on the number of processors used in the computation, decreasing from an ideal 100% as the number of processors increases (for a given number of unknowns). The decrease in efficiency is due to the necessary communications for either a parallel in-core solver or a parallel out-of-core solver. One can improve the performance of a parallel LU factorization by organizing the communication pattern (for example, changing process-grid shape). For an out-of-core solver, file I/O operations also incur some overhead, and it can be reduced, to some degree, by using RAID controller.

Once the matrix is factored into lower and upper triangular matrices, the procedure of the forward and backward substitutions is performed to complete the solution. Both of the substitution algorithms contain loops where calculations performed at each stage in one process depend on the previous calculations by other processes. A straightforward parallelization therefore becomes more involved. The parallel algorithm used essentially rewrites the solution vector to all the processes as the backward and the forward substitution progresses.

Parallel postprocessing. After the matrix equation is solved, the known current vector can now reside in each processor and the secondary calculations related to the observed quantities can be made. Though the time taken to perform this operation is quite insignificant to the time required to set up the problem and solve the matrix equations, it may be instructive to examine the scalability involved in the calculation of the fields.

In this section, the calculation of the radiated field will be distributed over the different processes as shown in Figure C.8, where *ntheta* and *nphi* denote the total number of angles in the θ and ϕ directions, respectively. *NumberofProcess* is the total number of processes and *thisID* is the rank of a process.

Parallel far-field calculation:

Do *itheta*=1, *ntheta*   !Loop over θ
Do *jphi*=1, *nphi*   !Loop over ϕ
If (mod ((*itheta*-1)*nphi +*jphi*−1, *NumberofProcess*) == *thisID*)   ! θ or $\phi \in$ *this process*
    Calculate far field on this process
Endif
Enddo
Enddo

Figure C.8. A parallel calculation of the far field.

For example, if we limit our discussion to only four processes, when *ntheta* is taken as 1, we can assign various angle ϕ to different processes as follows: 1,5,9,... to process 0, 2,6,10,... to process 1, 3,7,11,... to process 2 and 4,8,12,... to process 3. By distributing the computation of the fields at these angles over different processes in a cyclical pattern, good load balancing can be obtained. *Load balancing* refers to the practice of distributing work among the various tasks so that all the processors are kept busy all of the time. It can be considered as a minimization of the task idle time. Load balancing is important for parallel programs to obtain good performance. For example, if all tasks are subject to a barrier synchronization point, the slowest task will determine the quality of the overall performance.

The parallel MoM (FORTRAN) source code using this parallel postprocessing scheme is given in the next section. The RCS of a circular PEC cylinder, when it is illuminated by an incident field propagating along the +*x* axis, is given in Figure C.9. The result from the text by Peterson et al. [3] is also plotted simultaneously in Figure C.9.

The other way to parallelize the far-field calculations is to cyclically assign each process a portion of the current distribution and to calculate the fields for all the angles to ensure a proper load balance. A discussion on the distribution of the current coefficients can be found in the literature [4], where a MoM-PO (physical optics) hybrid method is discussed and the MoM current distribution is used as the excitation of the PO region and distributed over different processes.

The near field calculation can also be distributed to different processes in much the same way as discussed for the far-field calculation. This will also be incorporated in the source code presented in the next section. The magnitude (in V/m) of the total field inside a sampling window surrounding a PEC cylinder (*ka* = 1) is plotted in Figure C.10, where the numbers in the horizontal and vertical axes represent the value of the sample. The size of the sampling window is $8\lambda \times 8\lambda$.

MPI Finalizing. MPI must be "finalized" at the end of the program. Subroutine MPI_FINALIZE cleans up all the MPI states. Once this routine is called, no MPI routine (even MPI_INIT) can further be called. The user must ensure that all pending communications involving a process have been completed before the process calls MPI_FINALIZE.

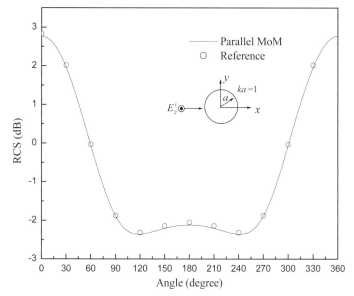

Figure C.9. Bistatic RCS of a PEC cylinder.

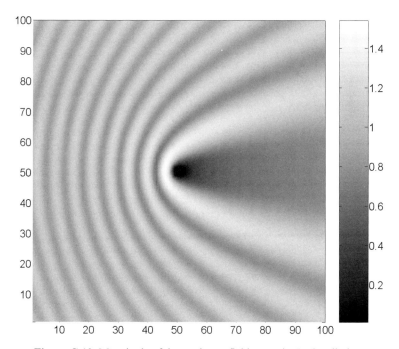

Figure C.10. Magnitude of the total near field around a PEC cylinder.

C.4.2 A Parallel MoM Source Code Using ScaLAPACK for the 2D Scattering Problem

```
C *-----------------------------------------------------------------------------------------
C * + Name
C *      PARALLEL_MOM
C *
C *      MoM Analysis of PEC Cylinders for TM-POL
C *      Plane-wave Excitation -- Program Finds the Current
C *      Density and RCS
C *      The E-field Integral Equation is Discretized with Pulse Basis
C *      Functions and Tested with Point-Matching
C *
C * + Description
C *   A Frame Code for Parallelizing MoM Using ScaLAPACK and MPICH2
C * + Note: Demonstration Purpose Only, Not Optimized for Performance!!!
C * + Author
C *      Yu Zhang
C * + Language
C *      Fortran77
C * + Last Update:
C *      April, 2008
C *      Source code is available from the authors
C * -----------------------------------------------------------------------------------------


C *-----------------------------------------------------------------
C *      MOM  V A R I A B L E S
C *-----------------------------------------------------------------


      MODULE MOM_VARIABLES
      IMPLICIT NONE
      DOUBLE PRECISION ,PARAMETER ::
     + E=2.7182818284590451,GAMA=1.78107
      COMPLEX*16 ,PARAMETER ::        CJ=CMPLX(0,1)
      DOUBLE PRECISION ,PARAMETER :: PI=3.14159265
      DOUBLE PRECISION WAVElength, A,B, WAVEK,DPHI,CN,DIST,
     + DOAINC
      INTEGER NUN
      COMPLEX*16, ALLOCATABLE :: Z_MN(:,:) , V_M(:)
      COMPLEX*16, ALLOCATABLE :: CWW(:) , CCWW(:)
      COMPLEX*16, ALLOCATABLE :: NEARF(:,:),SUM_NEARF(:,:)
      DOUBLE PRECISION ,ALLOCATABLE :: RCS(:),SUM_RCS(:)
      INTEGER   NPHI, NX, NY
      DOUBLE PRECISION  W
      INTEGER :: M,N, LCOLNUM,LROWNUM
```

```
C *----------------------------------------------------------------
C *        G E O M I T R I C A L   V A R I A B L E S
C *----------------------------------------------------------------
      DOUBLE PRECISION,ALLOCATABLE ::XN(:),YN(:),ARCLENGTH(:)
     +   ,XCN(:),YCN(:)
      END MODULE MOM_VARIABLES


C *----------------------------------------------------------------
C *            MPI  V A R I A B L E S
C *----------------------------------------------------------------

      MODULE MPI_VARIABLES
      INTEGER :: MY_MPI_FLOAT, MY_ID, NPROCESS
      INTEGER, PARAMETER :: SFP = SELECTED_REAL_KIND(6,37)
      INTEGER, PARAMETER :: DFP = SELECTED_REAL_KIND(13,99)
      INTEGER, PARAMETER :: RFP=DFP ! 32 BIT PRECISION
      INTEGER :: ISTAT, ERROR,IERR
      END MODULE MPI_VARIABLES


C *----------------------------------------------------------------
C *            ScaLAPACK  V A R I A B L E S
C *----------------------------------------------------------------

      MODULE SCALAPACK_VARIABLES
      INTEGER   DLEN_, IA, JA, IB, JB, MB, NB, RSRC,
     +        CSRC, MXLLDA, MXLLDB, NRHS, NBRHS, NOUT,
     +        MXLOCR, MXLOCC, MXRHSC
      PARAMETER      ( DLEN_ = 9,  NOUT = 6 )
      DOUBLE PRECISION   ONE
      PARAMETER      ( ONE = 1.0D+0 )
C *   .. LOCAL SCALARS ..
      INTEGER      ICTXT, INFO, MYCOL, MYROW, NPCOL, NPROW
      INTEGER      NPROC_ROWS
      INTEGER      NPROC_COLS
      INTEGER      NROW_BLOCK  !MB
      INTEGER      NCOL_BLOCK  !NB
      INTEGER ICROW
      INTEGER ICCOL
      INTEGER LOCAL_MAT_ROWS,LOCAL_MAT_COLS,B_LOCAL_SIZE
      INTEGER, ALLOCATABLE:: IPIV( : ), DESCA( :), DESCB( :)
      CHARACTER,PARAMETER ::TRANS='N'

C*    .. EXTERNAL FUNCTIONS ..
      INTEGER,EXTERNAL ::   INDXG2P,INDXG2L,INDXL2G,NUMROC
      END MODULE SCALAPACK_VARIABLES
```

```
C *----------------------------------------------------------------
C *               PARALLEL CODE STARTS HERE
C *----------------------------------------------------------------


      PROGRAM PARALLEL_MOM

C *----------------------------------------------------------------
C *               G L O B A L  V A R I A B L E S
C *----------------------------------------------------------------
      USE MPI_VARIABLES
      USE MOM_VARIABLES
      USE SCALAPACK_VARIABLES
      USE IFPORT
      IMPLICIT NONE
      INCLUDE 'MPIF.H'
      COMPLEX*16,EXTERNAL :: H02

C *----------------------------------------------------------------
C *               L O C A L  V A R I A B L E S
C *----------------------------------------------------------------
      INTEGER::I,J
      INTEGER ITEMP,JTEMP
      INTEGER STATUS(MPI_STATUS_SIZE)
      INTEGER :: III,JJJ

C *----------------------------------------------------------------
C *               MAIN PROGRAM STARTS HERE
C *----------------------------------------------------------------


C *----------------------------------------------------------------
C *               MPI INITIALIZING
C *----------------------------------------------------------------
      CALL MPI_INIT(IERR)
      CALL MPI_COMM_SIZE(MPI_COMM_WORLD, NPROCESS, IERR)
      CALL MPI_COMM_RANK(MPI_COMM_WORLD, MY_ID, IERR)

          IF (RFP==SFP) THEN
          MY_MPI_FLOAT = MPI_REAL
          ELSE IF (RFP==DFP) THEN
          MY_MPI_FLOAT = MPI_DOUBLE_PRECISION
          ELSE
          WRITE(*,*) 'FATAL ERROR! , MY_ID = ', MY_ID
          STOP
          END IF
```

```
C *-------------------------------------------------------------------
C              INPUT PROCESS GRID AND BLOCK SIZE
C *-------------------------------------------------------------------
          IF (MY_ID.EQ.0) THEN
          WRITE(*,*)'Input Number of Rows in the process grid,'
          WRITE(*,*)' Number of Columns in the process grid, '
          WRITE(*,*)' and Row and Column Block  size '

          READ(5,*) NPROC_ROWS, NPROC_COLS, NROW_BLOCK,
     +         NCOL_BLOCK
          IF (NPROCESS.LT.(NPROC_ROWS*NPROC_COLS)) THEN
             WRITE(6,250)  NPROCESS, NPROC_ROWS, NPROC_COLS
 250         FORMAT(' ','PROC ',' > NP = ',I2,', NPROC_ROWS = ',I2,
     +          ', NPROC_COLS = ',I2)
             WRITE(6,260)
 260      FORMAT(' ','NEED MORE PROCESSES!  QUITTING.')
          WRITE(*,*) 'ABNORMALLY QUIT!'
          CALL MPI_ABORT(MPI_COMM_WORLD, -1, IERR)
          END IF
          IF (NPROCESS.GT.(NPROC_ROWS*NPROC_COLS)) THEN
             WRITE(6,270)  NPROCESS, NPROC_ROWS, NPROC_COLS
 270         FORMAT(' ','PROC ',' < NP = ',I2,', NPROC_ROWS = ',I2,
     +          ', NPROC_COLS = ',I2)
             WRITE(6,280)
 280      FORMAT(' ','CHECK PROCESS GRID!  QUITTING.')
          WRITE(*,*) 'ABNORMALLY QUIT!'
          CALL MPI_ABORT(MPI_COMM_WORLD, -1, IERR)
          ENDIF
          IF (NROW_BLOCK .NE. NCOL_BLOCK) THEN
          WRITE(6,290)  NROW_BLOCK, NCOL_BLOCK
 290      FORMAT(' ', 'NROW_BLOCK = ',I2,',','NCOL_BLOCK= ',I2)
          WRITE(6,291)
 291        FORMAT(' ','NROW_BLOCK!= NPROC_COLS,  QUITTING.')
          WRITE(*,*) 'ABNORMALLY QUIT!'
          CALL MPI_ABORT(MPI_COMM_WORLD, -1, IERR)
          END IF
          END IF

       CALL MPI_BCAST(NPROC_ROWS,1,MPI_INTEGER,
     +       0,MPI_COMM_WORLD,IERR)
       CALL MPI_BCAST(NPROC_COLS,1,MPI_INTEGER,
     +       0,MPI_COMM_WORLD,IERR)
       CALL MPI_BCAST(NROW_BLOCK,1,MPI_INTEGER,
     +       0,MPI_COMM_WORLD, IERR)
       CALL MPI_BCAST(NCOL_BLOCK,1,MPI_INTEGER,
```

```
    +        0,MPI_COMM_WORLD, IERR)
     CALL MPI_BARRIER(MPI_COMM_WORLD,IERR)


C *------------------------------------------------------------------
C *             MOM INITIALIZING
C *------------------------------------------------------------------

     NUN=100;
     ALLOCATE (XN(1:NUN),YN(1:NUN),ARCLENGTH(1:NUN),
    + STAT=ISTAT)
     ALLOCATE (XCN(1:NUN),YCN(1:NUN), STAT=ISTAT)
     ALLOCATE (IPIV(1:NUN), STAT=ISTAT)
     ALLOCATE( CWW(1:NUN),CCWW(1:NUN), STAT=ISTAT )

     IF (MY_ID ==0) THEN
     CALL GEONWAVE_INITIALIZE
     ENDIF
     CALL MPI_BARRIER(MPI_COMM_WORLD,IERR)
     CALL GEONWAVE_PARA_INITIALIZE
     CALL MPI_BARRIER(MPI_COMM_WORLD,IERR)


C *------------------------------------------------------------------
C *             BLACS  INITIALIZING
C *------------------------------------------------------------------

     CALL BLACS_PINFO(MY_ID,NPROCESS)
     CALL BLACS_GET(-1, 0,ICTXT)
     CALL BLACS_GRIDINIT(ICTXT,'R',NPROC_ROWS,
    +           NPROC_COLS)
     CALL BLACS_GRIDINFO(ICTXT , NPROC_ROWS,
    + NPROC_COLS, MYROW, MYCOL )
        CALL BLACS_PCOORD(ICTXT, MY_ID, MYROW,
    +        MYCOL)


C *------------------------------------------------------------------
C * GET LOCAL MATRIX DIMENSION INFORMATION
C *------------------------------------------------------------------

     LOCAL_MAT_ROWS = NUMROC(NUN, NROW_BLOCK, MYROW,
    +                 0, NPROC_ROWS)

     LOCAL_MAT_COLS = NUMROC(NUN, NCOL_BLOCK, MYCOL,
    +                 0, NPROC_COLS)

     B_LOCAL_SIZE = NUMROC(NUN, NROW_BLOCK, MYROW,
    +                 0, NPROC_ROWS)
```

```
C *-------------------------------------------------------------------
C *   ALLOCATE  RAM FOR IMPEDANCE MATRIX AND VOLTAGE
C *   VECTOR ON EACH PROCESS, THEN INITIALIZING THE ARRAY
C *-------------------------------------------------------------------

    ALLOCATE( V_M(1:B_LOCAL_SIZE),  STAT=ISTAT )
    ALLOCATE( Z_MN(1:LOCAL_MAT_ROWS ,1:LOCAL_MAT_COLS ),
   +                 STAT=ISTAT)
        Z_MN(:,:)=0.0D0
        V_M(:)=0

C *-------------------------------------------------------------------
C *    INITIALIZE THE ARRAY DESCRIPTORS
C *-------------------------------------------------------------------

    ALLOCATE( DESCA( DLEN_ ), DESCB( DLEN_ ))
    IA = 1
    JA = 1
    IB = 1
    JB = 1
    RSRC=0
    CSRC=0
    NRHS=1
    NBRHS=1

C * -------------------------------------------------------------------
C * DESCINIT INITIALIZES THE DESCRIPTOR VECTOR.
C * EACH GLOBAL DATA OBJECT IS DESCRIBED BY AN ASSOCIATED
C * DESCRIPTION VECTOR DESCA . THIS VECTOR STORES THE
C * INFORMATION REQUIRED TO ESTABLISH  THE MAPPING
C * BETWEEN AN OBJECT ELEMENT AND ITS CORRESPONDING
C * PROCESS AND MEMORY LOCATION.
C * -------------------------------------------------------------------


    CALL DESCINIT( DESCA, NUN, NUN, NROW_BLOCK,NCOL_BLOCK,
   +         RSRC, CSRC, ICTXT, LOCAL_MAT_ROWS,INFO )

    CALL DESCINIT( DESCB, NUN, NRHS, NROW_BLOCK, NBRHS,
   +         RSRC, CSRC, ICTXT,B_LOCAL_SIZE, INFO )

    IF (MY_ID ==0)  WRITE(*,*) 'START TO FILL THE  MATRIX'

    CALL MATRIX_FILLING
    IF (MY_ID ==0) WRITE(*,*) 'FINISH MATRIX FILLING'
```

```
C *--------------------------------------------------------------------
C *     CALL  SCALAPACK ROUTINES TO
C *     SOLVE THE LINEAR SYSTEM [Z]*[I] = [V]
C *     USING LU DECOMPOSITION METHOD
C *--------------------------------------------------------------------

    IF (MY_ID ==0) WRITE(*,*) 'START TO SOLVE THE MATRIX
   + EQUATION'

    CALL PZGETRF(NUN,NUN,Z_MN,IA,JA, DESCA,IPIV,INFO)

    CALL PZGETRS(TRANS,NUN,NRHS,Z_MN,IA,JA,DESCA,IPIV,
   + V_M,IB,JB,DESCB, INFO )

    IF (MY_ID ==0) WRITE(*,*) 'FINISH SOLVING THE MATRIX
   + EQUATION'


C *--------------------------------------------------------------------
C *     COLLECT CURRENT COEFFICIENT FROM DIFFERENT
C *     PROCESS USING MPI_REDUCE
C *--------------------------------------------------------------------

    CWW(1:NUN)=0
    IF(MYCOL==0)  THEN
         DO III=1,B_LOCAL_SIZE
      JJJ=INDXL2G(III,NROW_BLOCK,MYROW,RSRC,NPROC_ROWS)
      CWW(JJJ)=V_M(III)
     ENDDO
    ENDIF

    CCWW(1:NUN)=0
    CALL MPI_REDUCE(CWW,CCWW,NUN,MPI_DOUBLE_COMPLEX,
   +   MPI_SUM,0,MPI_COMM_WORLD,IERR)

    CALL MPI_BCAST(CCWW,NUN,MPI_DOUBLE_COMPLEX,
   +             0,MPI_COMM_WORLD,IERR)


C *--------------------------------------------------------------------
C *     OUTPUT THE CURRENT COEFFICIENT
C *--------------------------------------------------------------------

    IF (MY_ID ==0) THEN
    OPEN ( UNIT=2,  FILE='CURRENT_PARALLEL.DAT' )
    DO III=1,NUN
    WRITE(2,*) ABS(CCWW(III))
```

```
C *   *120*PI  ! IF NORMALIZE TO |HINC|
C *    REAL(CCWW(III)), CIMAG(CCWW(III))
      ENDDO
      CLOSE(2)
      ENDIF

C *------------------------------------------------------------------
C *       POST PROCESSING TO CALCULATE THE RCS
C *------------------------------------------------------------------

      IF (MY_ID ==0) WRITE(*,*) 'START POST PROCESSING'
      CALL POST_PROCESSING
      IF (MY_ID ==0) WRITE(*,*) 'FINISH POST PROCESSING'


C *------------------------------------------------------------------
C *       EXIT THE CODE
C *------------------------------------------------------------------
      IF ( MY_ID .EQ. 0) WRITE(*,*) 'PROGRAM FINISHED PROPERLY'
      CALL MPI_FINALIZE(IERR)

          END PROGRAM PARALLEL_MOM

      SUBROUTINE POST_PROCESSING
C *------------------------------------------------------------------
C *       THE CODE HERE IS NOT OPTIMIZED FOR A MINIMUM
C *       REQUIREMENT OF RAM FOR PARALLEL POST PROCESSING
C **-----------------------------------------------------------------
          USE MOM_VARIABLES
          USE MPI_VARIABLES, ONLY: MY_ID, ISTAT, NPROCESS
          IMPLICIT NONE

          INCLUDE 'MPIF.H'
          INTEGER IX, IY,I,IPHI,IERR
          DOUBLE PRECISION XO,YO
          DOUBLE PRECISION PHI
          COMPLEX*16,EXTERNAL :: H02,EINC
          COMPLEX*16 SIG
          W=4*WAVElength
          NX=100
          NY=100
          NPHI=360

          ALLOCATE( NEARF(1:NX,1:NY),SUM_NEARF(1:NX,1:NY),
     +         STAT=ISTAT)
```

```
          ALLOCATE( RCS(1:NPHI),SUM_RCS(1:NPHI), STAT=ISTAT)

      NEARF(1:NX ,1:NY)=0
      SUM_NEARF(1:NX,1:NY)=0
      RCS(1:NPHI)=0
      SUM_RCS(1:NPHI)=0


C *----------------------------------------------------------------
C *    COMPUTE THE BISTATIC RCS
C * ----------------------------------------------------------------

      DO IPHI=1,NPHI
      IF ( MOD((IPHI-1),NPROCESS) == MY_ID) THEN
      PHI=IPHI*PI/180.0
      SIG = (0.,0.)
      DO I=1,NUN
       SIG = SIG + CCWW(I)*cdEXP(1.0*CJ*WAVEK*( XCN(I)*COS(PHI)+
      +          YCN(I)*SIN(PHI)        ) )*ARCLENGTH(I)
      ENDDO
      RCS(IPHI) = 10.*LOG10(CDABS(SIG)**2 * WAVEK*(120*PI)**2/4.0)

      ENDIF
      ENDDO
      CALL  MPI_REDUCE(RCS,SUM_RCS,NPHI,MPI_DOUBLE_PRECISION,
      +    MPI_SUM,0,MPI_COMM_WORLD,IERR)
      IF (MY_ID==0)    THEN

      OPEN(4,FILE='RCS_PARALLEL.DAT')
      WRITE(4,*) 'RCS IN DB-WAVELENGTH'
      WRITE(4,*) '    ANGLE          RCS'
      WRITE(4,*)
      DO IPHI=1,NPHI
      WRITE(4,*) IPHI,SUM_RCS(IPHI)
      ENDDO
      ENDIF

      DO  IX=1,NX
        XO = (2*((IX-1.0)/(NX-1))-1)*W;
      DO IY=1,NY
        YO = (2*((IY-1.0)/(NY-1))-1)*W;
      IF (  MOD((IX-1)*NX+IY-1,NPROCESS)==MY_ID) THEN
      DO N=1,NUN
      DIST = SQRT((XO-XCN(N))**2+(YO-YCN(N))**2);
      NEARF(IX,IY) =  NEARF(IX,IY)-120*PI/4.*CN*WAVEK
      +              *CCWW(N)*H02(WAVEK*DIST)
      ENDDO
```

```
      NEARF(IX,IY) =  NEARF(IX,IY)+EINC(XO,YO,DOAINC)
      ENDIF

      ENDDO
      ENDDO

      SUM_NEARF(1:NX,1:NY)=0

      CALL MPI_REDUCE(NEARF,SUM_NEARF,NX*NY,
    +    MPI_DOUBLE_COMPLEX,MPI_SUM,0,MPI_COMM_WORLD,IERR)

      IF (MY_ID==0)   THEN
      OPEN (3, FILE = "NEARF_PARALLEL.DAT")
      DO  IX=1,NX
      DO IY=1,NY
      WRITE(3,*)ABS(SUM_NEARF(IX,IY))

      ENDDO
      ENDDO
      CLOSE(3)
      ENDIF

      END SUBROUTINE POST_PROCESSING

      SUBROUTINE GEONWAVE_INITIALIZE
C *----------------------------------------------------------------------
C *     THIS SUBROUTINE WILL RETURN HOW MANY UNKNOWNS
C *     AND GEOMETIRC INFO FOR IMPEDANCE MATRIX AND
C *     VOLTAGE MATRIX OF MOM.
C *     THE FOLLOWING ARE THE GEOMETRIC COORDINATES (X,Y)
C *     OF EACH  SUBDIVISION ALONG THE BOUNDARY OF
C *     THE PEC CYLINDER
C *----------------------------------------------------------------------
         USE MOM_VARIABLES
         USE MPI_VARIABLES,ONLY:MY_ID,
    +        ISTAT,IERR,MY_MPI_FLOAT
         USE SCALAPACK_VARIABLES,ONLY: IPIV
         IMPLICIT NONE
         INCLUDE 'MPIF.H'
         INTEGER I

      WAVElength=1.0;
      WAVEK=2*PI/WAVElength
      A=1.0/WAVEK;
      B=1.0/WAVEK;
```

```
      DPHI=2.0*PI/NUN    !ANGLE OF EACH SEGMENT
      DOAINC=180.0            ! DIRECTION OF INCIDENT WAVE

      IF(MY_ID==0) THEN
      DO I=1,NUN+1
          XN(I)=A*DCOS((I-1)*DPHI)
          YN(I)=B*DSIN((I-1)*DPHI)
      ENDDO
      DO I=1,NUN
C *    X AND Y COORDINATES AT THE CENTER OF EACH SEGMENT
      XCN(I)=(XN(I)+XN(I+1)) /2
      YCN(I)=(YN(I)+YN(I+1)) /2
      ARCLENGTH(I) =DSQRT((XN(I+1)-XN(I))**2+(YN(I+1)-YN(I))**2);
C*      FOR A CIRCULAR CYLINDER, USING THE FOLLOWING
C*      ARCLENGTH(I)= A*DPHI
      ENDDO
      ENDIF
      END SUBROUTINE GEONWAVE_INITIALIZE

      SUBROUTINE GEONWAVE_PARA_INITIALIZE
C *-------------------------------------------------------------------
C *    THIS SUBROUTINE WILL BROADCAST THE GEOMETRIC
C *    COORDINATES TO EVERY PROCESS
C *-------------------------------------------------------------------

      USE MOM_VARIABLES
      USE MPI_VARIABLES,ONLY:MY_ID, ISTAT,IERR,MY_MPI_FLOAT
      USE SCALAPACK_VARIABLES,ONLY: IPIV
      IMPLICIT NONE
      INCLUDE 'MPIF.H'
      CALL MPI_BCAST(a ,1, MY_MPI_FLOAT,
     +      0,MPI_COMM_WORLD,IERR)
        CALL MPI_BCAST(b ,1, MY_MPI_FLOAT,
     +      0,MPI_COMM_WORLD,IERR)
      CALL MPI_BCAST(NUN,1,MPI_INTEGER,
     +      0,MPI_COMM_WORLD,IERR)
     CALL MPI_BCAST(XN  ,NUN, MY_MPI_FLOAT,
     +      0,MPI_COMM_WORLD,IERR)
     CALL MPI_BCAST(YN  ,NUN, MY_MPI_FLOAT,
     +      0,MPI_COMM_WORLD,IERR)
     CALL MPI_BCAST(XCN  ,NUN, MY_MPI_FLOAT,
     +      0,MPI_COMM_WORLD,IERR)
     CALL MPI_BCAST(YCN  ,NUN, MY_MPI_FLOAT,
     +      0,MPI_COMM_WORLD,IERR)
     CALL MPI_BCAST(ARCLENGTH ,NUN, MY_MPI_FLOAT,
     +      0,MPI_COMM_WORLD,IERR)
```

```
      CALL MPI_BCAST(DOAINC ,1, MY_MPI_FLOAT,
     +       0,MPI_COMM_WORLD,IERR)
      CALL MPI_BCAST(WAVElength ,1, MY_MPI_FLOAT,
     +       0,MPI_COMM_WORLD,IERR)
         CALL MPI_BCAST(WAVEK ,1, MY_MPI_FLOAT,
     +       0,MPI_COMM_WORLD,IERR)
      END SUBROUTINE GEONWAVE_PARA_INITIALIZE

      SUBROUTINE MATRIX_FILLING
C *----------------------------------------------------------------------
C *    THIS SUBROUTINE FILLS THE MATRIX OF MOM IN PARALLEL
C *----------------------------------------------------------------------
      USE MPI_VARIABLES
      USE MOM_VARIABLES
      USE SCALAPACK_VARIABLES
      USE IFPORT
      IMPLICIT NONE
      INCLUDE 'MPIF.H'
         INTEGER::I,J
      INTEGER ITEMP,JTEMP
      INTEGER STATUS(MPI_STATUS_SIZE)
      INTEGER :: III,JJJ
      COMPLEX*16,EXTERNAL :: EINC
      COMPLEX*16,EXTERNAL :: H02


C *----------------------------------------------------------------------
C *  FILLING IMPEDANCE MATRIX ON EACH PROCESS
C *----------------------------------------------------------------------

      DO M=1,NUN
      ICROW = INDXG2P( M,  NROW_BLOCK, 0,  0, NPROC_ROWS)
      IF(ICROW==MYROW) THEN
      LROWNUM=INDXG2L( M,  NROW_BLOCK, 0,  0, NPROC_ROWS)
      DO N=1,NUN
      ICCOL = INDXG2P( N,  NCOL_BLOCK, 0,   0, NPROC_COLS)
      IF (ICCOL==MYCOL) THEN
      LCOLNUM=INDXG2L( N,  NCOL_BLOCK, 0,   0, NPROC_COLS)
      IF(M==N)THEN
      CN=ARCLENGTH(N)
      Z_MN(LROWNUM,LCOLNUM)=120*PI/4.*CN*WAVEK
     +    *(1-CJ*2.0/PI*LOG(GAMA*WAVEK*CN/(4.0*E) ) )
      ELSE
      CN=ARCLENGTH(N)
      DIST=DSQRT((XCN(M)-XCN(N))**2+(YCN(M)-YCN(N))**2)
      Z_MN(LROWNUM,LCOLNUM)=120*PI/4.*CN*WAVEK
```

```
     +    *H02(WAVEK*DIST)
     ENDIF
     ENDIF
     ENDDO
     ENDIF
     ENDDO
C *------------------------------------------------------------------
C * OUTPUT THE IMPEDANCE MATRIX THAT WAS DISTRIBUTED TO
C * THE FIRST PROCESS
C *          -- FOR VALIDATION PURPOSE ONLY ----
C *------------------------------------------------------------------
     IF (MY_ID==0)  THEN
     OPEN ( UNIT=1,  FILE='MATRIX0.DAT' )
       DO III=1,LOCAL_MAT_ROWS
       DO JJJ=1, LOCAL_MAT_COLS
         WRITE(1,*)DREAL(Z_MN(III,JJJ)),DIMAG(Z_MN(III,JJJ))
       ENDDO
       ENDDO
     CLOSE(1)
     ENDIF


C *------------------------------------------------------------------
C * FILLING EXCITATION VECTOR ON EACH PROCESS
C *------------------------------------------------------------------
     DO M=1,NUN
     ICROW = INDXG2P( M,   NROW_BLOCK, 0,  0, NPROC_ROWS)
     IF(ICROW==MYROW .AND. MYCOL==0) THEN
     LROWNUM=INDXG2L( M,   NROW_BLOCK, 0,  0, NPROC_ROWS)
     V_M(LROWNUM)=EINC(XCN(M),YCN(M),DOAINC)
     ENDIF
     ENDDO

   END SUBROUTINE MATRIX_FILLING

   COMPLEX*16 FUNCTION EINC(X,Y,INCPHI)
C *------------------------------------------------------------------
C *   THIS FUNCTION RETURNS THE INCIDENT WAVE
C *------------------------------------------------------------------
     USE MOM_VARIABLES,ONLY: WAVEK,CJ
     IMPLICIT NONE
     DOUBLE PRECISION X,Y,INCPHI

     EINC=cdEXP(1.0*CJ*WAVEK*( X*COSd(INCPHI)+Y*SINd(INCPHI) ))

     RETURN
     END
```

```
      COMPLEX*16 FUNCTION H02(A)

C *-------------------------------------------------------------------
C * ZERO-ORDER HANKEL FUNCTION (2ND KIND) FROM [5]
C *-------------------------------------------------------------------

      DOUBLE PRECISION A
      DOUBLE PRECISION T0,F0,BY,BJ

      IF(A.GT.3.000) GOTO 5
      BJ=(A/3.00)**2
      BJ=1.0+BJ*(-2.2499997+BJ*(1.2656208+BJ*(-.3163866+BJ*
   +   (.0444479+BJ*(-.0039444+BJ*.00021)))))
      BY=(A/3.00)**2
      BY=2.0/3.1415926*DLOG(A/2.)*BJ+.36746691+BY*(.60559366+BY
   +   *(-.74350384+BY*(.25300117+BY*(-.04261214+BY*(.00427916-
   +   BY*.00024846)))))
      GOTO 10
    5 BJ=3.00/A
      F0=.79788456+BJ*(-.00000077+BJ*(-.00552740+BJ*(-.00009512
   +   +BJ*(.00137237+BJ*(-.00072805+BJ*.00014476)))))
      T0=A-.78539816+BJ*(-.04166397+BJ*(-.00003954+BJ*(.00262573
   +   +BJ*(-.00054125+BJ*(-.00029333+BJ*.00013558)))))
      BY=DSQRT(A)
      BJ=F0*DCOS(T0)/BY
      BY=F0*DSIN(T0)/BY
   10 H02=DCMPLX(BJ,-BY)
      RETURN

      END
```

C.5 COMPILATION AND EXECUTION OF THE PARALLEL CODE

Here, we compile and execute the parallel code on an IA-32 Windows Desktop PC. The following are the details of the procedure:

1. Install Intel FORTRAN compiler (e.g., Version 9.0 is installed in "C:\Program Files\Intel\Compiler\Fortran").
2. Install ScaLAPACK library package, as an example, here Intel Cluster Math Kernel Library (CMKL) package has been installed. (e.g., CMKL Version 8.1 is installed in "C:\Program Files\Intel\cmkl")
3. Install MPI software (available at: http://www.mcs.anl.gov/research/projects/mpich2/). (e.g. MPICH2 is installed in "C:\Program Files\MPICH2")

4.   Set environment variables for MPI and CMKL by executing

> set lib=C:\Program Files\MPICH2\lib; C:\Program Files\
> Intel\cmkl\8.1\ia32\lib;%lib%
> set include=C:\Program Files\MPICH2\include;%include%

The screenshot for this operation is given in Figure C.11.

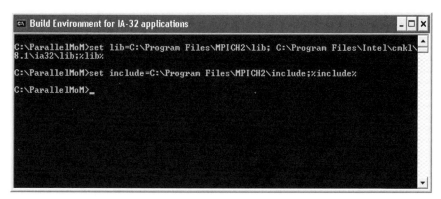

Figure C.11. A screenshot for setting up environment variables.

5.   Copy the code from the book (ParallelMoM.f), and place it inside a folder. (e.g. C:\ParallelMoM)
6.   Go to the directory "C:\ParallelMoM" in "Build Environment for IA-32 applications" window and execute

> ifort parallelMoM.f mkl_scalapack.lib mkl_blacs_mpich2.lib
> libguide.lib mkl_c.lib mpi.lib fmpich2.lib -o parallelMoM

The screenshot for the compilation of the code, using the ScaLAPACK (from Intel Math Kernel Library package) and MPI libraries, is shown in Figure C.12.

```
cv Build Environment for IA-32 applications                      _ □ x
C:\ParallelMoM>ifort parallelMoM.f  mkl_scalapack.lib mkl_blacs_mpich2.lib libgu
ide.lib mkl_c.lib mpi.lib fmpich2.lib  -o parallelMoM_
```

Figure C.12 A screenshot for compiling the parallel code.

7. Go to the command window and execute

"c:\Program Files\MPICH2\bin\mpiexec.exe" -n 4 parallelMoM.exe

The screenshot is given in Figure C.13, where "-n 4" implies that the number of processes is 4. In this example, the process grid used for the four processes is 2×2, and row and column block size are both 24. The numbers shown in Figure C.13 are the number of rows of the process grid, number of columns of the process grid, row block size, and column block size, respectively.

```
Build Environment for IA-32 applications - "c:\Program Files\MPICH2\bin\mpiexec.exe" -n ...

C:\ParallelMoM>"c:\Program Files\MPICH2\bin\mpiexec.exe" -n 4 parallelMoM.exe
 Input Number of Rows in the process grid,
 Number of Columns in the process grid,
 and Row and Column Block  size
2
2
24
24
```

Figure C.13 A screenshot for executing the parallel code.

C.6   CONCLUSION

In this appendix, the basic concept of the MoM technique is described. The MoM solution for a 2D PEC scattering problem is discussed in detail. The parallel implementation of the MoM code is given along with the source code in the hopes of providing some helpful guidelines for beginners.

REFERENCES

[1]   R. F. Harrington, *Field Computation by Moment Methods,* Macmillan, New York, 1968.

[2]   T. K. Sarkar, "A Note on the Choice Weighting Functions in the Method of Moments," *IEEE Transactions on Antennas and Propagation*, Vol. 33, No. 4, pp. 436–441, April 1985.

[3]   A. F. Peterson, S. L. Ray, and R. Mittra, *Computational Methods for Electromagnetics*, IEEE Press, 1998.

[4]   Y. Zhang, X. W. Zhao, M. Chen, and C. H. Liang, "An Efficient MPI Virtual Topology Based Parallel, Iterative MoM-PO Hybrid Method on PC Clusters," *Journal of Electromagnetic Waves and Applications*, Vol. 20, No. 5, pp. 661–676, 2006.

[5]   M. Abramowitz and I. A. Stegun, *Handbook of Mathematical Functions*, Dover Publications, New York, p. 392, 1972.

INDEX

A

Adaptive integral method (AIM), 6
Advanced micro devices corporation (AMD), 9
AMD Core math library (ACML), *see* Math library
Antenna
  coupling, 182, 241, 243, 244, 247
  dipole, 249
  impulse radiating antenna (IRA), 268
  isolation, 188, 244, 247
  monopole, 181, 237, 238, 241, 247
  network, 241
  wire antenna, 181, 182, 244
  Yagi-Uda, 264
Antenna array
  L-band, 271, 272
  Vivaldi, 141–143, 146–148, 162, 169, 216, 224–227, 248
  Yagi, 264–266
Architecture
  AMD64, 9, 11, 276, 279
  Intel architecture, 9, 10
    EM64T, 10, 11, 18, 141, 147, 187, 190, 276, 277, 279, 280, 282–284
    IA-32, 9, 11, 103, 147, 148, 276, 277, 331, 332
    IA-64, 10, 11, 141, 276

B

Basic linear algebra communication subroutines (BLACS), 13–15, 288, 293–295
Basic linear algebra subroutines (BLAS), 13–18, 31, 32, 46, 52, 288, 289

Basis function

Basis function
  edge basis function, 121–124
  higher-order, 117, 133, 177, 233, 234, 237, 240, 243
  node basis function, 118
  patch basis function, 121–123
  polynomial, 118, 128, 133, 137
  Rao–Wilton–Glisson (RWG), 6, 71, 74, 79, 190, 233, 234, 237, 240, 243
  segment basis function, 118
Bilinear surface, 116, 117, 119, 128, 130, 241
Block
  column, 44–54
  row, 44–46, 51
Block-cyclic distribution, *see* Data distribution
Block matrices, 39, 45–47
Block partitioned form, 45, 62
Block size, 36–39, 41, 46, 47, 53, 56, 61, 166, 171–173, 176, 201, 220, 222, 333
Buffer
  in-core buffer, 42, 98, 150, 153, 154, 158, 159, 170, 176, 186–188, 194–197, 204, 227, 230
  size of the in-core buffer for each process, 157. *See also* IASIZE

C

C (C++), *see* Compiler
Cache, 29–32
  cache miss, 30
  L1 cache, 30, 32
  L2 cache, 10, 30, 32, 277–280, 282–284
  on-chip cache, 30
  on-die cache, 30

primary cache, 30
Ceiling, 42, 159, 177
Chip-level multiprocessor (CMP), 10
Cluster
    Cluster-1, 11, 201, 202, 282–284
    Cluster-2, 11, 171–173, 201–204,
        254, 257, 259, 260, 266, 268, 282,
        283
    Cluster-3, 11, 201, 202, 282, 284
    Cluster-4, 11, 201–203, 282, 284
    Cluster-5, 11, 201–203, 282, 284
    Cluster-6, 11, 201, 202, 282, 284
    Cluster-7, 11, 174, 176, 284
    Cluster-8, 11, 175, 176, 284
Cluster Math Kernel Library (CMKL),
    see Math library
Compiler
    C/C++, 287, 288
    FORTRAN, 61, 220, 221, 287, 288,
        297, 316, 331
Composite structure, 110, 114, 141, 271
Computational electromagnetics (CEM),
    1, 2, 4, 5, 9, 141
Compute node, 97, 153, 158, 165, 168,
    177, 203, 226, 259, 268, 279, 280,
    283, 314
Computer
    CEM-1, 11, 92, 93, 136–139, 276,
        277
    CEM-2, 11, 93, 137–139, 161,
        276–278
    CEM-3, 11, 92, 93, 141–144, 146,
        248, 251, 276, 278
    CEM-4, 11, 93, 97, 98, 100,
        141–144, 146, 149, 150–154, 158,
        159, 161–165, 171, 188, 194, 195,
        197–199, 214, 225, 228, 244, 251,
        276, 279
    CEM-5, 11, 103, 148, 276, 279, 280
    CEM-6, 11, 102, 103, 141, 146–148,
        171, 172, 180, 276, 279, 280
    CEM-7, 11, 102, 103, 141, 146, 147,
        150, 152, 153, 171, 172, 177,
        180–183, 185, 186, 190, 276, 280,
        281
    CEM-8, 11, 190, 276, 280, 281
    CEM-9, 11, 169, 199, 200, 248, 276,
        280, 281
    CEM-10, 11, 187, 188, 276, 282
    personal computer (PC), 9
Conjugate gradient (CG), 6, 8, 207
    LU-CG, 215, 216

    parallel CG, 214, 216
    RCG, 210
    RPCG, 211
CPU, 3, 10, 11, 30, 276, 282, 284
    AMD, 9, 10–13, 146, 171, 275, 276,
        279
    dual-core, 10, 11, 137–139, 146–148,
        187, 252, 275–277, 279, 282–284
    Intel, 9, 10–13, 141, 146, 171,
        275–277, 279, 280, 282–284
    multicore, 3, 4, 10, 132, 136, 141,
        146, 154, 277
    quad-core, 10, 11, 146, 150, 152,
        169, 190, 248, 275, 276, 280,
        282–284
    single-core, 11, 136–138, 141,
        148–150, 152, 194, 196, 248,
        275–277, 284
Current
    distribution, 73, 101, 116, 121, 122,
        131, 139–141, 316
    electric current, 28, 72, 73, 113, 120,
        128
    expansion, 117, 119, 121, 131
    magnetic current, 28, 113, 128

D

Data decomposition, 42, 43, 86
Data distribution, 32, 38, 40, 171
    block-cyclic distribution, 36, 40
    physically based matrix distribution
        (PBMD), 16, 17, 38, 39, 41
Dielectric structure, 107, 110, 128, 162,
    233
Direction of arrival (DOA), 249, 251
Double data rate (DDR), 283, 284
Doublet, 118, 119, 122, 124
Dual-core, *see* CPU

E

Edge basis function, *see* Basis function
Efficiency, *see* Parallel
Electric field integral equation (EFIE),
    see Integral equation
Electromagnetic compatibility (EMC),
    1, 243, 244

Electromagnetic interference (EMI), 1, 268
EM64T, *see* Architecture
Equivalent model, 72, 109
Equivalent principle, 72, 108
Experiment, 271

F

Fast Fourier transform (FFT), 6, 213
Fast multipole method (FMM), 2. *See also* Frequency domain
Field
  electric field, 72–74, 81, 108–112, 126, 149, 305, 306
  magnetic field, 72, 81, 108–112, 126, 131
  near field, 316, 317
Finite-difference time-domain (FDTD), 3, 8, 9, 88. *See also* Time domain
Finite element method (FEM), 2, 3, 7, 88. *See also* Frequency domain
Finite-element tearing and interconnecting (FETI), 8. *See also* Frequency domain
Finite-element time-domain (FETD), 8, 9. *See also* Time domain
Floor, 38
FLOPS, 230
  GFLOPS, 230
  TFLOPS, 10
FMM-FFT, 6
FORTRAN, *see* Compiler
Frequency domain, 5
Frontside bus (FSB), 10, 157, 277, 279, 280, 282–284

G

Galerkin's method, 76, 124, 305
Ganglia, 158, 159, 177, 179
Gauss-Legendre, 127, 130
Gaussian quadrature, 80, 81
Generatrix, 114, 115, 117, 126
GFLOPS, *see* FLOPS
Gigabit Ethernet, *see* Switch
Gigabytes (GB), 7, 275
Green's function, 73, 80, 113, 125, 127, 131, 315

H

Hankel function, 306, 308
Hard disk, 30, 42, 48–50, 55, 62–64, 66, 86, 88, 134, 136, 165, 167–170, 275
  SAS, 169, 277, 279, 280, 283, 284
  SATA, 277, 279
  SCSI, 169, 277–279
High performance computing (HPC), 4, 12
High-power microwaves (HPM), 268
Higher-order basis function, *see* Basis function
Hybrid method, 5, 6, 7, 316

I

IASIZE, 157, 161, 176, 177
Impedance matrix, 28, 78, 82–84, 87, 126, 128–130, 132–136, 138, 216, 223, 309
Impulse radiating antenna (IRA), *see* Antenna
In-core buffer, *see* Buffer
In-core matrix filling, *see* Matrix filling
In-core solver, *see* Solver
Incomplete LU (ILU), *see* Preconditioner
Infiniband, *see* Switch
Instruction-level parallelism (ILP), 10
Integral equation (IE), 5, 9, 107, 112, 113, 306
  CFIE, 237
  EFIE, 71, 74, 76, 114, 124, 237, 238, 240–243, 303
  PMCHW, 107, 110, 114, 124, 224
  Surface integral equation (SIE), 107, 110
Intel architecture
  32 bit (IA-32), *see* Architecture
  64 bit (IA-64), *see* Architecture
Internet protocol (IP), 132

J

Junction, 118–122, 124, 141, 224, 238, 241, 244

K

Kilobytes (kB), 165, 166, 170, 188, 277–279

L

Left-looking algorithm, 44, 47–50, 53–55, 57, 59, 61, 65
Library package
    LAPACK, 9, 13–15, 31, 32, 52
    LINPACK, 29, 31
    PLAPACK, 13, 16–19, 27, 38, 40, 41, 61, 64, 65, 219, 220, 222, 287, 298
    ScaLAPACK, 13–15, 36, 40, 61, 287, 314, 331
Linux, 11
    Linux x86, 276
    Linux x86_64, 276
Load
    balance, 166, 203, 316
    balancing, 8, 9, 36, 86, 91, 133, 171, 202, 204, 221, 316
Local matrix index, 221
Locality, 29
LU factorization (decomposition), 7, 15, 28, 31, 32, 42–46, 50, 57, 60, 62, 64, 215

M

Managed message passing interface (MMPI), 16
Marching-on-in-degree (MOD), 9
Marching-on-in-time (MOT), 9. See also time domain
Math Kernel Library (MKL), see Math library
Math library
    ACML, 12, 13, 171
    CMKL, 16, 171, 287, 331
    MKL, 12, 13
Matrix equation, 28, 66, 78, 207, 307
Matrix filling
    in-core, 81–87, 132–134, 136, 220
    out-of-core, 86, 87, 134–136, 222, 223

Matrix solving, 95, 97, 98, 100–102, 104, 137–139, 149, 172, 173, 181, 228, 229, 315
Matrix–vector product (MVP), 2, 17, 213
Megabytes (MB), 182
Memory hierarchy, 29, 30
Message passing interface (MPI), 6, 12, 14, 16, 37, 288,
    HP-MPI, 12, 13,
    Intel MPI, 12, 13
Metallic, 110, 111, 114, 116, 254
Method of Moments (MoM), 2, 28, 78, 88, 92, 96, 124, 177, 190, 303, 305, 309, 312, 313, 318
Microsoft, 12
    Windows 2000, 276
    Windows Vista, 187, 276, 282
    Windows XP, 12, 191, 234, 276
MPI topology, 8
MPICH2, 13, 282, 287, 288, 332
Multicore, see CPU
Multilevel fast multipole algorithm (MLFMA), 2, 6, 234, 237, 238, 240–243. See also Frequency domain

N

Network, 4, 10, 11, 103, 132, 148, 158, 160, 173, 175, 177, 202
Node, 11. See also Compute node
Node basis function, see Basis function
Nonpivoted, see Pivot
Number of unknowns (NUN), 162, 237, 240, 242
Numerical electromagnetics code (NEC), 5–7

O

Operating system
    Linux, 11, 13, 97, 185, 187, 275, 276, 279, 280, 284, 287, 300. See also Linux
    Microsoft windows, 11, 234. See also Microsoft
Out-of-core matrix filling, see Matrix filling
Out-of-core solver (OCS), see Solver

P

Parallel, 3–5, 12, 32, 61, 64, 81, 86, 88, 157, 158, 161, 177, 190, 201, 213, 215, 220
  CG, 213
  code, 313, 318, 331
  computing, 3, 4, 10
  efficiency, 91, 92, 95, 96, 103, 139, 148, 161, 315
  LU-CG, 215
  scalability, 92, 98, 103, 141, 146, 147, 150, 152, 174, 175, 202, 204, 315
  speedup, 91–95, 132–134, 138, 139, 161, 202, 214
Parallel BLAS (PBLAS), 13–15
Parallel Linear algebra package (PLAPACK), *see* Library package
Parallel virtual machine (PVM), 6, 8
Patch basis function, *see* Basis function
PC, *see* Computer
Perfect electric conductor (PEC), 71–73, 88, 92, 93, 98, 110, 111, 114, 117, 124, 128, 136–141, 144, 161, 162, 213, 228, 234, 237, 241, 248, 300, 305, 309, 312, 316, 317
Physical memory, 3, 10, 42, 185, 186
Physical optics (PO), 6
Physically based matrix distribution (PBMD), *see* Data distribution
Pivot, 49, 53, 54, 56, 63, 65
  nonpivoted, 53
  pivoted, 46, 49, 53, 63–65
  pivoting, 28, 32, 36, 44, 45, 48, 49, 53, 55–60, 66, 315
Plane wave, 72, 88, 92, 98, 136–138, 144, 228, 234, 252, 257, 305, 306
Poggio–Miller–Chang–Harrington–Wu (PMCHW), *see* Integral equation
Portability, 12, 92, 95, 98, 103, 138, 139, 141, 146, 147, 150
Postprocessing, 160, 216, 234, 309, 315, 316
Potential functions, 74, 84, 87, 133, 136
  electric scalar potential, 73, 125, 126
  magnetic vector potential, 73, 74, 125, 126
Preconditioner, 213, 237, 238, 240, 243
  incomplete LU (ILU), 237, 238, 243

sparse approximate inverse (SPAI), 238, 240
Preconditioning operator, 211
Precorrected FFT, 6
Process, 13–15, 32–43, 58, 61, 62, 82–88, 91, 97, 98, 132–136, 159, 188, 214, 227, 230, 300, 316
  column, 37, 39, 220, 221
  coordinate, 39–41, 221
  grid, 15, 33, 36–41, 86, 161, 166, 170, 173–176, 197, 199–201, 220, 222, 333
  ID, 40
  rank, 39–41, 203, 204, 315
  row, 37, 39, 220, 221

Q

Quad-core, *see* CPU
Quadrilateral, 107, 116, 133, 234, 244

R

Radar cross section (RCS), 139, 252
  bistatic RCS, 140, 145, 154, 228, 234, 236, 253, 254, 258–260, 263, 317
  monostatic RCS, 254–256
Radome, 265, 271
Random access memory (RAM), 7, 29, 30, 42, 97, 98, 134, 147–150, 153, 158, 161, 177, 185–191, 194–197, 227, 275, 276, 282, 284
Rao–Wilton–Glisson (RWG), *see* Basis function
Redundant array of inexpensive (or in dependent) disks (RAID)
  RAID 0, 148, 154, 168–170, 279, 280, 283
  RAID 1, 169
  RAID 5, 169
  RAID 6, 169
  RAID 10, 169
  RAID controller, 169, 170, 196
Residual, 208, 209, 211–213, 215, 216
  of CG (RCG), *see* Conjugate gradient
  of preconditioned CG (RPCG), *see* Conjugate gradient

Revolutions per minute (rpm), 157, 279–280, 282–284
Right-hand side (RHS), 85, 133
Right-looking algorithm, 44, 45, 47, 49, 50, 55, 59, 60
Robert A. van de Geijn, 6, 18, 220

S

Scalability, *see* Parallel
Scalable file share (SFS), 282, 284
Scalable linear algebra package (ScaLAPACK), *see* Library package
Scattering, 71, 72, 92, 98, 108, 109, 114, 144, 150, 190, 192, 194, 202, 228, 234, 254, 303, 305
SCons, 287, 288
Script file, 288, 289, 292, 293, 296–298
Serial advanced technology attachment (SATA), *see* Hard disk
Serial attached SCSI (SAS), *see* Hard disk
Single-core, *see* CPU
Singleton, 119
Slab, 42, 43, 55–57, 59, 61, 62, 64–66, 86, 87, 135, 136, 158, 160, 177, 221–223, 227
Slab solver, 55
    one-slab solver, 55, 61
Small computer system interface (SCSI), *see* Hard disk
Solver
    in-core, 7, 27, 32, 92, 96, 97, 136, 139, 143, 144, 149, 150, 153, 161, 162, 164, 170, 177, 215, 225, 227, 228, 230, 248.
    out-of-core, 7, 27, 42, 65, 96–98, 101, 103, 136, 143, 144, 147–150, 153, 154, 158, 161–165, 167–171, 175–177, 180, 184–186, 194–197, 201, 215, 225, 227, 229, 230, 252, 266.
Sparse approximate inverse (SPAI), *see* Preconditioner
Speedup, *see* Parallel
Stability, 28, 45, 92, 259
Storage, 100, 151, 152, 161–163, 165–169, 174, 176, 201, 202, 275, 282, 284
Surface integral equation (SIE), *see* Integral equation

Swap space, 185, 186
Swapping, 185
Switch
    Gigabit Ethernet, 196, 279, 280
    Infiniband, 282–284

T

Target identification software (TIDES), 202, 204, 233–243, 248, 264, 271, 272
Task manager, 187, 190, 191
Terabytes (TB), 10, 201, 202, 203, 275
Testing function, 74, 76, 77, 126, 128–131, 304, 305, 307, 309
TFLOPS, *see* FLOPS
Time
    communication time, 91, 138, 146, 177
    CPU time, 3, 97, 98, 137, 143, 158, 160–164, 167, 168, 172, 230, 300
    execution time, 91, 97, 141, 144, 146, 147, 161, 162, 172, 175–177, 182, 234, 237, 242, 243
    idle time, 91, 316
    I/O time, 177
    matrix filling, 100, 102, 137, 174, 181, 183, 192, 227
    matrix solving, 100–102, 104, 172, 173, 191
    total time, 95, 97, 100, 102, 103, 137–139, 143, 146–151, 153, 160, 162, 163, 167, 172, 173, 181, 184, 191, 192, 196, 197, 199, 202, 204, 228, 229
    wall (clock) time, 4, 91, 100, 144, 151, 152, 162–164, 167, 168, 171–173, 175, 177,180–184, 186, 188–204, 214, 228, 248, 251
Time domain, 5
Time domain integral equation (TDIE), 5, 9
Topology, 14, 37. *See also* MPI topology
Translation look-aside buffer (TLB), 14, 32
Transverse magnetic (TM), 305, 306
Truncated cone, 114, 115, 117, 118, 126
Tuning the performance, 158, 176, 300

U

Ultrawideband (UWB), 141, 268
Uniform geometrical theory of
    diffraction (UTD), 6
Unmanned aerial vehicle (UAV), 249,
    250
User account control (UAC), 282

V

Virtual memory, 7, 10, 42, 184–186,
    193
Vista, *see* Microsoft
Vivaldi, *see* Antenna array
Voltage matrix, 28, 78, 85, 309, 315

W

Wall clock time, *see* Time
Weighted residual method, 2
Weighting function, 2, 304
Windows, *see* Microsoft
Wire, 114, 115, 117–119, 124, 125, 254,
    255, 264

X

XP, *see* Microsoft

Y

Yagi, *see* Antenna; Antenna array

Z

Zero-field region, 111

ANALYSIS OF MULTICONDUCTOR TRANSMISSION LINES • *Clayton R. Paul*

INTRODUCTION TO ELECTROMAGNETIC COMPATIBILITY, Second Edition • *Clayton R. Paul*

ADAPTIVE OPTICS FOR VISION SCIENCE: PRINCIPLES, PRACTICES, DESIGN AND APPLICATIONS • *Jason Porter, Hope Queener, Julianna Lin, Karen Thorn, and Abdul Awwal (eds.)*

ELECTROMAGNETIC OPTIMIZATION BY GENETIC ALGORITHMS • *Yahya Rahmat-Samii and Eric Michielssen (eds.)*

INTRODUCTION TO HIGH-SPEED ELECTRONICS AND OPTOELECTRONICS • *Leonard M. Riaziat*

NEW FRONTIERS IN MEDICAL DEVICE TECHNOLOGY • *Arye Rosen and Harel Rosen (eds.)*

ELECTROMAGNETIC PROPAGATION IN MULTI-MODE RANDOM MEDIA • *Harrison E. Rowe*

ELECTROMAGNETIC PROPAGATION IN ONE-DIMENSIONAL RANDOM MEDIA • *Harrison E. Rowe*

HISTORY OF WIRELESS • *Tapan K. Sarkar, Robert J. Mailloux, Arthur A. Oliner, Magdalena Salazar-Palma, and Dipak L. Sengupta*

PHYSICS OF MULTIANTENNA SYSTEMS AND BROADBAND PROCESSING • *Tapan K. Sarkar, Magdalena Salazar-Palma, and Eric L. Mokole*

SMART ANTENNAS • *Tapan K. Sarkar, Michael C. Wicks, Magdalena Salazar-Palma, and Robert J. Bonneau*

NONLINEAR OPTICS • *E. G. Sauter*

APPLIED ELECTROMAGNETICS AND ELECTROMAGNETIC COMPATIBILITY • *Dipak L. Sengupta and Valdis V. Liepa*

COPLANAR WAVEGUIDE CIRCUITS, COMPONENTS, AND SYSTEMS • *Rainee N. Simons*

ELECTROMAGNETIC FIELDS IN UNCONVENTIONAL MATERIALS AND STRUCTURES • *Onkar N. Singh and Akhlesh Lakhtakia (eds.)*

ANALYSIS AND DESIGN OF AUTONOMOUS MICROWAVE CIRCUITS • *Almudena Suárez*

ELECTRON BEAMS AND MICROWAVE VACUUM ELECTRONICS • *Shulim E. Tsimring*

FUNDAMENTALS OF GLOBAL POSITIONING SYSTEM RECEIVERS: A SOFTWARE APPROACH, Second Edition • *James Bao-yen Tsui*

RF/MICROWAVE INTERACTION WITH BIOLOGICAL TISSUES • *André Vander Vorst, Arye Rosen, and Youji Kotsuka*

InP-BASED MATERIALS AND DEVICES: PHYSICS AND TECHNOLOGY • *Osamu Wada and Hideki Hasegawa (eds.)*

COMPACT AND BROADBAND MICROSTRIP ANTENNAS • *Kin-Lu Wong*

DESIGN OF NONPLANAR MICROSTRIP ANTENNAS AND TRANSMISSION LINES • *Kin-Lu Wong*

PLANAR ANTENNAS FOR WIRELESS COMMUNICATIONS • *Kin-Lu Wong*

FREQUENCY SELECTIVE SURFACE AND GRID ARRAY • *T. K. Wu (ed.)*

ACTIVE AND QUASI-OPTICAL ARRAYS FOR SOLID-STATE POWER COMBINING • *Robert A. York and Zoya B. Popović (eds.)*

OPTICAL SIGNAL PROCESSING, COMPUTING AND NEURAL NETWORKS • *Francis T. S. Yu and Suganda Jutamulia*

SiGe, GaAs, AND InP HETEROJUNCTION BIPOLAR TRANSISTORS • *Jiann Yuan*

PARALLEL SOLUTION OF INTEGRAL EQUATION-BASED EM PROBLEMS • *Yu Zhang and Tapan K. Sarkar*

ELECTRODYNAMICS OF SOLIDS AND MICROWAVE SUPERCONDUCTIVITY • *Shu-Ang Zhou*